Gemstones

To Annie, Lucy, Clare and Peter

Gemstones

Michael O'Donoghue

London   New York
Chapman and Hall

First published in 1988 by Chapman and Hall Ltd
11 New Fetter Lane, London EC4P 4EE
Published in the USA by Chapman and Hall
29 West 35th Street, New York NY 10001

© M. O'Donoghue
Softcover reprint of the hardcover 1st edition 1988

ISBN-13:978-94-010-7030-0

British Library Cataloguing in Publication Data

O'Donoghue, Michael
    Gemstones.
    1. Precious stones
    I. Title
    553.8   QE392
    ISBN-13:978-94-010-7030-0   e-ISBN-13:978-94-009-1191-8
    DOI: 10.1007/978-94-009-1191-8

Library of Congress Cataloging in Publication Data

O'Donoghue, Michael.
    Gemstones/Michael O'Donoghue.
        p.   cm.
    Bibliography: p.
    Includes index.
    ISBN-13:978-94-010-7030-0
    1. Precious stones.   I. Title.
    TS752.03 1987   87–15746 CIP
    553.8—dc19

Contents

Contents   vii

Preface

Gemstones is the first attempt in English to bring together the geological, mineralogical and gemmological developments that have taken place during the last thirty years. Though there have been many gemstone books published in that time, most have been concerned, understandably and rightly, with the science of gem testing and have covered that area very well. Details of the geological occurrence of many of the classic gemstones, and of nearly all those which have only recently been discovered, have been less adequately dealt with. Coverage has been restricted to a number of papers in a wide variety of geological and mineralogical journals. *Gemstones* hopes to get the balance right. In the preparation of the book all the journals and monographs in the field have been consulted so that the book should stand for some years as the authority to which gemmologists and others turn in the first instance. Ease of reference and depth of coverage make *Gemstones* both a reference book and a bench book.

Acknowledgements

I am grateful to Brian Jackson of the Department of Geology, Royal Museums of Scotland who read the manuscript and painstakingly indicated places where amendment or alteration was needed. I am also grateful to the publishers of the books on which the line illustrations are based. The colour plates present in the book derive from a number of sources, which I would like to acknowledge as follows.

Plates 1, 4–15 and 17 Crown copyright reserved. Reproduced by courtesy of the Director, British Geological Survey.
Plates 2 and 3 Courtesy Sotheby's.
Plate 16 Photograph courtesy Gemological Institute of America. Copyright © GIA.
Plate 18 Copyright © P. O. Knischka.

Glossary

The glossary includes those terms which are not explained in the text.

absorption spectrum	dark lines or bands seen vertically crossing a complete spectrum obtained by passing white light through, or having it reflected from, certain coloured minerals. The position of the lines or bands is characteristic for a particular element.
accretion	growth by successive layers.
acicular	needle-like.
adularescence	refers to the characteristic moonstone effect arising from light reflected from layers of orthoclase and albite.
allochromatic minerals	colourless when chemically pure, but more commonly coloured due to small amounts of colouring elements.
anions	ions with a negative charge.
anisotropic	crystals in which the optical properties vary with direction.
annealing	the heating of a substance to a temperature below its melting point; the rutile in star corundum precipitates out in star-like rays when the crystal is heated to approximately 1500°C (the melting point of corundum is 2037°C).
arborescent	with a branching, tree-like form.

asterism	the star effect seen by reflected light (epiasterism) or by transmitted light (diasterism) in cabochon-cut stones with suitably oriented rod-like inclusions or cavities.
atom	the smallest particle into which an element can be divided and still keep its individuality.
atomic mass	the mass of an atom compared with an atom of carbon, the mass of which is arbitrarily assigned as 12.
birefringence	a crystal with more than one refractive index is said to show birefringence or double refraction.
bonding	the process by which atoms or ions are linked together.
byssolite fibres	fibres of a mineral belonging to the actinolite-tremolite group having the nature of an asbestos.
cabochon	a gemstone cut into a domed form whose height and slope may vary and which usually has a flat bottom.
carat	a unit of mass internationally accepted as the standard weight for gemstones: 0.2 g.
cations	ions with a positive charge.
chatoyancy	the 'cat's-eye' effect in which a sharp band of light changes position as the specimen (cut as a domed cabochon) is turned. It is caused by the reflection of light from minute parallel crystals or cavities.
culet	the small facet at the bottom of a cut stone; the name is also used for the sharp point in the same position when a facet is not polished there.
defect	an internal irregularity in a crystal.
dendritic	having a plant-like form.
dislocations	line defects in crystals.

dispersion   the splitting up of white light into its component spectrum colours.

dopant   a substance added to another for the purpose of altering its colour or other properties.

electron   a negatively charged particle outside the nucleus of an atom.

epitaxial growth   growth of one substance upon another of the same or of different composition.

etch figures   pits on crystal surfaces resulting from chemical attack; their shape reflects the underlying crystal structure.

flux   a substance in which the ingredients of a desired material are dissolved before crystal growth begins. The flux and its dissolved material have a lower melting point than that of the substance to be grown.

foiling   the enhancement of the colour of a stone by placing a coloured metallic layer beneath it, the whole being enclosed by the setting.

frequency   the number of complete waves of electromagnetic radiation such as light which pass a given point in a second.

hydrothermal   precipitable from a hot aqueous solution.

idiochromatic minerals having colour resulting from elements which are an integral part of their chemical composition.

ion   a positively or negatively charged particle formed when an atom loses or gains an electron(s).

isotropic   a transparent body in which light travels with the same velocity in any direction.

lamella   a thin plate or leaf, e.g. one of the units of a polysynthetically twinned mineral.

lapidary   cutter, polisher or engraver of gems.

magma   naturally-occurring mobile rock material capable of forming intrusions and from which igneous rocks are thought to derive.

metamict a mineral in which degenerative changes to the crystal lattice have taken place as a result of radioactive inclusions.

nanometre unit of length used for the measurement of the wavelength of light and for shorter wavelengths such as UV and X-rays. $1\,\text{nm} = 10^{-9}\,\text{m}$.

neutron a particle with no charge in the nucleus of an atom.

nucleation initiation of crystal growth.

nucleus the central portion of an atom, positively charged and containing most of the weight.

opalescence the milky effect seen in common opal and in some precious opal. Not to be confused with play of colour.

piezoelectricity the development of an electrical potential in certain crystals when mechanical strain is applied, or the development of a mechanical strain when an electrical potential is applied.

pneumatolytic description of the phase of magma solidification in which minerals can be formed or altered from gaseous emanations.

polysynthetic twinning repeated twinning of three or more individual crystals according to the same twin law and on parallel composition planes.

proton a positively charged particle forming part of the nucleus of an atom.

pyroelectricity the simultaneous development of opposite charges at opposite ends of a crystal upon changes in temperature. Crystals behaving in this way invariably lack a centre of symmetry.

re-entrant angle an angle between two plane surfaces on a crystalline solid in which the external angle is less than 180 degrees; V-shaped angles characteristic of twinned crystals.

refractive index a number expressing the ratio of the velocity of light *in vacuo* to its velocity in a transparent substance.

roiling swirling variations of the same colour occurring randomly in a stone, i.e. not according to direction.

schist foliated metamorphic rock presenting layers of different minerals and splitting into thin irregular plates.

sectile having a texture sufficiently tenacious to be sliced with a knife.

sintered fused powder.

specific gravity the weight of a substance compared with the weight of an equal volume of pure water at 4°C.

spectrophotometer an instrument for comparing the intensity of light in different parts of the spectrum as a function of wavelength.

substitution more correctly ionic substitution; the replacement of one or more ions in a crystal structure by others of similar size and charge.

thermal conductivity the ability of a substance to conduct heat away; crystals are much more efficient in this respect than glass and so feel cooler when touched by the tip of the tongue. Diamonds are particularly efficient conductors, and devices have been designed to exploit this property to distinguish them from simulants.

unit cell the fundamental parallelepiped that forms a crystal lattice by regular repetition in space.

valency the number expressing the combining power of an atom of any element.

1 Formation and occurrence of gemstones

Most gemstones are minerals and stand out from other minerals by their beauty, durability and comparative rarity. These terms are to some extent flexible: diamond for example is very rare indeed compared to salt or quartz, but not compared to ruby, the particular conditions needed for the formation of which are encountered hardly at all. Few entirely new gem mineral species have been found in the last few years; tanzanite, appearing on the gem market in 1967, is a fine blue transparent variety of a mineral which had previously appeared in a totally different form with no gem potential whatever. It is not likely that many entirely new mineral species with gem application will now be found, although minerals already known occasionally turn up with new colours or optical effects such as asterism (star formation).

The several conditions necessary for a mineral of gem quality to form occur together very rarely. The conditions required to produce gem-quality crystals are much rarer. For example, the red and green of ruby and emerald require a source of chromium; without it the corundum and beryl minerals, of which ruby and emerald are special varieties would be colourless or coloured less attractively by other elements. For gem crystals, cooling must take place slowly, as fast cooling (such as occurs on the ejection of lava from volcanoes) produces either glasses (whose atoms have no time to assemble into the regular structures of crystals) or very small crystals. Those crystals which form in gas pockets in rocks have time to grow slowly, to attain considerable size and to incorporate rarer elements which are among the last ones to come out of the original melt. Relatively large crystals are found in igneous rocks, which are formed through the crystallization of molten rock material (or magma) which originated at great depths and which has consolidated near or at the earth's crust at temperatures in the range 950 to 450°C. Large deep-seated masses deep below the surface are called plutons; veins and dykes run from these masses through fissures in the surrounding rocks. In these conditions the magma cools more slowly with formation of larger crystals.

Much primary magma arising within the crust has a basaltic composition – about 50% silica (a relatively low percentage) with alumina (Al_2O_3), lime (CaO), magnesia (MgO) and iron oxides (FeO and Fe_2O_3). As the magma begins to crystallize magnesian olivine (the mineral of which green peridot is a variety) is the first mineral to separate out in any quantity. Here the silica content is even lower than that of the magma as a whole. Diamondiferous pipes in South Africa and elsewhere contain this kind of rock.

When these early-forming minerals have crystallized the remainder of the melt is relatively rich in silica and minerals crystallizing later are likely to be more acid in consequence. A highly acid melt is so silica rich that much of it crystallizes as free silica in the form of quartz, as in a granite. Along with the increasing silica content other elements will concentrate in the melt; they include lithium, beryllium and boron which have not been used in earlier crystallization. Highly volatile constituents such as fluorine, chlorine and hydroxyl ions are also concentrated in the melt, lowering its viscosity and thus allowing larger crystals to grow in later stages. Pegmatites, rich in topaz, tourmaline and feldspar varieties, are often found in this late stage of activity.

Most unweathered rock is hard to exploit for any gems it may contain (apart from diamondiferous pipes). Hard gemstone crystals survive the vicissitudes of the host rock and can be more easily recovered from the weathered and softer product. Their hardness also ensures their survival during the process of travel from the decomposing host to the beds of rivers and streams from which they are eventually recovered. Gem crystals found in alluvial deposits such as these also include materials derived from metamorphic rocks. Magmatic emplacement among existing rock heats the latter and completely or partially recrystallizes it; this process is called contact metamorphism and can well be seen in some limestones which are recrystallized to marble – Burma rubies are formed in this type of rock. The ruby comes from corundum derived from clay impurities in the original limestone. Spinel, which forms in a similar way, derives its magnesium from original dolomite. Sometimes volatile constituents pass from the magma into the surrounding rock and help to form new minerals such as lapis-lazuli. Such volatile constituents may also affect the igneous body itself – olivine may be altered to serpentine and opal or agate may be deposited in cavities. Contact metamorphism is usually fairly local in extent but where the country rock is recrystallized over a much larger area the process is known as regional metamorphism. This produces the gem minerals nephrite, kyanite, staurolite and sillimanite. Many of these minerals, too, are recovered from alluvial deposits.

The third major class of rock, sedimentary rock, comprises sands and gravels as well as other material derived from weathering of existing rock. Sedimentary rocks can also be derived from the long accumulation of skeletal matter from marine organisms. Such soft rocks may be penetrated by hot mineral-bearing solutions which may allow opal and other gem materials to form.

1.1 FORMATION OF INCLUSIONS

The history of gemstone formation can be found inside the stones themselves; such information is of considerable importance as it allows geologists to predict further occurrences and gemmologists to say where a particular stone has come from – this information is now increasingly required for buyers of important stones. The study of inclusions is probably therefore the most important way of identifying specimens. It is also convenient since stones do not need to be free from their settings and the test is non-destructive, needing no instrument save a microscope with a suitable light source. Loose stones can be immersed in a liquid matching their mean refractive index as closely as possible but much can be seen even if this cannot be done without undue trouble. Most gemmologists have their own methods of examining stones under magnification but a microscope that can provide dark-field illumination is very useful since the inclusions can then be seen illuminated against a dark background. A similar effect can be obtained by manoeuvering the light source with skill and care.

Not only can characteristic inclusions identify a natural stone, distinguish it from its synthetic counterpart and provide valuable information on the formation conditions of an unknown specimen; they can also in many cases establish exactly where the stone came from – sometimes down to the actual mine. Today it is commonplace to see in sale catalogues issued by the major auction houses a note stating that a particular ruby is from Burma or a particular blue sapphire from Kashmir (both these places undoubtedly give a cachet to an item which can be exemplified in commercial terms). Such certification can only be given by a gemmologist familiar with inclusions and since attributions can be challenged in the courts it is all the more vital to study what Gübelin has called 'the internal world of gemstones'. Gübelin has, in fact, established himself as an outstanding authority on the topic and his book with the same title was published in 1979 (the second edition). Much of the material following is drawn from this source.

Gemstone inclusions fall into three classes:

1. *PROTOGENETIC INCLUSIONS* are those which are already formed before they are absorbed into the growing crystal; they are always minerals;
2. *SYNGENETIC INCLUSIONS* fall into five groups: (a) minerals (b)

liquid inclusions which may under some conditions also form gas bubbles and mineral crystallites as new phases (c) liquids in cavities which have been sealed by the host crystal growing over them (sometimes these are known as primary syngenetic inclusions) (d) liquids trapped in cracks and fractures which have been sealed by the growth of the host (secondary syngenetic liquid inclusions) (e) growth zoning and the formation of twin crystals; 3. *EPIGENETIC INCLUSIONS* develop after growth of the host crystal is complete. They comprise such events as the formation of stars, the development of cleavages within the stone resulting from accidental knocks, the gradual alteration of the stone due to included radioactive material (as in the case of metamict zircon) and the development of mineral inclusions from exsolution. Traces of radioactivity can also be described as epigenetic inclusions.

Protogenetic inclusions

Mineral protogenetic inclusions may originate in a rock-forming event much earlier than the formation of the host crystal. When a gemstone is formed by the metamorphic process, minerals from the original rock (which have remained unaltered) are enclosed by the crystals and appear as accidental intruders with no obvious relationship to the composition of the host. Sometimes conditions obtaining at the time of the growth of a mineral inclusion are such that the mineral forms at a lower rate than the host crystal, so that the inclusion is overgrown by the crystal. In such a case the inclusion and the host are often chemically related (sometimes they are the same mineral), and the process occurs at an earlier stage in crystallization than inclusion of the 'accidental intruders'.

Many protogenetic inclusions are corroded, rounded or etched, showing signs of an eventful history. Sometimes they are skeletal, the bulk of the crystal having been dissolved. On the other hand, some protogenetic crystals are notably well-shaped and in good condition with sharp edges and smooth faces, particularly if the inclusions have been formed from the same sequence of crystallization as the host but at an earlier stage of it. Spinel crystals in Burma ruby and olivine crystals in spinel are good examples.

Although there are many examples of protogenetic inclusions, the main ones of importance to the gemmologist are:

- actinolite crystals (resembling bamboo stalks) in some emerald, quartz and garnet
- anatase in rock crystal
- apatite in garnet, kornerupine, corundum and spinel
- chromite in emerald
- diamond in diamond

- epidote in rock crystal
- garnet in corundum
- hematite in topaz and euclase
- mica: biotite, phlogopite or fuchsite in quartz, emerald and corundum
- pyrite in corundum and emerald
- rutile in quartz
- spinel in ruby
- zircon in garnet, corundum and kornerupine

Many of these minerals also occur syngenetically or epigenetically and it is their age in relation to that of the host which is of importance. This can be inferred either from the appearance of the inclusion or from the type of deposit from which the stone comes.

Among the protogenetic inclusions are several whose frequency and appearance make further comment desirable. The crystals of apatite so often found in corundum and hessonite garnet are recognized by their well-formed appearance. This is because the crystals were formed early from the melt and were able to float freely in it. Their small size is accounted for by the small amount of phosphoric acid available in the melt, the result being that larger crystals inevitably overtake and envelop the apatites. The various types of mica form rocks known as mica schists; these are metamorphic rocks. Emeralds may be formed when the schists are penetrated by hot pegmatitic residual melts which recrystallize the mica, leaving corroded and etched fragments as inclusions in the emerald crystals. Mica crystals may also re-form after dissolution and remain in the host crystals as syngenetic inclusions. Pyrite is a common and easily recognizable inclusion (it displays a yellow, brassy lustre); since it is formed by so many different mechanisms it is not surprising that it appears as an inclusion in all three categories.

Syngenetic inclusions

Minerals Syngenetic mineral inclusions are minerals which arise from the solution from which both they and the host originate. These solutions may be complex, in that several components may be dissolved and more than one mineral can crystallize at the same time. In some instances the host and the inclusions may be intimately related without being the same substance. The close relationship can sometimes be spotted by the way in which crystal faces of both individuals echo or complement each other. When two individuals are placed adjacently and are of approximately equal dimensions it is known as 'ordered intergrowth'. More commonly minerals settle on a crystal face of the host in which the atomic intervals allow settling – this is known as 'epitaxial growth'. Where minerals have grown epitaxially and are once more overtaken by the growth of the host they become inclusions and reflect earlier phases of their host's development. Garnet may be seen in diamond crystals and examination shows that they conform to similar crystallographic directions as those of the host, resting on earlier surfaces. Syngenetic mineral inclusions often show very well-developed crystal forms as in the byssolite fibres in demantoid garnet. Many minerals which occur as protogenetic inclusions can also occur as syngenetic ones, the latter indicating a younger generation.

Mineral inclusions can indicate the way in which the host was formed; magmatic minerals are found in crystals of magmatic (igneous) origin such as olivine, pyrope and diopside in diamond. Pegmatite minerals such as apatite and quartz may be found in another pegmatitic mineral, aquamarine. A mineral from one type of formation cannot occur in a host from another; pyrope which is formed at great depths and high pressures will not be found in sapphire which is formed in the pneumatolytic phase of solidifying magma. The close relationship between inclusion and host enables stones from different areas to be distinguished from each other as their appearance and form reflect the supply and type of nutrient material in any particular location. For example, crystals as tiny red specks are found in blue sapphires only from Kampuchea (the inclusion is a uranium-bearing pyrochlore) and zircon crystals are common in Sri Lanka gemstones.

The more important syngenetic mineral inclusions are:

- apatite in a variety of hosts
- biotite mica in andalusite, chrysoberyl, kornerupine and Sri Lanka sapphire
- calcite in Burmese ruby and spinel; emerald from Muzo, Colombia
- chrome diopside in diamond
- chromite in peridot
- olivine in diamond
- pyrite in fluorite, sapphire and emerald
- pyrrhotite in corundum, diamond and emerald
- quartz in aquamarine, topaz and emerald
- rutile in corundum, quartz and andalusite
- sphene in ruby and spinel
- uraninite in Sri Lanka sapphire
- zircon in corundum.

Apatite is a characteristic mineral of metamorphism and is thus found in contact-metamorphic gemstones such as garnet, ruby, sapphire and spinel. Calcite which is found in emerald from Muzo and corundum and spinel from Mogok, Burma, shows that lime-bearing rocks were in the vicinity of the contact metamorphism which formed the gemstones. Crystals of tourmaline are frequent in rock crystal; black tourmaline is usually of pegmatitic origin and green tourmaline is formed hydrothermally so that the origin of the quartz host can be deduced. Hematite discs seen in rock crystal, and flakes of biotite in iolite show preferential accretion on the growth faces of the host.

Liquid inclusions Syngenetic liquid inclusions are found in cavities in the host, often accompanied by gas bubbles and mineral crystals. Stones which have crystallized from a melt free from water, as in the case of many magmas and lavas, do not contain liquid inclusions. Neither do diamond and pyrope which have crystallized at great depths and at high pressures; the absence of liquid inclusions helps to identify the stones. Those gemstones which crystallized from a mother-liquor which itself is made up of water and gases and which remains liquid at the temperatures on the earth's surface do contain liquid inclusions (which prove their mode of origin). If a crystal containing a two-phase (liquid and gas) inclusion is heated the inclusion will transform to a single phase, the gas bubble vanishing. The temperature of this homogenization varies depending on chemical composition. The temperature at which the gas phase disappears gives, with reference to simple fluid systems, an indication of the pressure, volume, temperature and composition state of the fluids at the time of trapping.

Liquids in cavities During the growth of a crystal cavities may form in which liquids and gases become trapped. Cavities arise when a local shortage of nutrient leaves gaps in the growing crystals which are covered over by fresh material when it becomes available once more. As, during the shortage of nutrient, the growing crystal concentrates on building up the more important edges and faces (which these are depends upon the crystal symmetry), other parts of the crystal are left behind and filled with a solution or gas other than the nutrient. The form of the cavities may echo that of the host crystal or they may take the shape of long growth tubes parallel to the growth direction and often tapering. The liquids filling the cavities are known as 'primary growth inclusions'.

Cavities may also be formed by mechanical means, as when the development of part of the crystal is prevented by the presence of another crystal or by a mineral inclusion. Hollows formed by these means are often noticeably regular in shape and they are bounded by plane faces. These 'hollow crystals' are known as negative crystals; they can, in fact, be mistaken for solid crystalline inclusions, but careful microscopic examination will identify gas bubbles which will give the cavity away.

Healing cracks Secondary liquid inclusions may also develop during the growth of the crystal. These occupy the cleavages and cracks which form while growth is proceeding, filling them up with nutrient which may then crystallize in the crack and 'heal' it, to form an inclusion known as a 'healing crack'. When healing is complete there is always a residue of liquid left in the former crack which forms networks and patterns. Healing cracks can originate around included mineral fragments as can be seen in aquamarine, garnet and corundum. When the healing cracks follow crystallographic directions they may appear, under magnification, as straight lines or planes. Other healing directions are independent of any crystallographic planes such as cleavages and take their own course through the host, often developing curves and twists.

Inside the cavities the liquids may contain tiny gas bubbles and small crystals. When all three phases are present the inclusion is known as a three-phase inclusion, familiar in Colombian emeralds. Two-phase inclusions contain liquid and gas; among the liquids found are water and brine, carbonic acid and even hydrocarbons such as oil (e.g. in some fluorite from Illinois, USA). Among the large number of gases are methane, ethane, propane and sulphur dioxide.

Water, however, is the commonest cavity filling and often has salts dissolved in it. Sometimes water will co-exist in a cavity with another liquid with which it will not mix (immiscible solutions); water and carbonic acid are common associates.

Secondary syngenetic inclusions are generally formed at lower temperatures than primary ones and they occupy large parts of the interior of a crystal. When they consist of residual mother-liquor the cavity filling is described as 'isophysical'. When the liquid is foreign to the crystal the filling is said to be 'xenophysical'. If cavities are filled entirely with foreign liquids it shows that the crystal containing the cavity had stopped growing, as the mother-liquor would have penetrated the cavity preferentially. In one sense, therefore, xenophysical fillings are epigenetic rather than syngenetic inclusions. It is only possible to determine what a liquid was made of when crystals have formed from it.

When a cavity is healed the healing is hard to see if the healing agent is the original nutrient (i.e. when the healing takes place while the crystal is still growing). However there is usually some liquid which has not been used up in the healing process and this can be observed under magnification. On entry of the healing liquid into the cavity the healing process operates over the whole area at once. Any liquid remaining which does not fit the crystal lattice of the host is sealed by the healing process to form a liquid inclusion. This liquid is split into individual droplets which take up positions amongst the healed portions of the host. The droplets may take the form of negative crystals and the vessels within which the secondary liquid inclusions are sealed also comprise a variety of shapes. These may be long or short, angular or rounded; the precise factors influencing the shape are not known, although accident seems unlikely. To understand this process, cracks were artificially induced in synthetic rubies which were then returned to their mother-liquor for healing to take place; from many months of observation the stages were found to be (a) penetration of the mother liquor, (b) crystallizing out of the filling material in the form of dendrites, (c) separation of the undigested liquid into a series of individual droplets and (d) the transformation of the droplets into two-phase inclusions through contraction and formation of a gas bubble in each inclusion. This work was reported by W. F. Eppler (1966).

Zoning and twinning Gas bubbles other than those encountered in two- or three-phase inclusions are a sure sign of a synthetic material or a glass. Their bold outline shows that the refractive index of the contents differs markedly from that of the host. Generally gas bubbles in synthetic crystals occur in groups whereas in both natural and artificial glasses they are isolated. The occurrence of twinning and growth banding assists the gemmologist by giving clues about the nature of the material; often no further tests are necessary. Examples of twinning can be seen in rubies from Thailand and in the feldspar gemstones. Zones of growth arise through fluctuations in the supply of nutrient, through changes in pressure during growth or alterations in the chemical composition of the mother liquor. In all natural gemstones growth banding is angular compared to the curved banding seen in the Verneuil-grown synthetic stones (see below). Irregular distribution of colour such as can be seen in hessonite, garnet, corundum and emerald also ranks as a syngenetic inclusion.

Epigenetic inclusions

Epigenetic inclusions are the result of: (a) the internal recrystallization of the host, sometimes with the production of new crystals by the process of exsolution; (b) the leaching out of cracks and cleavages and their re-filling or partial filling by new epigenetic material; (c) damage by radioactivity emanating from included mineral matter. The formation of stars (asterism) is an epigenetic event; it is the way in which internal recrystallization is most easily seen and is its most beautiful manifestation. The most important star stones are members of the corundum family and in many countries star rubies and sapphires are highly prized. The mineral responsible is the titanium oxide rutile, TiO_2 which at normal temperatures is randomly distributed in the crystal. On annealing of the crystal the rutile precipitates out at right angles to the vertical crystal axis and the crystal needs only to be cut into the cabochon shape for the star to be visible. Garnets and some other gemstones may show four-rayed stars; those displayed by corundum have six rays.

Minerals with cleavages such as topaz or moonstone may develop internal cleavages (splitting along crystallographic directions – see below) at any time. The cleavage can be detected by the presence of interference colours; directions are straight whereas those taken by fracture or tension cracks may not be. Some of the cleavages and cracks may be filled with xenophysical liquids which sometimes give a dendritic or arborescent patterning, as in moss agate and similar materials.

The presence of radioactive material in a crystal may lead to the gradual breakdown of its crystal lattice; this is particularly well seen in some green and brown zircon from Sri Lanka. The process is called 'metamictization' and the stones known as metamict stones. The zircon crystals contain some uranium and/or thorium which radiate alpha–particles; the appearance of the inclusion changes through time as its volume changes slightly, causing mechanical stress around it. The stress marks thus produced form characteristic haloes; they surround the inclusion radially and can easily be seen under low magnification. In time (up to 800 million years) the radiation destroys the inclusion's crystal structure, and renders the stones virtually amorphous.

1.2 INCLUSIONS IN THE MAJOR GEMSTONES

Diamond

It is fortunate that diamond contains so wide a variety of mineral inclusions since they are excellent aids to identification. There is no possibility of liquid inclusions in diamond because of formation at high temperature and pressure.

Protogenetic inclusions in diamond are almost invariably crystals of diamond itself and even these can be thought of as being at least partly syngenetic. Diamonds form and re-form in the parent rock and crystals formed at one time may act as nuclei for new crystallization later. Diamond in diamond inclusions which show signs of damage (the breaking-off of an edge or point from an octahedron, for example) are sure to be protogenetic.

Most inclusions in diamond, however, are syngenetic and the commonest ones are olivine, garnet, diamond, chrome diopside, chrome enstatite, pyrrhotite, pyrite, pentlandite, ilmenite, rutile and coesite. In recent years Gübelin has made the exciting discovery that ruby can be added to this list (Gübelin, 1986). Some of these inclusions show oddities of growth: some garnet crystals, for example, are elongated and olivine may behave in a similar way. The latter, in fact, has frequently been mis-identified as zircon because the crystals appeared to be tetragonal rather than orthorhombic.

Epigenetic inclusions which often enter the diamond crystal through cracks include rutile, calcite, goethite, quartz and hematite. These are frequently deposited from xenophysical solutions. Some syngenetic inclusions change to epigenetic ones through weathering; garnet crystals may alter (in the main) to chlorite and olivine may alter to serpentine or to chlorite. Olivine inclusions appear as transparent crystallites which are too pale in general to evoke the normal olivine colour response; some contain grains of chromite. The crystallites are found either as individuals or as groups. They grow on the octahedral faces of the diamond and are then enclosed by further growth. In composition the olivines are about 94% forsterite, the colourless magnesium-rich olivine, and this accounts for their pale colour. Garnet crystals range in colour from brown to yellow, orange, red and purple and may be 0.5 mm or larger. Some, mostly chromium-rich pyrope garnets, reach the surface via fractures. Picotite, the chromian spinel, appears reddish-brown and is found in the form of distorted octahedra. Diamond itself can be recognized by its adamantine lustre. Chrome diopside and chrome enstatite are difficult to distinguish from one another as they are both emerald-green and the crystallites of both minerals are well formed. Graphite flakes are quite common as inclusions, as are ores such as pyrrhotite. Sometimes the pyrrhotite is intergrown with other sulphur minerals including pyrite and pentlandite.

It is possible, although difficult, to locate the origin of some diamonds from their inclusions. Fortunately this type of identification is not yet stipulated for important stones, as now required for ruby and emerald. A few clues are available, however: chrome spinel seems to be common in Siberian diamonds and South African stones are rich in garnets. A structural defect like a Maltese cross has often been found in diamonds from India.

Corundum

The most valuable corundum gemstones are rubies and sapphires. Inclusions in corundum serve not only to identify the stone but in some cases to pinpoint its place of origin. All types of inclusion are represented. Corundum crystallizes rapidly and thus encloses a variety of syngenetic minerals. The quick crystallization also encourages cracks to develop through stress and therefore healing cracks and liquid inclusions are especially prominent. The most important inclusion in corundum is rutile which occurs epigenetically to form stars and also to form fine networks of needles which intersect at 120 degrees to form a highly characteristic pattern, traditionally known as 'silk'. Apatite, pyrrhotite and zircon also occur, as syngenetic inclusions. Primary liquid inclusions contain two-phase contents and sometimes three-phase ones with aqueous mother liquor and carbonic acid in liquid form. Cavities frequently take the form of negative crystals. Healing cracks give rise to complex patterns, especially in sapphires from Sri Lanka.

Burmese ruby contains short rutile needles, calcite rhombs, spinel octahedra, apatite prisms and a mixing of colour traditionally known as 'roiling'. Rutile in Sri Lanka ruby is finer and forms longer crystals; liquid inclusions are quite rare showing that the stones crystallized from a melt deficient in liquid. Sri Lanka ruby shows the rutile crystals in addition to flakes of biotite, healing cracks and zircon crystals with haloes. Siam rubies, as the stones from what is now Thailand are still known (the name is often given to any ruby with a dark red colour inclining to brown), are characterized by mineral grains surrounded by liquid films along the contact planes of the twin crystals which form the rubies. Rutile is rarely found but the lamellae of polysynthetic twinning show as parallel lines. Almandine garnet, apatite and pyrrhotite crystals are found. Healing cracks with a script-like pattern are characteristic, as are plagioclase crystals indicating pegmatitic origin. Ruby from Tanzania shows no rutile but twinning lamellae are common. Mineral grains in liquid films are not found. Rubies from Longido have zoisite and spinel inclusions. Rubies from the Hunza valley deposits of Pakistan show cracks, parting and polysynthetic twin planes and swirl marks like those seen in Burmese stones. Calcite is the commonest mineral inclusion and dolomite is also found. Red–brown flakes of phlogopite are prominent. No rutile silk has been seen, although individual crystals occur, together with apatite, pyrrhotite and spinel.

Sapphire from Burma contains healing cracks resembling crumpled flags which are characteristic. Short rutile needles, the individual needles appearing sectioned, and pyrrhotite crystals, make up the only common mineral inclusions. Sri Lanka sapphire shows inclusions similar to ruby from the same country. Kashmir sapphires have a cloudiness caused by minute fissures arranged in three directions intersecting at 120 degrees. Zoning of colour is also characteristic. Montana sapphires contain hexagonal mineral inclusions surrounded by liquid sprays and in general the inclusions resemble those found in Siam rubies. Sapphires from the Pailin gem field in Kampuchea are characterized by tiny red uranium pyrochlore crystals. Siam sapphires contain plagioclase crystals but no rutile, indicating a pegmatitic origin. In sapphires from Tanzania dislocation lamellae with fringes are common, with apatite, zircon, graphite and pyrrhotite crystals.

Emerald

Inclusions in emerald are useful indicators of a stone's provenance and some specimens can even be assigned to individual mines. Emeralds are formed in two distinct ways. Some are formed during the pneumatolytic phase of mineral formation: as solutions at high temperature and pressure alter existing rocks, some of the mineral components of the rocks become protogenetic inclusions in the emeralds. Most of the rocks thus altered are crystalline schists from which mica, tremolite and actinolite are captured by the emerald. Other emeralds, including those from Colombia, are formed at lower temperatures and arise from the hydrothermal phase of mineralization. In this process water at high temperature and pressure, with some liquid carbonic acid, allows emeralds to grow from solutions containing salt. In consequence these emeralds show fewer mineral inclusions, those which do occur being formed syngenetically. The solid phase is halite salt. A para-epigenetic inclusion may be the correct way in which to describe traces of oil inside an emerald which has been immersed to improve colour and clarity. Immersion minimizes reflection from the surface which hinders observation of what is inside the stone.

Colombian emeralds are especially rich in three-phase inclusions with jagged edges. They contain a halite crystal, a liquid and a gas bubble. Stones from the Chivor mines also contain very small plates of albite and crystals of pyrite. The latter can be recognized by its brassy lustre and colour. Growth bands are characteristic of emeralds from Gachalá which also contain growth tubes and albite platelets. Peculiar to Muzo emeralds which are found in calcareous shales, are calcite and yellow to brown crystals of rhombs of the crystal parisite. Both Muzo and Chivor produce the interesting and attractive trapiche emerald, described in the section on individual stones. In Muzo stones the core is black, indicating a carbonaceous material.

Emeralds from Brazil are usually full of thin liquid films and there are relatively few mineral inclusions. Dolomite, biotite mica and talc are found. Emeralds from the mica schist deposits of the Habachtal in Austria contain tremolite, tourmaline and biotite; the tremolite crystals are acicular (needle-like) and fairly broad, always straight. Indian stones are remarkable for two-phase inclusions which occur as negative crystals, resembling commas or hockey-sticks. Minerals included are fuchsite mica and chlorapatite. The Sandawana emeralds of Zimbabwe show tremolite crystals which are more hair-like than those of the Habachtal emeralds. The tremolite is also curved rather than straight. Transvaal emeralds have tiny corroded flakes of mica from the mica schist in which they were formed. Emeralds from the Urals in the USSR contain bamboo-like crystals of actinolite and as they were formed by the alteration of mica schists and hornblende their contents resemble those of Sandawana and Habachtal stones which were similarly formed.

Inclusions in emerald from Pakistan resemble those seen in Brazilian stones. Care should be taken as they could be mistaken for the smoke- or veil-like inclusions of flux-grown stones. Well-formed crystals of calcite and dolomite have been observed with jagged inclusions arranged parallel to the c-axis (the vertical crystal axis) of the crystals. These arise, as at Muzo, from the syngenetic growth defects which have trapped the hydrothermal solution during growth. Isolated minute crystals may also appear as a third phase. Some crystals may have growth tubes extending from them and could, under the right conditions, give rise to chatoyancy (cat's eye effect – see glossary). Very thin films are arranged parallel to basal planes and drops of solution are found in partially healed fractures; these drops may form hose-like structures. Tourmaline has been found as an inclusion in Zambian emerald; it is found throughout the schist in which the emeralds occur.

Aquamarine and other beryls

It is quite hard to find a cut aquamarine (or other beryls except emerald) which is interestingly included. Crystals commonly grow large enough for clear gems to be cut from sections if not from the whole. However, before cutting the commonest inclusion is a series of parallel channels to which the name 'rain' has been given (presumably the tropical rather than the English variety as the channels are long rather than drop-like). The channels are growth tubes which are voids containing crystallites and liquid and gas vesicles. They are negative crystals. Aquamarine also contains groups of two-phase inclusions which give a mosaic pattern characteristic of the gem. Ilmenite skeletons and foils of biotite may also be found, together with red hematite tablets.

Gübelin (1981b) has described some inclusions found more recently in beryl. Beryl of greenish and bluish colours, as well as some colourless material, emanating from widespread deposits, all shared at least one inclusion, either a large pseudohexagonal colourless or brown crystal and/or a brownish-orange crystal of the cubic system. In a Brazilian colourless beryl, particularly prominent was a large brown mineral inclusion surrounded by a whitish coating with a silvery lustre and microscopic flakes resembling fish scales emanating from it. Microprobe examination of one of the flakes, fortuitously protruding from the surface, showed that it was muscovite mica. The large brown crystal was found to be phlogopite mica. Black grains also found in the stone turned out to be niobite. The association of these three minerals in a single host is a newly observed paragenesis. Niobite crystals have also been noticed in a goshenite (colourless beryl) from Brazil and in the same crystal a manganese apatite and phlogopite crystals were found. Spessartine crystals have been found in light blue aquamarine from Nuristan, Afghanistan; the garnet appears as a number of brownish crystals. In a greenish-blue aquamarine from Kenya, albite, spessartine and euhedral green tourmaline crystals were found.

Feldspars

Feldspars, particularly moonstone, show characteristic fissure systems on their surfaces which have been named 'Chinese aeroplanes' or 'centipedes'. They are presumably caused by the stress imparted by the intergrowth of the albite and orthoclase varieties of feldspar.

Garnet

In the garnet group demantoid shows the diagnostic inclusions of byssolite fibres which grow from a chromite crystal. Grossular garnet in its hessonite variety contains well-shaped apatite crystals and has a granular appearance which is easily recognizable. The green, vanadium-bearing African grossular also shows apatite crystals and some chromium-bearing stones from Tanzania contain actinolite crystals. In spessartine, flags and shreds of liquid distinguish orange-brown stones from the similarly-coloured hessonite and from other orange stones such as fire opal. The pyrope–almandine series contains apatite, mica, quartz, spinel and zircon as well as rutile which may give a four-rayed star, being arranged parallel to the edges of rhombic dodecahedral faces.

Peridot

In peridot the well-known 'lotus leaves' are formed by chromite crystals surrounded by liquid films. Peridot can also contain glassy drops, especially in Hawaiian stones; the drops may contain chromite crystals.

Quartz

Inclusions in rock crystal are often so prominent that they create a special variety such as tourmalinated quartz or the rutile-filled Venus's hair stone, in which the rutile crystals are golden yellow. Attractive patterns are made by healing cracks, particularly in Alpine quartz. Some rock crystal displays a pseudo-asterism when goethite needles are arranged in the rhombohedron. In amethyst, goethite crystals provide the 'beetle legs' which are highly characteristic. Quartz has a wide variety of cavity fillings, all three phases being possible. Carbonic acid is commonly included. Fuchsite crystals are the cause of green aventurine quartz, fuchsite being the chrome mica. Yellow to brown goethite crystals can fill a rock crystal, giving a resemblance to tiger's-eye. 'Zebra stripes' in amethyst are caused by stripes of liquid inclusions alternating with negative crystallites.

Spinel

Spinel itself is the commonest inclusion in spinel, the tiny octahedra being arranged in rows. The iron spinel, hercynite, occurs in red spinel from Thailand; calcite crystals are found in Burmese spinel and zircon crystals in the variously coloured varieties (shades of lilac or blue) from Sri Lanka spinel. Occasional four- or six-rayed stars are caused by blades of the mineral sphene which have come out of solution within the spinel.

Topaz

Topaz, formed hydrothermally, shows characteristic liquid inclusions. More than one liquid is typically involved and the separate liquids will not mix; an aqueous solution and liquefied carbon dioxide are the commonest fillings of negative crystallites; they are arranged in rows aligned with the main crystallographic axis. Healing fissures are also found in topaz, lying as flat surfaces parallel to the direction of cleavage (parallel to the basal pinacoid) or crossing the crystal as feathers. Included minerals are apatite, mica, fluorite, goethite, calcite and hornblendes.

Tourmaline

In all tourmalines the mother-liquor is found in profuse liquid inclusions in hair-like tubes known as trichites. Patches of liquid are linked by these fine tubes, some of the tubes being so closely packed that asterism is possible.

Zircon

One of zircon's most familiar internal features is the number of epigenetic cracks following the cleavage direction. Film residues of yellow to brown liquid mark them out. The metamict stones are cloudy but display angular patterns which are caused by stress; the angles mark the positions of former crystal faces which have disappeared through metamictization.

Note: Inclusions in man-made stones and in glass are covered in the sections devoted to these materials.

1.3   PHOTOGRAPHING INCLUSIONS

Photography of inclusions is of great value both in the identification of specimens (the photograph can be passed to others to evaluate) and in the recording of particularly interesting phenomena. Especially important stones such as large diamonds can have distinguishing features recorded as a security measure.

In setting up the apparatus and specimen, care should be taken to ensure both are as clean as possible. The appropriate cleaning agent should be used for lenses (which should *not* be rubbed with a dry cloth). With pyroelectric specimens such as tourmaline, the heat from the light source may stimulate their trapping of dust; a cooled source such as that provided by a fibre-optic attachment is useful here. Camera shake must be avoided; inclusions are small and can easily get out of focus. In general a faster film gives less good results than improving the lighting of the subject because high speed films give a grainy image, making the enlargement quality indifferent. A film of 50 ASA should give good results.

Dark-field illumination is essential and although it can be contrived in microscopes not carrying a facility for it, the specially-designed instruments are quicker and more convenient. Inclusions are seen bright against a dark background; crystals, cleavages and some fluid inclusions respond best. Especial care should be taken to eliminate dust since specks can simulate crystallites. The use of polarized light can show up internal strain and if a colour film is used the effects are quite startling. If an included crystal has a refractive index close to that of its host (so that its outline is indistinct) the use of polarized light will show it up, if it is birefringent since rotation will change the interference colours. As the light available from a polarizing source is limited the exposure time needs to be longer than usual.

Transmitted light is best for the examination of large liquid inclusions but ensure that small, faint, solid inclusions and growth lines are not overwhelmed. Colour zoning also responds well to this technique. The shortest exposure times should be used and specks of dust are less of a problem since the amount of light used stops their being so prominent. The way in which the light reaches the subject can be experimented with, direct overhead illumination not always being the most desirable, save for the observation of stars and similar features. Thin films and liquid inclusions show at their best when illuminated at angles between the horizontal and the perpendicular. The use of ultra-violet light in photography requires very long exposure times and is not usually necessary for gem testing. The use of immersion liquids reduces image quality because there are more media for the light to traverse. As microscopists know, liquids shake, too. With time, many liquids used for immersion will darken and this militates against their frequent use; darkening may even be hastened by the use of bright photographic lamps. Dust needs to be filtered out and kept out. Liquids are handy, however, in improving a scratched surface; a drop of a liquid which closely matches the specimen in refractive index will, when placed on the surface to be photographed, allow the observer's eye and the camera to penetrate beyond it into the stone.

2 The nature of gemstones

All matter is made up of atoms which possess all the characteristics of the element but which are themselves made up of smaller particles. Niels Bohr in 1912 likened the relationship of the particles to a solar system with a central nucleus surrounded by circling electrons at different distances or energy levels from the nucleus. The nucleus, except in hydrogen atoms, is composed of protons with a positive electrical charge and neutrons which are electrically neutral. The orbiting electrons carry a negative charge and the whole atom is neutral; the number of electrons and protons are equal. The mass of the atom is in the nucleus as the proton is 1834 times heavier than the electron.

Orbits taken by the electrons around the nucleus are called shells and an atom may have up to seven of them, labelled alphabetically from K to Q outwards. Each shell has a limit on the number of electrons it can contain and the light elements with up to 18 protons in their nuclei fill their shells systematically outwards. Although the electron-swarm around nuclei is very small the nuclei of adjacent atoms do not collide and amalgamate to form larger atoms. Collision is prevented by mutual repulsion of the electrical charges on protons and electrons. Where atoms do link – the process called bonding – it is by the interaction of the outermost electrons. Hydrogen has one nuclear proton with a single orbiting electron in its K shell; helium with two protons has a complete K shell with two electrons; lithium with three protons has a full K shell and a single electron in the L shell. This pattern of shell-filling continues up to argon with 18 protons. After this many elements permit electrons to occupy outer shells before the inner ones are completely filled. The inert (or noble) gases – helium, neon, argon, krypton, xenon and radon have full outer shells. Apart from helium they have eight electrons in this outermost shell, filling it and thus making combination with other elements to form compounds very difficult.

Atoms link (bond) by sharing electrons so that their electron orbits approach completeness as far as possible (i.e. resemble those of inert gases). This movement of the electrons takes place in two stages. First, the atoms of an element gain or lose electrons till they reach the configuration of the nearest inert gas; they gain an electrical charge in so doing and are then known as ions. Those ions with gained electrons are negatively charged and are called anions; those which lose electrons are positively charged and are called cations.

2.1 VALENCY

The number of electrons gained or lost by an atom in its attempt to reach the electronic configuration of the nearest inert gas is called its valency. The elements sodium (Na) and fluorine (F) have, respectively, one more and one less electron than the inert gas neon (Ne). Their ions are written: Na^+ and F^-. These are monovalent ions, sodium a cation and fluorine an anion. Magnesium (Mg) has two more electrons than neon and by losing both of them it becomes Mg^{2+}; this is a divalent cation. Other elements such as carbon (C) and silicon (Si) have atomic structures midway between two inert gases so that carbon, for example, can either gain four electrons to become C^{4-} and resemble neon, or lose four electrons, becoming C^{4+} to resemble helium. Carbon may thus be either an anion or a cation. In cases where atoms have incomplete inner electron shells the element can form ions with several valencies. Iron (Fe), for example, may be divalent or trivalent in different circumstances.

2.2 BONDING

The ionic bond, the commonest bond type found in minerals, can easily occur between two elements that can form ions of equal and opposite valency. The spare electrons of one atom become detached from it and fill vacancies in the outermost shell of the other atom. The pair of ions formed in this way are held together by electrical force. Groups of ions can be linked in the same way, as in a divalent cation with two monovalent anions. In covalently bonded substances two or more atoms come close together by pooling their electrons so that both have eight in their outermost shells. Covalent and ionic bonds are strong; the former are fixed in specific directions around an atom and the latter are non-directional. Bonds can also be intermediate between the two extremes. Solids with ionic or covalent bonding are usually fairly rigid and fracture rather than deform at a blow. Metals form metallic bonds in which ions of the element pack closely together and leave their outermost loosely-bonded electrons to move about independently. It is these independent electrons which enable metals to be beaten out into thin sheets and which confer electrical conductivity.

When two or more atoms are bonded they form a compound, which can be distinguished from a mixture by its electrical neutrality, achievable only when the component atoms are combined in a fixed ratio. Some sub-groups of atoms are linked strongly in a predominantly covalently bonded complex ion or radical. The tetrahedral arrangement of four oxygen atoms round a single atom of silicon gives a complex anion which is written $(SiO_4)^{4-}$. Such atomic groups will bond ionically with single atoms or with other groups to give a wide variety of mineral structures. The $(SiO_4)^{4-}$ complex anion must be bonded to a cation or cations with a total valency of 4 so that an electrically balanced compound is achieved. Two atoms of Mg (divalent) would be suitable and the mineral forsterite contains the compound Mg_2SiO_4.

2.3   THE BUILDING BLOCKS OF CRYSTALS

The ions which are the fundamental building blocks of crystals are more or less spherical and vary widely in size. Ions of different sizes combine into small groups and if these groups are imagined as solids with flat faces, these are the 'coordination polyhedra' which in turn can be packed together to form the crystal. The limited number of shapes that the polyhedra can take can be imagined if oranges (oxygen) touching a pea (carbon), so that all are in contact, are taken to represent the triangular carbonate ion $(CO_3)^{2-}$. The triangle is obtained by linking the centres of oranges and pea. If the pea is made larger four oranges will be needed to stack around it and these make the tetrahedron which is the shape of the silicate anion, $(SiO_4)^{4-}$. Further enlargement of the pea necessitates the introduction of six oranges to form the octahedron and when pea and oranges are the same size a cube results.

If ions are equal in size they can stack together without forming polyhedra. The various ways in which this can be achieved falls under the description of 'close packing'. Coordination polyhedra and close packing occur particularly in crystals with metallic or ionic bonding as these links are non-directional.

2.4   POLYMORPHISM AND ISOMORPHISM

Some chemical elements or compounds occur in two or more structural forms – diamond and graphite are different forms of carbon. X-ray crystallography can detect these differences in internal structures. The phenomenon of a chemical element or compound existing in two distinct physical forms is called dimorphism or polymorphism, the latter term being used both loosely and specifically in cases where two or more modifications are found. Isomorphism describes the converse situation, where the crystal structures of substances may be the same although their chemical composition is different. This arises from the substitution of ions which are similar in size so that no change to the structure occurs. However the substitution is worked out the crystal must maintain its electrically neutral character.

Some isomorphous pairs of minerals are related in a particular continuous way: the minerals forsterite and fayalite are Mg_2SiO_4 and Fe_2SiO_4 respectively. If we melt a mixture of the two and allow it to cool the crystals formed are an olivine containing both Mg and Fe. It is as if the two had dissolved in one another and the name solid solution is used. Between the two there is a complete series of intermediate olivines and this is known as an 'isomorphous series'. Such series are common in nature and among the gem minerals the garnet family displays several different series, that between pyrope and almandine being the most obvious.

Exsolution

Solid solutions are affected by changes in temperature. Atoms vibrating at high temperatures cause bonding to loosen and the crystal can then accommodate ions which would be too large for the sites in the more compact crystal at lower temperatures. When a high-temperature solution in which this has happened is cooled, the over-sized ions which cannot be accommodated at lower temperatures are forced out and the crystal unmixes into two separate species, usually closely associated as grains or layers within one another. This process is called exsolution.

2.5 THE CLASSIFICATION OF MINERALS

The chemical classification of minerals forms the basis of most serious books on the subject. The system in current use groups minerals by their anions (simple or complex). The mineral groups from which most gemstones are formed include silicates, oxides and carbonates.

In silicates, silicon combines with oxygen to form the tetrahedral complex anion $(SiO_4)^{4-}$. The tetrahedra can join to form rings, chains, sheets, frameworks or other structures. In the olivine group and in the garnets the tetrahedra are independent and the SiO_4 groups are linked by interspersed cations. Crystals with this structure are usually dense and hard. Silicates with pairs of tetrahedra are rare but ring silicates are commoner; in beryl, $Be_3Al_2Si_6O_{18}$, there are six tetrahedra of SiO_4 in the rings. The tetrahedra share oxygen anions at the touching corners, the rings stacking one above another to produce a hard and dense crystal. The framework silicates have their tetrahedra linked at all four corners to give a rigid three-dimensional network; feldspars and quartz are members of this group.

The density and hardness of oxides make good gemstones. In oxides the anion is oxygen alone; when the sole anion is the complex $(OH)^-$ the mineral is a hydroxide. In many oxides the cations are much smaller than the O^{2-} anion and form dense minerals with close packing; corundum is a good example.

The carbonate group contains a number of softer gemstones and has the complex anion $(CO_3)^{2-}$.

2.6 CRYSTAL STRUCTURE

It is ironic that to many people the word 'crystal' means an elaborately-fashioned glass ornament, for glass is one of the very few substances known which is not crystalline! The regular faces on cut glass and cut gemstones are produced artificially; those regular faces on the natural objects which are true crystals arise from the regular internal arrangement of the atoms which compose the elements forming the crystal. There is always an 'ideal' arrangement of faces for a crystal of a given composition although the real-life crystal may appear to the eye as a shapeless mass. In such cases the atomic arrangement and the type of the crystal can be worked out by X-ray diffraction. Substances with an irregular internal atomic arrangement are known as amorphous (without shape).

The growth of crystals

A crystal can only grow when a nucleus of a critical size acts as a starter. Even grains of dust (as crystal growers find to their cost) can act as a nucleus provided that the size is above the critical size. If there is no nucleus of this size growth will not begin until groups of atoms come together long enough to form a tiny nucleus. The chance of such an accident happening is much greater when temperature is reduced. Fastest growth takes place when fresh material arriving at the surface generates sites suitable for still more material to grow. This type of growth takes place when a crystal contains a screw dislocation, one of several types of defect in atomic arrangement that a crystal can undergo during formation. The dislocation can be seen on the crystal face and is characteristic of synthetic quartz where the rhombohedral faces show the spiral when moved under the light. Screw dislocations allow crystal growth to take place at lower energy levels. Crystal growth by layers can take place at surfaces but needs more energy.

As a crystal grows other substances not of the same chemical composition but with similar ionic size to one of the components may move into sites left vacant.

The substitution of chromium for aluminium in ruby and emerald is an example. The mineral does not change species as the amount of impurity added is very small. The process is known as isomorphous replacement.

Crystal symmetry

In any discussion of crystals it is much easier to take ideal examples than to illustrate the broken, distorted, massive or fragmentary specimens that are found in the field. The crystals of the gem minerals are among the easier ones to study since they are relatively large, are often well-formed and are transparent. Even gem crystals, however, are subject to accidents of growth. Discussions of crystallographic symmetry for the purpose of gemstone study can and probably should be limited to basic concepts; for this reason some of the less common types of symmetry, together with systems of notation, are omitted; they can be found in any text book of crystallography, some of which are listed in the bibliography.

An examination of a well-shaped crystal will show that some of the faces are so arranged that their edges are parallel. These sets of plane faces are called zones and the common edge direction is known as the zone axis. Faces of the same size and shape on opposite sides of the crystal are also noticeable and where a crystal has like faces and edges on opposite sides it is said to have a centre of symmetry. No centre of symmetry can exist if a face is opposite to a point. Crystals which can be divided into two or more parts, each of which is the mirror image of the other are said to have planes of symmetry which divide these parts. The more regular the crystal the more planes of symmetry it will have – a cube has nine. An axis of symmetry is the third element of symmetry. When a crystal is rotated about such an axis it assumes a congruent position n times during a complete revolution where n is the degree of the axis; n may be 2, 3, 4 or 6. A cube, in addition to its nine planes of symmetry, will have 13 axes of symmetry, three four-fold (tetrad), four three-fold (triad) and six two-fold (diad). Similar symmetry is found in shapes derived from (modifying) the cube, such as the octahedron, rhombic dodecahedron and icositetrahedron. This is a simple example but in all crystals the symmetry elements interact so that the plane of symmetry will repeat an axis of symmetry inclined to it at any angle save a right angle.

Form and habit

Although the faces of a cube are all squares and those of the rhombic dodecahedron are all rhombuses many crystals have faces of several different shapes. The corners of a cube may be modified to give small equilaterally triangular faces. The symmetry of the cube dictates that if one such corner is so modified the others must be too. The original six cube faces and the eight triangular faces are two different 'forms'. Though this word can be used loosely to describe general shape, the crystallographic sense of the term is defined as the group of faces made essential by the symmetry when one face is given. Looking once more at the modification of the cube corners by the small equilateral triangular faces, if more of the corners are cut off so that the original square faces are replaced by eight triangular faces, we have the form known as the octahedron. But if we do not cut quite so much of the corner the new faces will meet in a new set of edges. The various stages between cube and octahedron show the 'habit' of the crystal, the habit being the general shape taken by the crystal according to the ways in which the various forms are developed.

The variety of habits shown by crystals can be confusing since the same substance may occur as crystals of differing habits. The reason for this lies in minute variations of composition or of growth conditions such as fluctuations of temperature or pressure. This can be particularly well observed when growing crystals in the laboratory; growth of ruby from a certain solvent yields crystals of a platy (flat) habit, too thin for gemstones to be cut from them. Alteration of the flux (the compound in which the ingredients of ruby are dissolved before melting) alters the crystals to another, thicker habit. The habit however, refers to the outward shape only – the underlying symmetry cannot be changed however the habit is varied.

In 1669 Steno discovered that in all crystals of the same substance the angles between corresponding faces have a constant value – the 'law of constant angles'. Steno is here referring to interfacial angles, which can be measured by a simple instrument known as the contact goniometer. Crystals of gem materials will not normally need to be tested in this way since their habits are well-known and they are fairly large as crystals go. (Optical goniometers are more elaborate and can be used for experiments and measurements involving light.)

Crystallographic axes

Crystals attain their external regularities from the regularity of the 'building blocks' from which they are constructed. The Abbé Haüy is credited with the discovery that calcite crystals were built up of a *unit cell*, the smallest single unit whose contents determine the composition of and whose given symmetry and dimensions determine the symmetry and morphology of the whole crystal. Providing that the unit cells will stack together without leaving spaces in between a variety of shapes is possible. Haüy postulated, on the basis of his work on the unit cell, his law of rational indices, the most important concept in crystallography. This law involves the use of crystallographic axes, a set of axes to which the geometry of the crystal can be referred (they are often known as axes of reference). Faces make intercepts on these axes and various notation systems have been devised to describe the slope of faces and the relationships of the various faces to each other.

The axes of reference are any three straight lines parallel to actual or possible edges of the crystal. They are not in the same plane and may not always be at right angles to one another. Once these have been chosen the next choice involves a unit plane which defines the units of measurement used when measuring along each of the three axes of reference. This is known as the parametral plane as it defines the crystal's parameters. It will be parallel to any face on the crystal but it must cut all three axes which may be produced to accomplish this.

Crystallographers recognize seven crystal systems which are defined in terms of axes of reference (crystallographic axes).

1. *CUBIC* (isometric): three axes of equal length and intersecting at right angles (Fig. 2.1).
2. *TETRAGONAL*: three axes, two equal in length, the third longer or shorter, meeting at right angles (Fig. 2.2).
3. *ORTHORHOMBIC*: three axes, none equal in length, meeting at right angles (Fig. 2.3).
4. *MONOCLINIC*: three axes, none equal in length, one being at right angles to the other two which do not meet at right angles (Fig. 2.4).
5. *TRICLINIC* (anorthic): three axes, none equal in length and none meeting at right angles (Fig. 2.5).
6. *HEXAGONAL*: four axes, three equal in length and meeting at

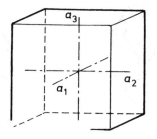

Figure 2.1 Cube.

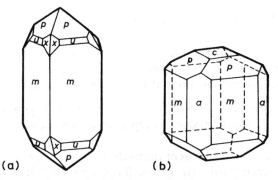

Figure 2.2 Tetragonal crystals: (a) Zircon; (b) Idocrase.

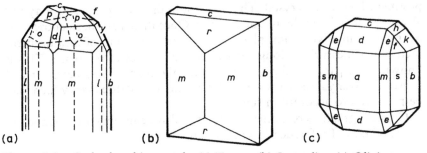

Figure 2.3   Orthorhombic crystals: (a) Topaz; (b) Staurolite; (c) Olivine.

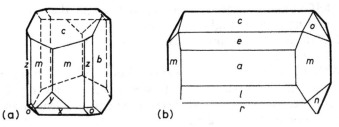

Figure 2.4   Monoclinic crystals: (a) Orthoclase; (b) Epidote.

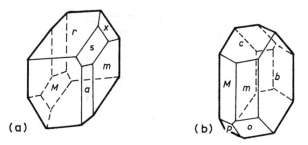

Figure 2.5   Triclinic crystals: (a) Axinite; (b) Albite.

60 degrees; the fourth at right angles to the three and longer or shorter (Fig. 2.6).

7. *TRIGONAL:* similar to the hexagonal, but this system has only three planes of symmetry and one trigonal and three digonal axes of symmetry; the hexagonal system has seven of both (Figs 2.7 and 2.8).

The seven systems can be divided further into 32 crystal classes, which can be subdivided into 230 space groups, not all of which are represented by a known crystal.

Although crystals of the gem minerals do not always show the highest symmetry of their particular system (holosymmetry) it is convenient to discuss the systems further as if all crystals were in fact holosymmetric.

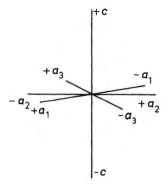

Figure 2.6 Hexagonal crystal axes.

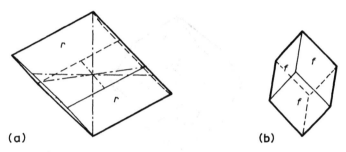

(a) **(b)**

Figure 2.7 Rhombohedrons.

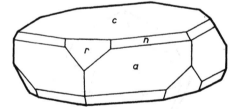

Figure 2.8 Corundum crystal (tabular: typical of ruby).

When a crystal consists entirely of one form (octahedron, cube) it is said to have a closed form; when two or more forms are needed to complete the crystal it has an open form. Closed forms are most common in the cubic system where the cube, octahedron (Fig. 2.9), rhombic dodecahedron (Fig. 2.10) and icositetrahedron (Fig. 2.11) (with combinations) are seen most often.

In the tetragonal system the prism form (with faces meeting in parallel edges) gives the tetragonal or square prism. Both need to be completed by another form, either that with the plane normal to the vertical axis and

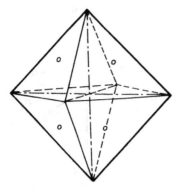

Figure 2.9   Octahedron.

Figure 2.10   Rhombic dodecahedron.

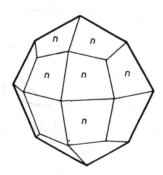

Figure 2.11   Icositetrahedron.

parallel to the lateral axes (the pinacoid, or the basal pinacoid when forming
the bottom of the crystal) or a pointed end (termination) consisting of four
triangular faces, the pyramid form. Another common tetragonal form is
the tetragonal bipyramid, resembling the octahedron but with one longer
or shorter dimension. If a horizontal crystallographic axis emerges from the
centre of each of the four prism faces the form is known as a second order

prism; in the first order prism the axes emerge at the centres of the vertical edges. Another possible form is the ditetragonal bipyramid in which there are eight pyramid faces above and below the centre.

Special forms of the orthorhombic system include the dome (which intersects two of the crystal axes but is parallel to the third – a chisel-like face with four sides often indicates a dome) and the pinacoid. In topaz the prominent basal pinacoid is a direction of easy cleavage. The orthorhombic system contains the second greatest number of crystals, after the monoclinic system. In the monoclinic system the general form is a prism though pinacoids are also found. About half of the total of known mineral species belong to this class. The triclinic system, with the lowest degree of symmetry of any of the systems, has no special forms, the general form being the pinacoid (three representatives at least must be present on a crystal since each form consists of a pair of parallel faces). In the hexagonal system the special forms are the pinacoid, prism and hexagonal bipyramid and in the trigonal system similar forms are also found with the addition of rhombohedra. The ditrigonal bipyramidal class has as its only member the gemstone benitoite. Scalenohedra are found in some trigonal minerals.

Twinning

Many crystals are composed of two related individuals, each being the reflection of the other across a central plane which is known as the 'twinning plane'. Other individuals may be brought into a parallel relationship with each other by a rotation of 180 degrees about an axis called the twinning axis, which may be an edge or the perpendicular to a face. The composite crystals are known as twins (twinned diamonds are known as macles). The twinning plane is at right angles to the twinning axis and is usually the same as the composition plane, the plane common to the two individuals. Some crystals obey more than one twinning law; calcite has two common and two less common directions of twinning. Diamond and spinel form common twins in which the two individuals in contact may resemble the wings of a butterfly before being rotated to form an octahedron. Figure 2.12 shows a number of examples of twinned crystals.

Crystals which protrude into each other are known as interpenetrant twins. Minerals such as fluorite show cubes protruding from other cubes but interpenetrant twins are not always so easy to see; some quartz crystals, for example, are twinned but show no signs of it on the surface. The gem mineral chrysoberyl shows repeated twinning giving the impression of a higher degree of symmetry than the one it really has. Twinned orthorhombic crystals arranged to give an overall hexagonal shape are familiar in chrysoberyl (pseudo-hexagonal twinning) and only the

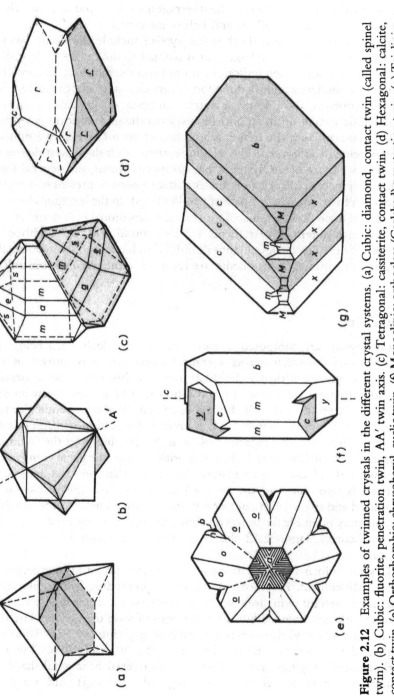

Figure 2.12 Examples of twinned crystals in the different crystal systems. (a) Cubic: diamond, contact twin. (b) Cubic: fluorite, penetration twin, AA' twin axis. (c) Tetragonal: cassiterite, contact twin. (d) Hexagonal: calcite, contact twin. (e) Orthorhombic: chrysoberyl, cyclic twin. (f) Monoclinic: orthoclase (Carlsbad) penetration twin. (g) Triclinic: albite, polysynthetic contact twins.

presence of re-entrant angles, resembling notches on the outer edge, give the true situation away. Re-entrant angles in a crystal are a sure sign of twinning. Where the individuals composing the twin are parallel thin lamellae, and are repeated many times, the twinning is known as polysynthetic and often manifests itself as a series of parallel striations on the surface. This effect is common in ruby.

Although the individual crystals forming a twin may look as if they had grown and rotated at the same time this is not really the case; the individuals have grown in this way from the start, the twinning arising at some stage from a change in growth circumstances. Due to pressure some individual portions of the twin may take an inverted structure where one portion is reversed with respect to the other; this is known as secondary twinning since in this particular case movement has taken place since the original growth. Crystals of this kind are likely to split along the planes of the lamellae and are said to have a plane of parting. This is seen prominently in corundum in which the plane of parting is parallel to the basal face. The splitting of boules of Verneuil-grown synthetic corundum arises from other causes connected with cooling and with the composition of the feed powder.

X-ray diffraction in crystals

The structure of crystals is routinely established using what is known as the Bragg law, following on the work of Max von Laue who discovered in 1912 that X-rays could be used to study crystals. X-ray beams passed through a crystal will be diffracted from atomic planes but act as though they were reflected. The reflections are not continuous but obey Bragg's law which states that $n\lambda = 2d \sin\theta$, where n is a whole number, λ is the wavelength, d is the distance between successive parallel lattice planes and θ is the angle at which the X-ray beam strikes a given atomic plane. When the beam enters the crystal it reflects from a number of parallel planes; the reflections needed for recording have to reinforce each other and thus be in phase. Various techniques are used in X-ray diffraction but they all involve rotation of the specimen so that all atomic planes are able to show reflections. The pattern seen is a function of the position of the atoms in the crystals. Some results are recorded and processed electronically and in other instances the reflections show up as curved bands on a film. These are obtained with the powder method in which small scrapings are taken from the specimen and are then placed in the centre of a cylindrical camera. Interpretation of results for all methods involves comparison with known standards.

2.7   PROPERTIES OF CRYSTALS

A number of properties shown by gemstones depend directly on their crystal structure. Some are useful in field identification (hardness, lustre) and others need to be investigated by instruments, sometimes after a face has been polished or the specimen otherwise prepared.

Cleavage

Cleavage is the way some crystals split along planes related to the molecular structure of the mineral and parallel to atomic planes (possible crystal faces). Some crystals have more than one direction of cleavage: diamond has four, parallel to the faces of the octahedron (octahedral cleavage); orthoclase has two at right angles to each other; sphalerite has six. Lapidaries and diamond cutters may find cleavage a help if unsightly inclusions can be removed by cleaving, as in diamond; alternatively, the cleavage may make cutting hazardous, as in kunzite. Cleavage may be described as perfect if the surfaces left are almost plane; other qualifications being good, imperfect, etc.

In the field, cleavage is one of the features that help in identification – it can be indicated by interference colours, rainbow-like markings showing that cracks or incipient cleavages exist at the surface or inside the stone if transparent. Some crystals show parting rather than cleavage; this takes place along planes of weakness rather than along significant crystallo-graphic directions. In corundum parting occurs along directions of lamellar twinning. Parting occurs along discrete twin planes with a fair distance between them whereas cleavage planes are spaced only interatomic distances apart.

Fracture

If a mineral breaks other than in cleavage or parting directions it is said to 'fracture'. Most minerals show a conchoidal (shell-like) fracture; the glasses show this type of fracture particularly well. Other types of fracture are described as splintery (the jade minerals), even, uneven or rough.

Electrical conductivity

The electrical conductivity of a crystal is related to its type of chemical bonding. The gem minerals are almost always poor electrical conductors (but see under diamond) since they do not possess metallic bonding. Crystals with no centre of symmetry and with polar axes are pyroelectric, that is a temperature change induces an electrical charge of opposite sign at either end. Tourmaline, which shows this property strongly, attracts dust in brightly lit shop windows as a result of the heat and subsequent induction of polarity. Piezoelectricity, when the same effect is produced by increase in pressure, is shown by tourmaline and by quartz; thus, quartz has become important in the electronics industry, for example in watch manufacture. A few gem materials develop frictional electricity when rubbed. Amber (and, unfortunately, most of its imitations) will pick up small fragments of paper when rubbed.

I have already said that crystals do not always show all the symmetry desired of their system and class if they have not grown in ideal conditions. Even where this ideal growth is attainable may minerals never show outward crystal form since they are composed of a multitude of single crystals too small for any type of magnification to detect them individually. These are called crypto-crystalline minerals and their habit can be described as massive. Their structure can be determined by X-rays. Among the more important examples of this phenomenon are:

- Turquoise, which has been found in minute triclinic crystals at Lynch, Virginia, USA (but masses are far more common).
- Quartz: rose quartz single crystals are known but masses are far more common; the granular varieties of quartz such as the jaspers are distinct from the chalcedony group, the members of which tend to be fibrous.

2.8 GEM CRYSTALS

Crystals of some of the gem minerals are described here for the purpose of identification in the field. Diamond usually takes the form of octahedra, cubes or twins but few people are likely to find diamonds for themselves; even the purchase of rough crystals is difficult sometimes. Crystals of the other gemstones are more easily obtained and studied because many of them are not of cutting quality, but still display the crystalline features shared by their more expensive fellows. Before leaving diamond, however, it is worth noting that a student, when presented with two colourless octahedra and having only a lens with which to examine the crystals, will find that the octahedron faces of the diamond show characteristic triangular markings known as trigons, the points taking the reverse direction to the point of the octahedron. A similar feature can be seen on some flux-grown synthetic corundrum. Octahedra of other minerals will not show these markings.

Corundum: the habit is most often the hexagonal prism but may also be steeply pyramidal, the bipyramid being a common habit of both ruby and sapphire. Burma ruby is rhombohedral and Montana sapphire is tabular. Ruby from East Africa is bipyramidal and Sri Lanka sapphire is bipyramidal. Etch figures when found reflect trigonal symmetry as do the rhombohedra seen on tabular crystals. Horizontal striations are found on pyramidal and prism faces; these are due to oscillations between forms.

Beryl: occurs most commonly in hexagonal prisms, the faces vertically striated. The first-order prism and basal pinacoid are the only forms. Some aquamarine or other beryl crystals that have been able to grow into space have pyramidal terminations and may show a number of faces.

Topaz: prismatic crystals with terminations consisting of pyramid and dome forms are common; the basal pinacoid is a cleavage plane, so doubly terminated crystals are rare. Both first- and second-order prisms may be present.

Chrysoberyl: simple crystals, usually tabular or short prismatic, but more commonly pseudo-hexagonal twins known as 'trillings'. They are usually flattened at right angles to the composition plane.

Spinel: the habit is commonly octahedral and twinning by the spinel law where the individuals are flattened parallel to the composition plane, is very common.

Tourmaline: vertically striated prismatic crystals with a rounded triangular cross-section are the commonest habit. The cross-section is formed from three faces of the trigonal prism and six faces of the second-order hexagonal prism. No other mineral shows this shape of cross-section. Tourmaline does not have a centre of symmetry so that both ends of the crystal can be differently terminated. Crystals are pyroelectric for this reason.

Quartz: crystals belong to a low symmetry class of the trigonal system and have neither planes nor a centre of symmetry. Prism and rhombohedral forms are found; the prism faces being horizontally striated (a useful identification feature) and terminated by a combination of positive and negative rhombohedrons which may be equally developed so that the crystal may be mistaken for a hexagonal dipyramid. The tetrahedral unit cells of the quartz are arranged in a helical way, spiralling to the left in some crystals and to the right in others. Quartz is almost always twinned according to two different laws: the Brazil twin in which left- and right-handed crystals interpenetrate and the Dauphiné twin in which two right- or left-handed crystals interpenetrate. Attractive crystals of quartz, sometimes in two varieties, may be found in Japan twins in which the two individuals form a V shape.

Zircon: crystals show prismatic (usually second-order) and pyramid forms. Some reddish-brown crystals from Nigeria have attractive pyramidal terminations with a number of faces.

Peridot: a variety of the mineral olivine, peridot crystals occur as forms showing the three pinacoids, three prisms and a dipyramid (see page 26). They are sometimes flattened.

Garnet: the garnet group shows the common forms of rhombic dodecahedra and icositetrahedra and very frequently a combination of the two. Cubes are much less common, being most frequently seen in the black varieties.

Feldspar: orthoclase and microcline are monoclinic and triclinic respectively and may resemble each other superficially. Crystals are twinned very frequently. Orthoclase is usually short prismatic with some elongation and flattening. Twinning shows up best under crossed polars when the specimen is transparent or in thin section. The plagioclase feldspars are triclinic and show polysynthetic twinning. Rough crystals show close striations on the basal cleavage.

3 Gem testing

3.1 HARDNESS

Hardness is the power of resistance offered by a substance to scratching. A table of relative scratch hardness was devised in 1824 by Friedrich Mohs; the table is not quantitative but will indicate when a mineral can be scratched by a sharp edge or point of others above it in the scale. Diamond is placed at 10 and the test is diagnostic only for this species.

10 Diamond
 9 Corundum
 8 Topaz
 7 Quartz
 6 Feldspar (orthoclase)
 5 Apatite
 4 Fluorite
 3 Calcite
 2 Gypsum
 1 Talc

A mineral can scatch any below it in the scale, and two of the same hardness can scratch each other. On this scale a finger-nail would be placed near 2.5, a common type of coin at about 3, a penknife blade just over 5 and window glass about 5.5. Substances placed lower than 7 are subject to abrasion by airborne dust which consists largely of silica grains; opal (non-crystalline silica), for example, used as a ring stone, will inevitably need repolishing after years of wear.

Hardness testing should never be carried out on cut stones, only on large objects where a tiny mark on the base will not be noticed, and, of course, on mineral specimens in the field. Marking is done with a hardness pencil, a piece of wood with a sharp piece of a known hardness set in it. The point is drawn across the specimen and the mark examined with a lens or under the microscope to make sure that a true scratch has been made. The surface around the mark should be wiped first.

Some crystals can show directional hardness; kyanite, for example, is 5 on the scale in one direction and 7 in a direction at right angles to the first. The variation in the hardness of diamond (planes parallel to the cube face being less hard than those parallel to an octahedron face) is important to polishers. Those who require a linear scale of hardness must turn to one of the other hardness tables. Many of these are based on indentation rather than scratch resistance. In such tables diamond can be seen to be four times harder than corundum, which itself is four times harder than quartz. Many minerals show a considerable degree of toughness even when they are not of great hardness. This property is notable in the jade minerals, jadeite and nephrite; it is called 'tenacity' but on the whole it is not significant in gem testing.

3.2 GEMSTONES AND LIGHT

Lustre and diaphaneity

The importance of a gemstone's response to light is self-evident and both transmitted and reflected light have their part to play in establishing the final appearance of the stone. The lustre is a function of light reflected from the surface and is more important when crystals are being tested in the field as the lustre of most transparent faceted stones does not vary much. Diamond crystals show a particular 'greasy' lustre but the cut diamond is traditionally described as adamantine. A glassy lustre (vitreous) character-izes most of the remainder of faceted stones. Zircon and sphalerite may be described as resinous amd cleavage faces show a pearly lustre. Light transmitted through a stone shows it to be transparent when the outline of an object is distinctly seen through it, and translucent when light is transmitted but nothing can be detected other than the light. Opaque stones pass no light at all. In many materials light will not pass through the centre but will pass through a thin edge.

Reflection and refraction

When a ray of white light passes from an optically rarer to an optically denser medium (as from air to a gemstone) it obeys two laws. The laws of reflection state that the angle of incidence equals the angle of reflection and that the incident ray, the reflected ray and the normal to the point of incidence lie in the same plane. That part of the incident ray which passes into the stone is refracted towards the normal to the surface making an angle of refraction which is always smaller than the angle of incidence; the amount of refraction is a function of the difference in optical density of stone and surrounding medium (air in most cases but see below, the refractometer) and of the obliquity of the incident ray.

Refractive index The laws of refraction were first published by Descartes but were discovered by Snell (1591–1626). The law states that the refracted ray lies in the plane containing the incident ray and the normal to the plane surface separating the two media and that the angle r which the refracted ray makes with the normal is related to the angle i which the incident ray makes with the same direction by the equation

$$n \sin i = n' \sin r$$

where n and n' are constants for the two media which are known as the refractive indices (RI). Figure 3.1 shows the refraction of light on entering and leaving a gemstone.

Light travels with a velocity of about 299 776 km s^{-1} in a vacuum and its velocity is reduced appreciably when passing through a transparent medium, the amount of reduction depending on the optical density of that medium. Thus, the RI can be expressed as the ratio between the velocity of light in a vacuum and the velocity in the denser medium, or

$$RI = \frac{\text{velocity of light in vacuum (taken as 1)}}{\text{velocity of light in stone}}$$

The velocity of light in air is so close to that in a vacuum (it is 0.997 in air) that for most purposes it is taken as unity. Measured in air, therefore, RI = $1/v$ where v is the velocity (so the RI is the reciprocal of the velocity).

The phenomenon of dispersion not only gives colourless transparent

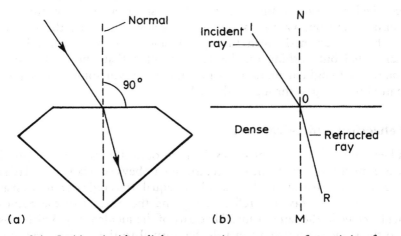

Figure 3.1 In (a) an incident light ray entering a gemstone from air is refracted towards the normal. A ray leaving the gemstone and passing into air will be refracted away from normal. The general case is shown in (b), where the refractive index of an optically dense material in a less dense medium (such as air) can be calculated by dividing the sine of angle ION by the sine of angle MOR.

stones their play of spectrum colour, or fire, it also affects the way in which their RI is measured. Where white light is used to measure RI we find that each wavelength making up white light is refracted by a different amount so that a standard wavelength has to be established for measurements. This is the wavelength of sodium light at 589.3 nm (1 nm $= 10^{-9}$ m) and instruments measuring RI are calibrated for this wavelength.

Total internal reflection Light travelling from an optically denser to an optically rarer medium as from a transparent gemstone into air is refracted away from the normal to the interface between the two media. Part of the light is reflected at the interface and the angle made by the incident light with the normal which allows the refracted ray to pass along the interface is called the critical angle. Rays reaching the interface at angles greater than the critical angle cannot escape from the denser to the rarer medium but instead are reflected back into the denser medium. This is known as total internal reflection and is of great importance both in the appearance of stones and in gem testing.

The refractometer The refractometer depends upon total internal reflection and the critical angle is used as the basis for measuring the RI of a stone. The instrument uses a truncated prism of a dense lead glass (RI 1.962) upon the upper surface of which the table facet of the gemstone to be tested is placed (Figs 3.2 and 3.3). Optical contact is made by a liquid whose RI must be higher than that of the stone tested but lower than that of the glass. Light passes through lenses via the glass and stone to fall upon a scale. Part of the light is reflected and part refracted, provided that the angles are less than the critical angle for the specimen. Those rays which arrive at the interface between stone and glass at angles greater than the critical angle pass back into the instrument and illuminate the scale so that there is a cut-off between the higher (light) part of the scale and the lower (dark) part. The cut-off is read off directly as the RI of the stone. Provided that the RI of the glass and its critical angle are known the RI of the stone can easily be established as

$$\text{RI (stone)} = \text{RI (glass)} \times \sin \text{critical angle (glass)}$$

If the critical angle of a stone is 58° 48′, the sine of this angle is 0.855; if the RI of the glass is taken as 1.90, the RI of the stone is 0.855 × 1.90 = 1.625. If white light is used for testing, the RI place on the scale will be shown by a band of spectrum colours: the RI should be read off at the point at which the yellow and green meet. If monochromatic light is used (sodium light) the RI will be seen as a fairly sharp shadow-edge dividing the scale into light and dark portions. Many gemmologists find that a test with white light first is useful since it is easier to see a spectrum than a change in intensity. When the position of the RI is known approximately from the

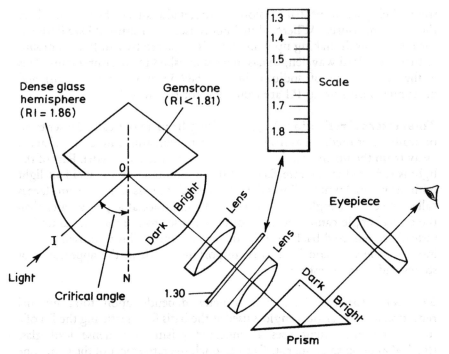

Figure 3.2 Components and ray path for the critical angle refractometer.

spectrum that area can then be searched carefully using sodium light for greater accuracy. Stones with high dispersion will show a less broad spectrum on a glass refractometer than stones of low dispersion. Using spinel (synthetic white material) instead of glass shortens the scale but increases resolution, also giving a broadish spectrum for highly dispersive stones. The scale on the refractometer using dense glass ranges from 1.30 to 1.86 and that of the spinel instrument from 1.30 to 1.68. Obviously, stones with RI higher than these figures cannot be tested on the glass or spinel instruments and refractometers using sphalerite or diamond have been made experimentally and commercially. The nature of the contact liquid limits the usefulness of such high-reading instruments as liquids with high RI are dangerous or unpleasant to use. The contact liquid in normal use (RI = 1.81) is made up of methylene iodide with tetraiodoethylene and dissolved sulphur. A faint line can be seen at the 1.81 position on the refractometer when liquid is on the glass.

To test a stone on the refractometer a small drop of the contact liquid is placed on the polished upper side of the glass and the specimen gently slid over the drop to make optical contact. The lid is closed and the shadow-edge observed through the focusing eyepiece. Care should be taken to clean instrument and specimen before and after use since the contact liquid is

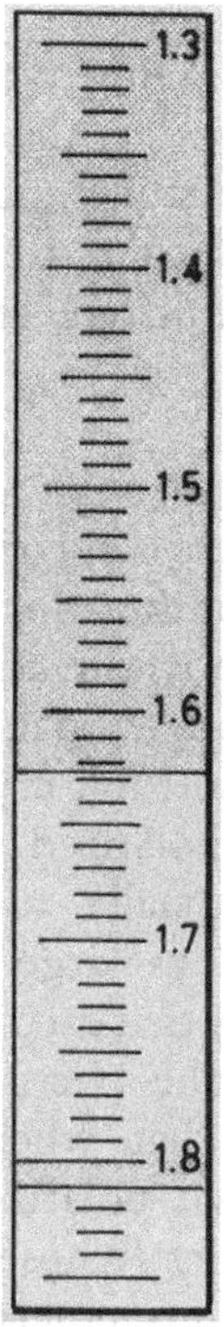

Figure 3.3 Scale as seen through the eyepiece of a refractometer with shadow edge at 1.625. The faint edge at 1.81 indicates the refractive index of the liquid.

corrosive. Most types of refractometer have a scale which can easily be read, but a model made in the United Kingdom uses a calibrated dial on the side of the instrument to make life easier.

The relatively high cost and limited availability of the standard sodium light source used for the refractometer has made the use of alternatives an important matter. One of the most recent is a set of LEDs. One of the drawbacks to the use of such simple devices as a yellow filter is that the transmission loss makes the use of a high-powered light source necessary and to some extent this is also a problem with the LED source. A type tested by Read (1983) was the Vitality, type CM4-582B which gave an emission peak centred on 585 nm with a bandwidth from 560–640 nm. Powered from the mains the LEDs give an adequate light though battery operation is also possible provided several are connected in parallel.

Double refraction So far we have considered (by implication) only those stones which show a single shadow-edge on the refractometer. These stones belong to the cubic system where there is no splitting of the incident ray or are amorphous. They are isotropic and include liquids and gases as well as glass and opal. Stones belonging to the other crystal systems split the incident ray into two when it enters them, the two rays proceeding at different velocities through the stone and therefore being refracted by different amounts. This group of stones are anisotropic. The splitting up of the incident ray into two by anisotropic crystals is called birefringence or double refraction (DR).

Polarized light Light passing through an anisotropic stone is not only split into two rays but each of these is polarized in a direction at right angles to the direction of polarization of the other. Polarized light vibrates in one direction only at right angles to its direction of travel; unpolarized light vibrates in all directions at right angles to its direction of travel. If one of the two rays passing through an anisotropic crystal can be eliminated the other will leave the crystal as plane polarized light (i.e. vibrating in a single plane). The Nicol prism, originally used to obtain polarized light, employed Iceland spar (calcite in large clear cleavage masses) which exhibits birefringence so strongly that each ray gives a separate image when a spot is viewed through it. The Nicol prism is made by sawing an elongated cleavage fragment so that, when the sections are rejoined with Canada balsam, one of the two polarized rays is totally reflected on reaching the balsam and the other (whose RI is very close to that of the balsam) passes through the prism to emerge as plane polarized light. The first ray is known as the ordinary ray since its RI is invariable however the crystal is rotated; the RI of the second ray varies with direction and is known as the extraordinary ray.

Polarized light can also be obtained from crystals which absorb light differentially. Tourmaline absorbs light very strongly in a direction parallel to the vertical crystal axis so that most light which passes through a dark tourmaline crystal is plane polarized. Nowadays polarized light is produced by the sheet polarizer Polaroid® in which crystals aligned on an acetate base absorb most of the light in one direction but transmit nearly all the wavelengths in the other direction as plane polarized light. Reflected light is partially polarized, particularly when reflected from a smooth non-metallic surface but this is not particularly useful in gem testing.

When anisotropic crystals are rotated between two pieces of polaroid with their vibration directions at right angles to each other, alternate light and dark bands are seen four times in a complete revolution. Isotropic stones remain dark during a complete rotation. The polariscope is a simple instrument which makes use of these properties; a light source is placed beneath the lower polaroid (the polarizer) and the specimen is viewed through the upper polaroid (the analyser). In some instruments either or both polaroids may be rotated. Many stones show anomalous double refraction, a striped effect often known as 'tabby extinction'; this effect is particularly noticeable in synthetic Verneuil–grown spinel but some garnets and diamonds also show it.

Optic axes Anisotropic crystals may be divided into two groups on the basis of their optic axes. An optic axis is a direction of single refraction in an otherwise birefringent material and is always parallel to the vertical crystal axis in the tetragonal, hexagonal and trigonal systems; in these there is only one optic axis and their group is known as uniaxial. Crystals of the other three systems have two optic axes which are not parallel to any crystallographic axes.

When light passes through a uniaxial crystal in any other direction than that of the optic axis it is split into two rays with different velocities: the ordinary ray (ω) has a constant velocity and always vibrates parallel to the basal plane. The other, the extraordinary ray (ϵ) vibrates at right angles to the ordinary ray in a plane that includes the optic axis. The greatest difference in velocity occurs when light moves at right angles to the optic axis with the extraordinary ray vibrating parallel to the vertical (c) axis. As an RI is the reciprocal of the velocity, this position gives the maximum difference between the RIs of the two rays. If the extraordinary ray reaches a higher arithmetical value than the ordinary ray the stone is said to be optically positive. Conversely, in optically negative stones, the ordinary ray has the higher value.

When light passes through a biaxial crystal it creates three mutually perpendicular vibrational directions and thus three refractive indices, which may all be different in value from each other. The lowest value is the α index, the higher value is the γ index, and the β index is an intermediate value. In optically positive crystals the β index has a value closer to the value of the α index; in negative crystals the β index is closer to the γ index. Birefringence is given as γ–α.

Behaviour of anisotropic stones on the refractometer While iso-tropic stones give a single shadow–edge, anisotropic specimens will show two in most circumstances, as already mentioned. Stones tested on the refractometers should always be rotated (or a piece of polaroid used as a filter and rotated in the eyepiece) since a chance observation along an optic axis may indicate that a stone is isotropic when it is in fact anisotropic.

When the table facet (the flat top of a cut stone) of a uniaxial stone is cut so that it is at right angles to the optic axis, the two edges representing the ordinary and extraordinary rays will remain stationary at their maximum separation throughout a complete turn. Where the optic axis lies in the same plane as the table facet and this direction lines up with the direction light takes through the refractometer then a single edge will be seen. To obtain full birefringence the stone needs to be turned away from this position, when the other edge will move to the full birefringence position twice in a complete turn. In practice stones may be cut randomly to either of the two directions described; thus, although two edges will be seen they will never coincide though moving to the full birefringence position twice in a turn.

With biaxial stones each shadow edge should be observed in turn while the stone is rotated, taking the upper edge to its highest position and the lower edge to the lowest position, subtracting the lower from the upper to get the DR. The intermediate value, β, then corresponds to either the lower value for the highest edge or the highest value for the lower edge. If the facet tested lies in the plane of one or both of the optical axes a single edge corresponding to the β index will be seen in some positions. It is important to remember that the full birefringence can be read from any facet tested in all anisotropic stones.

Gemstones which are crypto-crystalline will usually show a single shadow-edge since their individual crystals are in random orientation; some of the crypto-crystalline types of quartz such as onyx show a DR of about 0.006. This does not alter as the stone is rotated and has been called 'form birefringence'. It may be due to small fibres of quartz being embedded in material with a different RI.

The spot method of obtaining RI Cabochons and stones whose table facets are not completely flat (as is often the case with glass) can be tested by placing a very small drop of the contact liquid on the stone and then placing the stone on the refractometer glass in such a way that the drop makes the required optical contact. The scale is read from about 18 inches away: a bubble is seen, bisected horizontally with a dark portion above the bisection and a light portion below. The bisection indicates the RI value and the head should be moved up and down until the exact bisection can be seen clearly. Keeping the position of the bisection in mind, the eye is brought to the eyepiece to read the scale, which is out of focus from 18 inches.

When it proves difficult to take readings, remove the lid of the refractometer, block the window at the back of the instrument through which light normally enters, and adjust a lamp to give grazing incidence on the specimen when on the refractometer glass. (A grazing incidence is achieved by positioning the light source so that the beam passes along or across the place where specimen and instrument meet.) Bright lines should then appear at the RI position; the stone may need to be rotated for the effect to show at its best and, since the edges may move if the light is moved, a control stone should be used while setting up the experiment. Crypto-crystalline materials are made up of many minute crystals in random orientation. This structure causes the crystal to behave as an anisotropic material (which quartz is anyway, of course) but with a lower DR than a single crystal of the same material would show. Stones with an RI higher than that of the contact liquid or the glass are said to give a negative reading. Some stones stubbornly refuse to show satisfactory refractometer readings. This may be due to some eccentricity of polishing – coated stones are unlikely to give a response and this coating (which has a different RI from that of the main body of the stone) can often be removed with jeweller's rouge (powdered hematite).

Real and apparent depths Where a microscope with calibrated fine focus adjustment is available the direct measurement of refractive index is possible. The microscope should be monocular. The clean specimen is placed on the slide (with the aid of plasticine) and then viewed directly under a high-power objective which permits sharp focusing. The surface of the table is brought into focus (using polishing marks or dust particles) and a reading taken. The focus is then lowered to bring the culet into focus looking through the stone and the focal point is measured again. Then the stone is moved aside so that the surface of the slide can be brought into focus and a third reading obtained. The formula

$$RI = \frac{\text{real depth}}{\text{apparent depth}}$$

is used, the apparent depth being through the stone. Results may be accurate within 0.02 but only the ordinary ray of uniaxial stones or the single index of isotropic stones can be accurately read. Anderson mentions that the depth of an inclusion below the facet of a stone can be estimated by focusing first on the facet surface, then on the inclusion, and multiplying the result by the RI of the stone.

Refractive indices are conventionally quoted to three significant figures where appropriate, as are figures for double refraction. Stones which give a single shadow edge in the range 1.50 to 1.70 are very likely to be glass since only pollucite and rhodizite among natural transparent crystalline materials of gem quality are singly refractive in this range. Stones of a high dispersion are likely to show negative readings due to their high RI but a broad band of spectrum colours on a spinel refractometer suggests glass.

Immersion methods

Stones which for various reasons cannot be tested on the refractometer or by the real and apparent depth method may be immersed in a liquid of known RI and the behaviour of the outline and the facet edges observed. If a stone is immersed in a liquid of closely matching RI the stone will virtually disappear like ice in water, as with fire opal immersed in carbon tetrachloride or beryl in benzyl benzoate. The liquids with most applicability are probably bromoform with an RI of 1.59 and methylene iodide with an RI of 1.745. The immersion method is particularly useful for jewellery set with many small stones (perhaps diamonds) into which a synthetic spinel has intruded. The stone should be unmounted and immersed in a liquid over a sheet of white paper, the whole arrangement being illuminated by a single overhead light (preferably in a dark room). If the stones have a higher RI than the liquid they cast a dark-rimmed shadow on the paper; if the liquid has the higher RI the outline will be bright and the facet edges dark. The breadth of the outline gives some idea of how different the RIs of stone and liquid are. If there is a coloured margin the stone and liquid have closely matched RIs. Various devices have been made to view these effects from below.

The immersion contact photograph is a useful permanent record of particularly important stones. It is important to remember that negative film records the reverse effect to that outlined above: some of the effects give particularly interesting and attractive photographs; a bonus may be the appearance of curved growth lines in flame-fusion synthetic stones, although these are not always visible to the unaided eye.

Large objects can be tested by this method if a small fragment can be obtained without damaging the specimen. The fragment can then be immersed and the relative refractive indices of the liquid and fragment determined with the aid of the microscope. The piece is placed on the microscope slide and a drop of the chosen liquid placed on the piece. The relief can be observed as already described but further use can be made of the microscope provided it has a sub-stage condenser (in fact this is not vital but its presence helps). When the condenser is available it is lowered and the microscope focused on the edge of the fragment. The focus is then raised and lowered above and below this position while the edge of the specimen is closely observed. If the focus is raised a bright line of light passes into the medium of higher refractive index. If the focus is lowered the bright line passes into the medium of lower refractive index. The effect is not very easy to see and again it is better to work in a darkened room if possible.

Independently of each other a number of workers have found that a slight modification of this technique gives good results. The sub-stage condenser is lowered with the diaphragm partially closed. As the focus is lowered towards the stone the facet edges either appear first white and then dark or the other way round. If the facet edges are white first then the stone has the higher refractive index. The stone should be examined with the table facet down and the pavilion junction facets should be the ones observed. Anderson (1983), who gives an excellent account of all these tests suggests that ruby and red spinel can easily be distinguished from one another when immersed in methylene iodide (ruby has an RI of approximately 1.76 compared to spinel's 1.72).

This test is often known as the Becke line test and although it is more often used in the identification of minerals than gemstone fragments it can be used to determine whether a fair-sized mineral inclusion in a gemstone has a higher or lower RI than its host. In any case, immersion tells the gemmologist a good deal about a specimen and the lapidary benefits from the colour distribution showing up so well.

Minimum deviation method

The most accurate method for establishing the refractive index of a gemstone is by the method of minimum deviation, using the table spectrometer (goniometer). This instrument has a fixed collimator, the slit of which is adjusted to produce as fine a beam of parallel light as possible by focusing the slit on the cross-wires of the eyepiece of a radially mounted telescope. The stone, which must have facets which behave as the two faces of a prism, is placed on a platform so that the prism edges are exactly vertical and so that the light from the collimator falls on these adjacent faces. The angle between the faces is measured by turning the telescope until the image of the collimator slit, reflected from one of the prism faces, is centred in the cross wires. The telescope position is read on a vernier to scale (V). The telescope is then turned to focus the image from the other prism face and a second reading (W) is taken. The prism angle (A) is half the difference between V and W

$$A = \frac{V-W}{2}$$

The stone is now taken off the platform and a direct reading (X) of the collimator slit as seen through the telescope is taken. The stone is then replaced so that only one of its prism faces takes the beam from the collimator. The prism refracts the beam and the telescope is moved so that it picks up the refracted image (the slit will be seen as a spectrum and the cross wires are focused on its red component). The stone is now rotated until the refracted spectrum moves towards the line of light from the collimator; this image is followed by rotating the telescope until the image appears to stop moving and then reverse its direction. At the position at which the image ceases to move the telescope is adjusted so that the cross-wires once again coincide with the red part of the spectrum; the telescope vernier is read to give the value Y. The angle of minimum deviation is found by subtracting X from Y to give B. The refractive index is calculated using the formula

$$RI = \frac{\sin 0.5\,(A+B)}{\sin 0.5\,A}.$$

It is normally possible to obtain only the ordinary ray RI of uniaxial stones. With biaxial stones the reading may lie anywhere between the α and γ readings. Dispersion is measured with this instrument using light of appropriate colours or filters to obtain the RI's at two different wavelengths (686.7 and 430.8 nm). It is not possible to use a refractometer to obtain dispersion measurements because it is calibrated for sodium light only.

Dichroism and pleochroism

When the two rays emerging from a coloured anisotropic crystal appear to be of different colours the crystal is said to be dichroic. Strictly speaking, the term dichroism should be reserved for uniaxial stones and pleochroism used for biaxial stones, since more than two colours can be displayed in the latter case (if the specimen is turned). In fact, the term pleochroism is used for all examples. Each of the emergent rays is plane-polarized with respect to the other and this property, particularly as displayed by dark brown tourmaline, has been utilized in gem testing. Dark brown tourmaline is virtually opaque to the ordinary ray save in very thin sections. If a plate is cut parallel to the main crystallographic axis it will cause light transmitted through it to become plane-polarized. Until the Nicol prism was invented this was the only way in which polarized light could be obtained. Many gemstones show such pronounced pleochroism that it can easily be detected by the unaided eye; to observe the property in other coloured anisotropic stones the dichroscope is needed or, alternatively, the polarizing microscope can be called into service.

The dichroscope The dichroscope is a short metal or plastic tube with an eyepiece at one end and a small window at the other. Light entering the window passes to the eye through a rhomb of optical calcite (Iceland spar) to each end of which a glass prism is cemented (Fig. 3.4). The calcite causes the image of the window to be doubled so that two windows are seen side by side. In some instruments there is an attachment at the window end for holding and rotating the stone, which must be done so that the possibility of a chance observation along an optic axis does not cause a serious error of identification.

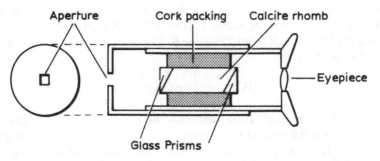

Figure 3.4 Diagram showing the construction of the prism-type dichroscope.

The light passing through the instrument to the observer has vibration directions of its own and only when these coincide with those from the specimen can the two rays be seen in their true colours. When the two sets of vibration directions, from the specimen and from the light forming the two images of the window, are at 45 degrees to each other the two rays will appear as the same colour. As long as the observer is aware of this danger there will be little trouble but it should also be borne in mind that only fairly deep-coloured stones should be tested since differences in the dichroic colours of pale stones may be lost or misinterpreted due to inexperience or to the effects of the polarization of daylight.

Operating the dichroscope is quite easy, although bright daylight or a strong single source of light is needed. Examining the plate of tourmaline mentioned above and rotating it, two positions are seen in which the images of the window are of the same colour. At the two alternate positions there are two contrasting tints side by side (light and dark of the same colour in the case of tourmaline). If the plate is held at an angle to the dichroscope the colour corresponding to the ordinary ray will be unchanged but the other colour will now show as a shade intermediate between its own colour and that of the ordinary ray. In biaxial stones there are three possible colours corresponding to the three principal optical directions.

Although dichroism cannot be said to be a determinative test, it is nonetheless a very valuable guide to the gemmologist. Many, if not most, synthetic rubies for example are cut with their table facet parallel to their optic axis and thus show dichroism through their table. Most natural rubies are cut at right angles to this direction and therefore show no dichroism through the table. The three colours of alexandrite (the stones from the USSR or the flux-grown synthetics) show three colours – red, orange and green – when appropriately rotated. The blue transparent tanzanite and blue iolite are extremely pleochroic and since the amount of pleochroism can be linked with treatment (since treatment may diminish dichroism) its observation can be of some importance; both stones are biaxial. Blue sapphire shows dark and light blue and aquamarine also shows a dark (Oxford) and a light (Cambridge) blue; aquamarine's finest blue belongs to the extraordinary ray and thus can only be seen through the dichroscope or similar apparatus.

It is possible to make a simple dichroscope by cutting a Polaroid sheet so that two or four alternating sections show the differing dichroic colours. On the polarizing microscope the polarizer can be swung into place and the stage rotated to give similar effects.

The 10 × lens

The hand lens used by an expert can identify a surprising number of gemstones. Most gemstones show enough of their inclusions under a magnification of $10 \times$ for their identity or quality to be determined. This degree of magnification is in fact taken as the international standard for the clarity grading of diamond. It is important to choose a good quality lens, because even at this magnification image distortion and colour fringing may occur in a cheaper lens which is unlikely to be made up of several components. Older lenses may show quite noticeable distortion at the edge of the field. Distortion arises from the curvature of the lens, the focus of the rays passing through the lens edge being closer to the lens than the focus of the rays which pass through its centre. Colour fringes arise from the dispersion of the lens. A triplet lens is free from distortion (aplanatic) by using lenses of different radii, and is free from colour fringing (achromatic) by using two sections of different dispersion with one bi-convex and one bi-concave lens.

Apochromatic lenses (those corrected for both colour fringing and distortion) are still a little difficult to use because their focus is critical and the field of view small. With lenses of still higher magnification (up to $25 \times$) both these factors make use really difficult and in most circumstances a microscope is preferred. Gemmologists make a habit of keeping their hands together while using a lens (the stone being held in tongs) to prevent loss of focus. The lens should be close to the eye or in contact with spectacles where necessary (magnifying attachments to spectacles are available). The light should be positioned so as to illuminate the side of the stone and not shine into the eyes. It is now possible to obtain lenses with wide fields of vision. The $10 \times$ lens will show the characteristic features of various types of pearl and its imitations; diamond can be recognized from its polishing features and inclusions (and the efficacy of the cutting assessed); glass imitations can be recognized as can such natural stones as hessonite, demantoid, peridot, amethyst, tourmaline, sphene and zircon, the marked birefringence of the latter showing up well. The advantage of using the lens to spot the doubling of the back facet edges seen in highly birefringent stones is that the specimen is easier to manoeuvre than under the microscope.

The microscope

There is no doubt that the microscope is the single instrument that a gemmologist would take to the proverbial desert island. With some adaptation it can be used for a variety of purposes other than straight magnification. It is vital for identifying synthetic materials.

The most important feature to look for when choosing a microscope is the working distance – the space between stage and lower lens (objective) (Fig. 3.5). If this is very small, large stones will not fit and there will be no possibility of looking at crystals. Prospective buyers should remember that

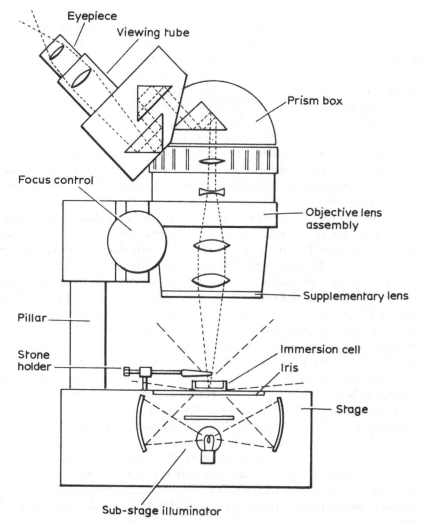

Eyepiece

Viewing tube

Prism box

Focus control

Objective lens assembly

Supplementary lens

Pillar

Stone holder

Immersion cell

Iris

Stage

Sub-stage illuminator

Figure 3.5 Sketch showing the components and ray path for one half of a typical stereo microscope. The illuminator in the sub-stage assembly is set for dark-field work. A stone holder and an immersion cell are shown on the stage.

the higher the magnification the smaller the working distance and the field of view. There is very little need for high magnification with gem testing, 15 × to 30 × being a perfectly adequate range. In the diamond trade 10 × is the statutory magnification (so that a description of a stone by one dealer to another will be taken automatically to be under 10 × magnification). Diamond microscopes with 10 × magnification are used specifically for diamond grading.

Most gemmologists use one of two kinds of binocular microscope: one type has a single objective lens system with the image shared between the two eyepieces (oculars); the other type has two objective lens systems each linked with one of the oculars (Fig. 3.5). This gives the stereo effect and is very useful in practice, although the microscopes are expensive. Some have zoom lenses but they are by no means vital.

The magnification of a microscope is calculated by multiplying the magnification of the ocular by that of the objective (those values are shown on the lens attachment). If a supplementary lens is added then the total magnification is multiplied by the magnification of that lens. Increasing the power of the objective reduces the field of view and the working distance. Changes in the power of the ocular will also reduce the field of view but not the working distance.

Various lighting arrangements are available but dark-field illumination is so useful for the identification of inclusions (which show up bright against a dark background) that it should be stipulated if an expensive instrument is to be purchased. Other attachments are useful: after dark-field illumination the facility to immerse the specimen is most desirable. This minimizes reflections from the surface, particularly at high magnifications (reflections prevent sufficient light getting into the specimen to show up its internal features). Some microscopes employ co-axial illumination through the lenses via a beam splitter but immersion is simpler and cheaper. The liquids bromoform with an RI of 1.59, monobromonaphthalene (RI 1.66) or methylene iodide (RI 1.74) are useful for immersing specimens (as for specific gravity determinations). Immersion cells can be fixed on to a horizontal microscope. The specimen can be rotated in the liquid by tongs and the height of the cell is sufficient to accommodate large specimens (this is not easy with the upright or sloping types of microscope).

A simple microscope with a calibrated fine adjustment can be used to measure refractive index by the real and apparent depth method. The petrological microscope with a rotating stage and fitted with polarizer and analyser can be used as polariscope or dichroscope (with polarizer used alone). A petrological microscope is designed primarily for the examination of rocks in thin section, so remember that the working distance needs to be larger if it is used for other purposes.

In operating the microscope be careful not to collide the objective and the specimen (or liquid). This can be obviated by first lowering the focus while observing from the side and then racking the body upwards while using the ocular.

Photomicrography is a useful way of testing stones and of providing a record of important features in major stones. Using a 35 mm camera it is not necessary to remove the ocular; the subject is brought into focus and the camera set at infinity and a wide stop, f/4 for example, is used. A single-lens reflex camera has its lens removed and can be mounted on the microscope tube by one of a variety of available attachments.

Fluorescence

When some materials are irradiated with energy of a certain wavelength, the molecules of that material are elevated to a higher energy level. One of the ways in which they can return to the 'ground level' is to emit the excess energy as light. This is termed fluorescence. Some gemstones exhibit this phenomenon in response to heat or other light of a certain wavelength. Although easy to observe, it is not always possible to identify a gemstone by fluorescence alone; it is valuable as a secondary or confirmatory test. Probably the best known fluorescent mineral is fluorite, which gives the term its name.

Crossed filters A simple method of exciting fluorescence is to examine a chromium-bearing specimen such as ruby in a blue light (such as that obtained by passing white light through a flask of copper sulphate); the stone, irradiated by the blue light, is observed through a red filter. Fluorescent light has a longer wavelength than incident light (and therefore less energy). Blue light has more energy and a shorter wavelength than red light. If a bright red glow is observed through the red filter it must have arisen by fluorescence, since the incident light had no red component (the red filter absorbs the blue light to make viewing easier). This simple but sensitive test was devised by G. G. Stokes in 1852 to detect the presence of chromium, and is known as the 'crossed filters' technique.

Any stone containing chromium (ruby, red spinel, emerald, alexandrite and pink topaz) will show a red glow by this method, as will the synthetic vanadium-bearing corundum simulating alexandrite, most synthetic blue spinels and synthetic green sapphires. Ruby and synthetic ruby show two fluorescence lines (a doublet) at very close wavelengths 692.8 and 694.2 nm, so that they appear as a single intense line. Natural red spinel shows a group of fluorescence lines, two of equal intensity being in the middle. Synthetic Verneuil red spinel shows a single intense line near 685 nm against a background of fainter lines.

Alexandrite and emerald may show red or be virtually inert according to the amount of iron present. A strong red response in an emerald suggests a man-made origin; South African and Indian stones show very little if any red. Pink topaz shows a reddish response and, surprisingly, some yellow Brazilian topaz glows reddish but not so strongly. Even more unexpected may be the response of some diamonds which will glow a faint red – as will strontium titanate, a diamond simulant. Natural black pearls will glow a dim red, which distinguishes them from those stained black with silver nitrate. The presence of rare-earth metals can also give rise to a red glow. The combination of blue incident light and a yellow filter enables the gemmologist to see orange and yellow-fluorescing stones, as with yellow sapphire from Sri Lanka. Passing white light through a copper sulphate solution with added ammonia gives a green light.

Long-wave ultraviolet Long-wave ultraviolet (UV) is obtained by passing the light from a high-pressure mercury discharge lamp through a filter, most commonly made of Wood's glass which contains cobalt and some nickel. This allows radiation at 365 nm to pass through which stimulates fluorescence. Note that it is dangerous to shine UV light into the eyes.

The crossed filter technique is especially sensitive for those stones which show a less decided fluorescence under UV, such as emerald and alexandrite, but for red spinel and ruby, long-wave UV gives a fine red fluorescence. Some green and purple specimens give a purple glow activated by the UV present in daylight (the Blue John variety does not respond however). Fine rose-pink colours are shown by aragonite crystals from Girgenti, Sicily, when under long-wave UV. When the UV is turned off the crystals phosphoresce green for a few seconds (phosphorescence is radiation emitted after the existing source has been removed). Yellow scapolite from Ontario which is sometimes cut into cabochons shows a strong yellow fluorescence, and willemite from Franklin, New Jersey, glows a bright green, from manganese impurities. Combined with red-fluorescing calcite a fine mixture of colours is given and the rock is sometimes cut into spheres which can be set to revolve in UV at gem shows to give a startling effect.

Another material with a trace of manganese, synthetic yellow-green spinel gives a bright green, somewhat lurid glow. An orange colour is shown by kunzite and some Brazilian yellow topaz (kunzite may be altered in colour by exposure) and some white zircon (the same proviso applies) and yellow apatite fluoresce yellow. Yellow and some off-white Sri Lanka sapphires glow an apricot-yellow which is characteristic. Synthetic orange sapphire glows red but only when chromium has been added for colouring purposes.

Diamond shows a variety of fluorescence effects and they cannot be accurately predicted. Most show a sky-blue colour under long-wave UV; those in the Cape series fluoresce this colour together with a yellow phosphorescence (for which the eye needs to be dark-adapted). This combination of blue fluorescence with yellow phosphorescence is only seen in diamond. Some at least of the blue fluorescence can be correlated with certain wavelengths at which emission bands can be observed; here the key band is at 415.5 nm (the position of the most significant absorption band in diamond). Broad regions of continuous emission can be seen extending to the red end of the spectrum. When a fluorescing diamond is examined with a spectrophotometer peaks representing lines or bands can be seen in the vicinity of 504 and 415.5 nm, areas where absorption can be seen with the hand spectroscope in some specimens. Reports of emission (bright) lines in fluorescing diamond are scattered throughout the literature. In some diamonds emission lines in the yellow and red can be seen and these are less sensitive to temperature, being visible at room temperature. Sir William Crookes (1909) recorded bright lines when a diamond was subjected to cathode rays; the lines were at 500, 513 and 537 nm, the stone fluorescing pale green. In another diamond lines in the yellow were seen. Anderson (1983) found that certain pink diamonds showing a strong orange or apricot-yellow fluorescence under all types of irradiation gave a fluorescence spectrum with a strong narrow line at 575 nm with other lines towards the red at 587, 589 and 617 nm, the strength of the lines decreasing as the red end of the spectrum was approached. Some stones from Sierra Leone showed a very sharp though weak line at 575 nm under blue-violet light or long-wave UV. The line was accompanied by another at 537 nm. Anderson found that a true canary-yellow diamond examined after cooling with ice showed a series of fine lines while fluorescing yellow.

Short-wave ultraviolet Short-wave UV of 253.7 nm is obtained by passing light from a low-pressure vapour lamp in a quartz tube or envelope through a Chance OX7 filter. Filters used for short-wave UV deteriorate after 100 h of use and may need replacement but otherwise the equipment is convenient – light to carry and easy to use. The lamps too may need changing after the first 100 h as their efficiency is likely to diminish by up to 20% in this time.

Short-wave UV is particularly useful for testing suspected synthetic white spinels which glow a strong bluish-white. Benitoite glows a bright blue and scheelite a bright whitish-blue. Synthetic rubies show a brighter glow than most natural rubies and most synthetic blue sapphires show a greenish glow which appears to be on the surface; sometimes this glow is more whitish. Some Sri Lanka sapphires may show a green fluorescence when virtually free from iron. When synthetic Verneuil-type sapphires are fluorescing their curved structure bands are quite easy to see. In general synthetic stones are more transparent to short-wave UV than their natural counterparts. This property can be tested by placing the stones on a contact paper immersed in water; a short-wave lamp about 15 inches above the dish can be switched on for a few seconds, the resulting print showing the various degrees of transparency to the rays. Control stones are needed for this experiment as the exposure time is critical.

X-rays Fluorescence can also be stimulated by X-rays, although such equipment must be treated with due respect: the operator must be properly shielded from the radiation. Under X-rays the effects are similar to those shown under long-wave and short-wave UV. Synthetic rubies show phosphorescence (eyes need to be dark-adapted) which may persist for 10 s or more. Orange, yellow and colourless sapphires may behave in the same way, because of chromium present. The fluorescence of diamond under X-rays is more uniform than under UV. Most specimens show a chalky blue colour; some which glow green under UV will behave similarly under X-rays. Synthetic colourless spinels will show a bright greenish fluorescence followed by a prolonged phosphorescence, while diamond will not usually phosphoresce under X-rays when its fluorescence is blue.

Hydrogrossular garnet fluoresces orange under X-rays and cultured pearls show a yellowish fluorescence. Dark-adapted eyes may be able to see the glow given off by natural salt-water pearls. The glow in cultured pearls arises from traces of manganese. The glow is at the pearl surface in natural pearls and inside in cultured specimens. Freshwater cultured pearls from Lake Biwa show a bright fluorescence with phosphorescence.

It should be remembered that all tests involving X-rays and even UV are likely to cause discolouration of the specimen which may be permanent. Although the mineral prospector will tend to use short-wave UV more than long-wave, the gemmologist will find crossed filters useful for observations at the red end of the spectrum, long-wave UV appropriate for stones with green or blue fluorescence and short-wave UV or X-rays useful for particularly elusive effects.

The sadly-defunct journal *The Gemmologist* carried a series of articles by Webster on fluorescence from June 1953 to April 1954; much of the detail is, however, to be found in his *Gems*, the fourth edition of which was published in 1983.

3.3 SPECIFIC GRAVITY

Providing that a gemstone is unset or, if rough, has no foreign material adhering to it, a specific gravity (SG) test can be useful in identification where opacity or high refractive index makes optical tests difficult. Both methods of specific gravity determination are rather slow and inclined to system error but if an average of several results is taken (especially with the hydrostatic method) then the test can give definite results. For the hydrostatic method the SG of a substance is a ratio between its weight and the weight of an equal volume of water at 4°C. It can be expressed as:

$$SG = \frac{A}{A - W} \times C$$

where A is its mass in air; W is its mass when totally immersed in water; C is a constant indicating the SG of the liquid actually used in the determination. If this is water, $C = 1$. For accurate determinations the temperature should be taken and adjustments made accordingly by consulting the appropriate tables.

If no chemical or diamond balance is available to measure value A it is not too difficult to contrive an acceptable substitute; several mineral books show how this can be done (see O'Donoghue, 1976a). Value W is obtained by weighing the stone again while it is totally immersed in water or other liquid (Fig. 3.6). The immersion can be achieved by placing the stone in a counter-weighted wire cage so placed so that it hangs in a container of the liquid without touching sides or bottom. Surface tension should be eliminated either by the use of a liquid other than water (care should be taken – some of them have dangerous properties) or by using distilled water, in which case a few drops of a wetting agent should be added. Bubbles, less likely to be found in distilled than in tap water, should be removed from cage and specimen by brushing. Clearly there are limitations to the hydrostatic method; small stones will fall out of the cage and very large ones will not fit in it. While the latter can be accommodated by using a large container (even a clean dustbin) and weighing in air and in liquid as before, small stones need to be tested by other methods.

Specimens which cannot be weighed by the hydrostatic method can be introduced into liquids of known density and their behaviour observed. Stones with a higher SG than the liquid will sink at a rate depending on the difference between the SG of stone and liquid; stones of lower SG will float on the surface. All stones which float should be pushed firmly below the surface with a glass rod as surface tension may give incorrect results. Stones which remain suspended without perceptible upward or downward movement match the liquid in SG. Liquids in common use include bromoform, $CHBr_3$ with an SG of 2.8 and methylene iodide (diodomethane), CH_2I_2 with an SG of 3.3. Acetone will dilute these liquids to values at

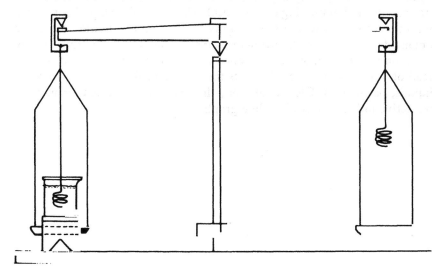

Figure 3.6 Sketch of a beam balance, showing a beaker of water supported over the left-hand pan with a gemstone holder immersed in water. The counterpoise is attached to the right-hand pan hanger.

which a specimen of known SG (indicator) will remain suspended, thus giving a useful quick reference liquid. For higher values of SG the available liquids are limited and often unpleasant to handle. Of these, the commonest is Clerici solution, an aqueous solution of thallium formate and thallium malonate in equal parts, with an SG of 4.15. It must be diluted with distilled water only. Its great advantage is its ability to float corundum at normal temperatures. Useful tests of a liquid's SG may be carried out with the assistance of a piece of rock crystal (SG 2.65) of known weight. The SG of the liquid will be

$$\frac{2.65 \times \text{loss of mass in the liquid}}{\text{mass of stone}}.$$

The SG of a variety of liquids at various temperatures are given in mineralogical texts and books of tables.

The specific gravity of a liquid can be found by using the pycnometer or specific gravity bottle which has a closely-fitting glass stopper which is pierced with a capillary opening. The mass of water contained in the bottle when completely full at a given temperature is engraved on the side – 5, 10, 25 or 50 g. The pycnometer is filled and the stopper replaced, forcing some water to escape through the capillary opening; when this process is complete the bottle is dried and checked for air bubbles. The dry mass subtracted from the mass when full gives the mass of the contained water. To find the specific gravity of an unknown liquid the now-calibrated pycnometer (completely dry) is filled via a small funnel as before and the stopper inserted, removing excess liquid. The mass of the liquid compared with that of water gives its specific gravity. The SG of a liquid can also be obtained by the use of a hydrometer (a glass weighted bulb with a narrow stem marked with an SG scale). When immersed in a liquid the SG can be read off on the scale where it meets the surface of the liquid. With some liquids, particularly Clerici solution, the SG can be plotted against the refractive index on a straight line graph.

4 Recent developments in gem testing

4.1 THERMAL CONDUCTIVITY

Thermal conductivity is a test used mainly for diamond and its simulants. At room temperature the thermal conductivity of a single crystal diamond varies from $1000 \text{ W m}^{-1}{}^{\circ}\text{C}^{-1}$ for Type I stones to $2600 \text{ W m}^{-1}{}^{\circ}\text{C}^{-1}$ for Type IIa stones. For comparison with simulants of diamond, cubic zirconia has the value $10 \text{ W m}^{-1}{}^{\circ}\text{C}^{-1}$; corundum has $40 \text{ W m}^{-1}{}^{\circ}\text{C}^{-1}$ (the highest value shown by any of the diamond simulants).

One of the early examples of the thermal conductivity tester was the Ceres Diamond Probe manufactured by the Ceres Electronics Corporation. The pen-like probe has a copper tip of 0.55 mm diameter which is in thermal contact with two miniature thermistors. (A thermistor is an element whose electrical resistance varies inversely with temperature; it is usually made of ceramic material.) One thermistor is fed with pulses of current at intervals of 1 s and provides the heating. The other is used for temperature compensation. In the 1 s intervals when no heat pulse is applied the resistance of the first thermistor is monitored to sample the temperature of the copper tip of the probe. When this is surrounded by air its temperature rises a few degrees above ambient temperature, as little heat is being conducted away. When in contact with a facet of a diamond each pulse of heat is conducted away and the temperature falls. When held against a poor heat conductor the temperature rises to an intermediate level higher than that reached by diamond.

The temperature reached by the tip is indicated on a dial which in this instrument is divided into three zones each with its own colour. The broad left-hand (red) zone indicates a simulant, the green right-hand zone indicates diamond. In between an amber zone acts as a buffer keeping the indication from jumping from simulant to diamond or, more importantly, from staying on the edge between the two. Coloured lamps flash to match the meter indications. The tip should be held as closely as possible to perpendicular on the facet and for at least 3 s so that the probe can make several sample readings while the probe temperature settles down. The instrument is calibrated for a temperature range of 12–32°C. All in all this is a useful and sensitive instrument, with which stones down to 0.3 ct can be measured. The Rayner Diamond Tester is another well-known instrument which acts through a thermistor.

The Presidium Gem Tester uses a thermocouple rather than a thermistor; it is a copper–constantan thermocouple, the two junctions separated by 20 mm. Heat is applied half-way between the junctions by a resistor fed from a constant-voltage source. The output voltage from the thermocouple is amplified and drives the needle of the meter. The central copper section of the thermocouple is extended at one of the junctions to form the tip of the probe. When both junctions are at the same temperature they generate the same voltage and as they are in series opposition the resulting voltage is zero. If heat is drawn from the probe junction when a diamond is tested the voltage developed by this section falls and the overall output voltage rises causing the needle to swing into the (green) diamond zone.

Many simple instruments measuring thermal conductivity distinguish only between diamond and its simulants without stating which these are. The Gemmologist (made by Gemtek Gemmological Instruments) will identify a number of the simulants and some other gemstones as well. The thermal conductivity of stones other than diamond covers a narrow range between 10 and 40 $W\ m^{-1}°C^{-1}$, thus the instrument has to be made quite sensitive, and the room temperature must be kept constant. The stone must be at body temperature. One useful application of such an instrument is to distinguish flux-melt emerald from natural emerald.

4.2   MEASUREMENT OF REFRACTIVE INDEX

The air-boundary refractometer devised by Yu and Healey endeavours to measure the RI of stones which cannot be measured on the refractometer in common use due to limitations posed by the contact liquid and glass. Total internal reflection in a faceted stone is the function measured in this experimental instrument. The pavilion angle θ is measured first by placing the stone table down on the platform and shining a beam of light at a pavilion facet. The operator views the light through an aperture in the cursor and moves the latter until a mirror image of the light can be seen in the pavilion facet nearest to the light source. The angle θ is then read from the lower cursor scale. Next a light source immediately beneath the stone is switched on and the cursor moved until its aperture is immediately above the stone. Where the RI of the stone is less than 2.37 the pavilion will show bright through the aperture; above 2.37 it will appear dark. For stones with an RI below 2.37 the cursor is moved toward the left-hand end of the scale until the lower girdle facets viewed through the aperture begin to turn dark (do not wait until they are completely dark). At this point the RI is read off from the position of the cursor against the appropriate section of the scale. For stones with an RI above 2.37 the cursor is moved to the right until the lower girdle facets just begin to turn bright. The accuracy of this instrument is limited to ± 0.02 compared to ± 0.01 for the standard refractometer; it cannot be used on fancy cut stones with pavilion facet angles outside the range of 37–45° or on cabochons.

A refractometer utilizing the Brewster angle has been devised by Peter Read and is shown in his book (1983). Brewster's law states that 'complete polarization of a ray of light reflected from the surface of an optically denser medium takes place when it is normal to the associated refracted ray within the medium'. Where the Brewster angle of polarization is *A* then the RI of the reflecting medium is tan*A*. This instrument allows the reflected rays to pass through a polarizing filter appropriately placed, the image being seen as a dot on a translucent screen. When a gemstone is placed over the test aperture the angle of the incident beam is adjusted for extinction of the reflected dot of light, the extinction being a direct measurement of the RI. This instrument will not detect birefringence but is not adversely affected by the quality of the test surface which often upsets reflectivity meter readings.

A simpler instrument based on the same phenomenon was developed by Yu. Here a graduated transparent scale is illuminated by a lamp whose light is diffused by a strip of translucent white plastic. Light from the scale is reflected from the gemstone's surface and viewed through an orientated polarizing filter. As the various points on the scale subtend different angles to the reflecting surface the viewer sees a dark band on the evenly illuminated scale, this band occurring at angles of viewing close to the Brewster angle for the stone.

Read (1981) describes the Riplus extended-range refractometer developed by the Swiss firm Siber and Siber and manufactured by Krüss of Hamburg. The prism is made of strontium titanate (RI 2.41) and a new contact fluid with an RI of 2.22 is made by heating a viscous non–toxic brown paste to around 40°C to make it free-flowing. The high dispersion of strontium titanate makes it necessary to use a monochromatic light source to avoid colour fringes. The paste needs to be removed from the specimen after testing with a swab of cotton wool soaked in a solvent such as methylene iodide, which can also be used to clean the prism. The working range of refractive indices extends from 1.79 to 2.21.

4.3 MEASUREMENT OF REFLECTIVITY

The limitations of the standard refractometer operating with dense glass and a contact liquid have led to the development of alternative simple methods of testing. Of these probably the most used are devices which measure the reflectivity of the surface of the specimen (rather than the refractive index). L. C. Trumper constructed a prototype instrument in 1959 but this was not developed at the time. The electronic reflectivity meters on the market today depend on the measurement of the ratio between an incident ray shone on to the surface of the stone and its reflected counterpart. A simplified version of Fresnel's equation describing the reflectivity (R) of a transparent isotropic mineral in air is used: this states that for normal incidence

$$R = \frac{I}{I_o} = \frac{(n-a)^2}{(n+a)^2}$$

I is the intensity of the reflected beam, I_o is the intensity of the incident beam, n is the RI of the specimen and a is the RI of the surrounding medium (normally air). Multiplying this value by 100 gives the percentage of perpendicular incident light that the surface of the specimen reflects back. For example, if n equals 2.42 (the RI of diamond) or 1.55 (quartz) where a equals 1, the percentages are 17 and 4.6 respectively. In practice the angles of incidence and reflection are offset from the normal since it is hard to illuminate the surface and measure the reflectivity at the same time where normal incidence is used. The state of the surface of the specimen, its birefringence where appropriate and its absorptive powers all affect the result and are not allowed for in the equation so that an actual RI is not obtainable. Furthermore, the light source operates at 930 nm, whereas normal RI's are measured at 589.3 nm. The use of the infra-red wavelength of 930 nm allows the separation of readings for diamond and for strontium titanate (the latter having the higher dispersion and thus the lower reading).

One of the first reflectivity meters introduced was the Jeweler's Eye which consists of a light emitting diode (LED) of gallium arsenide giving a beam at 930 nm, a photo-diode which detects the beam reflected from the surface of the specimen and a meter to give an indication of the reflectivity. Two scales are used for stones of high and low readings, the upper scale having places for quartz up to rutile, the lower scale for glass to garnet. Spinel and diamond are used as calibration standards. Power comes from a 9 V battery and the stone while being tested is placed under a plastic cap. The top range is the more accurate and useful (as with all dual range meters). The Diamond Eye works in the same way but has only one scale; it is calibrated in arbitrary values of lustre which are not generally accepted; the Jeweler's Eye gives the names of stones.

The Gemlusta 400X produced by Gemtronics/Partech Electronics Ltd uses both a test platform and a hand-held probe. The scale is expanded to give greater discrimination between stones with similar reflective powers and shows a difference between the reflectivity of synthetic and natural corundum; it has been found, however that these differences are due to the differences in polish or lustre between machine-cut synthetic stones and hand-polished natural ones.

The Rayner DiamondScan (Gemmological Instruments Ltd) operates with a 6 V lithium manganese dioxide battery and resembles a pen in size and shape. The length is 155 mm and diameter 18 mm. Ten red LED's are used in place of a meter. The probe tip is placed perpendicularly on the facet and the angle of the stone adjusted so that the highest possible reading is obtained (rutile, nearest the probe tip, is 9, diamond is 8). In this case the Fresnel formula is closely approximated since the probe with its reflected ray is virtually normal to the test surface. Some false high readings may occur if there is reflection from the metal settings of stones.

Some instruments have combined the functions of a reflectivity meter and thermal conductivity meter; one such is the Duogemmeter gem tester made by Krüss of Hamburg. The probe can be removed for replacement when necessary and is gold tipped, minimizing the possibility of corrosion. The measuring range is said to include stones as small as 0.015 ct (this may refer to the thermal conductivity test rather than to the reflectivity test since the latter usually accommodates stones of 0.05 to 0.1 ct at best).

The Eickhorst Gemlyzer became available during 1984; a reflectivity meter, it uses a 30-element bar graph display instead of an analogue-type meter indicator, the bar graph display being directly calibrated with the names of ten possible diamond simulants. It covers a reflectivity range from quartz to rutile in a single span. The test platform, too, represents an improvement over previous instruments in that it is a plastic cap rather than a metal surface so that no scratching can be done by the stone tested. The cap can be unscrewed to remove accumulations of dust.

The GemPro Digital Analyser made by LCE Ltd of Worthington, Minnesota, USA, computes the 'total optical activity' (TOA) of the stone tested. It can give the optical character as well as quantifying fluorescence, phosphorescence, depth of colour and transparency at visible and UV wavelengths. The makers claim that the 'total intensity of radiation from a specimen as detected in a narrow spectrum resulting from the interaction of the specimen with a source of electromagnetic energy over a very broad spectrum will differ significantly between natural and synthetic specimens of the same species'. The transparency and luminescent properties of the stone are analysed by the instrument which then generates an analogue signal proportional to the intensity of the integrated response of the stone. This gives a digital TOA reading as well as an identification symbol; this is crossed mining hammers for a natural stone and a pestle and mortar for synthetic stones. Indeterminate results are indicated by a balance arm fulcrum. The test platform is in a light-proof compartment.

After calibration with a 3 mm diameter test sphere testing can begin. Pleochroism and polarization can be measured (the use of several filters allows quite faint pleochroism to be detected); freedom from strain is clearly assessed. The cut of a stone and its brilliance are also quantifiable by measuring the effectiveness of the total internal reflection in the specimen. Peter Read's assessment of this instrument (Read, 1984) is that stones with an iron content may give readings in which natural stones can easily be confused with synthetic specimens.

4.4   SURFACE ASSESSMENT

For many years it has been known that the surface of a diamond is related to the underlying crystal structure in such a way that each face gives a unique picture when examined by the appropriate instrument. It has been suggested that this phenomenon can be used to give a 'fingerprint' of important stones, at least. The Gemprint machine, devised by the Weizmann Institute of Science, of Rehovat, Israel and manufactured by Kulso Ltd of Haifa was the first such device to come on to the market. The intention is to build up a central register of stones in North America.

In operation, a laser beam is used to produce a pattern of spots by reflection and refraction, the pattern being peculiar to each stone tested. Minute differences in polish and symmetry are shown. Laser light passes through a polarizing filter and a shutter and is then focused by a lens on to a pinhole screen. The beam passes through the hole to a second lens which gives a parallel light beam. When the specimen is placed in this beam reflections and refractions from the facets return through the lens and appear on the front side of the screen. It is necessary to place the specimen so that reflection from the table facet reaches the pinhole. There is only one position for each stone in which this can be achieved and so each subsequent test of the same stone will show the same features. If, however, the stone is recut (or at least the table repolished) the pattern of the reflected light will be altered.

The Nomarski interference contrast phenomenon can be used to obtain an unchanging record of a diamond's surface and this was utilized by the London–based Independent Gemstone Testing Laboratory in the 1970s under Roy Huddlestone who devised the method. The surface contrast of a facet is magnified until both polishing and structural features can be seen (beautiful colour pictures can be obtained by using polarized light). These structures cannot be altered by subsequent polishing.

The Gem Diamond Pen and other similar instruments work by spreading a liquid on the surface of the stone. The liquid is viscous and non–drying so that its behaviour can be studied. The surface is first cleaned (it should also be flat for best results) and the pen held vertically above it, drawing a line with gentle pressure. On diamond the line should be straight with a slight spreading; on other stones the line will appear as a series of beads. Some inks have a blue dye added to aid visibility.

4.5 X-RAY TOPOGRAPHY

The use of X-ray topography to record internal defects in the crystal lattice of a diamond has been studied by Lang and Woods of the University of Bristol. A polished stone can be identified with the crystal from which it was cut and polished stones can have their features permanently recorded, these being unaffected by subsequent re-cutting of the stone. At present the technique is too expensive to be commercially viable; it employs a Bristol-Lang Topography Camera which has a spectrometer-like table on which the angular positions of the specimen and the film holder can be adjusted relative to the source; this is usually a vertical ribbon-like beam of collimated X-rays, 10 to 100 micrometres in width. These impinge upon the specimen, giving either a section topograph or a projection topograph. In the first case the ribbon of X-rays penetrates a slice of the diamond which is so positioned that the X-rays are diffracted by the atomic layers and emerge at an angle to the beam. These diffracted X-rays fall on to a photographic plate near to the specimen but shielded from the direct beam. Defects in the crystal lattice deflect the beam more strongly and show up darker on the photograph. In the projection topograph the specimen and the film holder move backwards and forwards together through the beam of X-rays and the image produced is a parallel-beam projection of the lattice defects.

As the RI of diamond is approximately unity at the wavelength of the X-rays used in this technique, few problems are caused by surface roughness or irregularities of shape. If the technique is applied to species with a lower RI than that of diamond, the wavelength of the X-rays needs to be adjusted. The method has been used to distinguish natural from synthetic emeralds.

4.6 THE ELECTRON MICROSCOPE

The scanning electron microscope (SEM) uses a focused beam of electrons to bombard the specimen. The beam causes secondary electrons to be emitted from the sample surface. The pattern made by the positioning and intensity of these electrons is shown on a video display. As the electrons have a much shorter wavelength than visible light much smaller objects can be detected and magnifications up to 200 000 × are possible. The SEM uses an electron gun with a directly heated tungsten filament in a vacuum chamber. A high negative potential is applied to the filament and the electrons thus produced are focused on to the sample. Focusing is by means of magnetic lenses. The X-rays are sampled by a spectrometer detector and displayed on the screen as an emission spectrum recording the elements in the specimen. X-ray data is routinely analysed by a computer using known standards for the various elements as in the microprobe.

4.7 THE ELECTRON MICROPROBE

A useful description of the work of the electron microprobe in the testing of minerals (including gem minerals) is given by Dunn (1977). The ARL–SEMQ microprobe has an electron gun, a spectrophotometer, an electronic recording system and a microscope. The electron gun is placed in a cylindrical vacuum chamber and contains a tungsten filament. A negative 15–20 kV potential is applied to the filament and the filament then acts as a directly heated cathode giving off electrons which pass through a control aperture and are focused by electromagnetic lenses on to the specimen. Focusing can be achieved down to a diameter of 0.001 mm. The specimen is placed under the electron beam and observed with the microscope. X-radiation is emitted as the electrons strike the specimen's surface, each element in the specimen giving off a characteristic radiation which is detected and measured by spectrophotometers which transmit the information as digital pulses to a terminal or other apparatus.

The spectrophotometers are set to record the wavelengths for iron, magnesium and silicon. Then the X-ray intensities for each of the standard elements in the known sample are determined and the intensities of the X-rays from the unknown are recorded under identical conditions. Measurements take from 10 to 100 s on each point of a sample and from 10 to 15 points may be needed for an accurate analysis. A final check is made by re-measuring the X-ray intensities of the elements in the known sample. The spectrophotometers are then de-tuned from the wavelengths of the elements being analysed so that any background radiation from scattered X-rays can be measured. These are subtracted from the intensity of both known and unknown samples and the amounts of the constituent elements in the unknown sample are calculated using the ratios of the two sets of intensity readings. Much of the data obtained is rough and needs refinement to take into account backscatter, fluorescence and absorption due to other elements in the sample. Data is analysed by comparing emission spectral data of natural or synthetic materials of established composition (this is now often done by computer). For example, if a peridot is to be analysed the standards for magnesium and silicon are chosen. With a low iron and high magnesium content, peridot is compared with an already tested mineral with comparable composition as a standard.

Preparation of the sample is easier when it can be ground down to a small size; where this is not possible, as in the case of inclusion studies, much more time is taken in preparation. One limitation (other than time and cost) of the microprobe is that it cannot detect elements with an atomic number lower than 9. This rules out beryllium, boron, lithium and oxygen.

The microprobe at the National Museum of Natural History, Smithsonian Institution, Washington DC, is capable of analysing nine elements simultaneously. Six of the nine spectrophotometers are set permanently to analyse for silicon, aluminium, iron, magnesium, calcium and potassium.

4.8 ELECTRON PARAMAGNETIC RESONANCE

Electron paramagnetic resonance (EPR) alternatively named electron spin resonance (ESR) is useful as a non-destructive test using microwaves which are orders of magnitude weaker than light waves. The sample, between 1 mm and 1 cm in width, is placed in a cavity within a strong magnetic field. Observed transitions correspond to energy differences much smaller than those observable with optical spectra. The method detects unpaired electrons such as those in transition metals. EPR spectra will yield information on the nature of colour centres, sometimes showing their position in the crystal lattice. It was with EPR that the nature of the colour centres in Maxixe and Maxixe-type blue beryl was determined, different impurity ions being needed in each case with CO_3^- for the Maxixe-type and NO_3 for the natural Maxixe material.

4.9 ENERGY DISPERSIVE X-RAY SPECTROPHOTOMETRY

Energy dispersive X-ray spectrophotometry will detect all elements from sodium to uranium in the periodic table. The technique is preferable to X-ray fluorescence analysis as the latter is expensive to operate and some gem materials can be damaged by the X-rays (colours may change irreversibly, for example). There are two techniques: wavelength dispersive and energy dispersive spectrometry. The sample is first excited either by electron/ion bombardment or by primary X-rays or gamma rays. The excited secondary X-rays are of different wavelengths and energies corresponding to the chemical elements in the sample. The X-rays need to be split up by a dispersing medium which is either a crystal with a known dispersive value (d-value) or (with energy dispersive spectrometry) a solid state lithium-drifted silicon detector with a dispersing power of about 150 eV. Results are relative and have to be compared (for an unknown) with those of standard reference materials. An on-line computer is needed to process results.

Gemmological applications include the separation of the jade minerals, identification and separation of turquoise from its substitutes, the detection of diamond substitutes in lots containing diamonds and imitations, the identification of stones which are hard to test by other means, such as mounted and cabochon-cut stones, the characterization of garnets and in some cases the identification of synthetic stones. Further details are given in a paper in the *Journal of gemmology*, **18**, 4, 1982.

4.10   UV SPECTROPHOTOMETRY

Ultraviolet spectrophotometry has been used to distinguish natural from synthetic rubies which transmit differentially in the UV region. The method used is described by Bosshart (1982). Among interesting phenomena found by this method are weak absorption lines on a band between 345 and 326 nm and a further band at 314 nm, showing that traces of iron are present in otherwise pure chromium-rich stones. Thai rubies show an iron band at 450 nm. A narrow band near 266 nm, with another band in the visible, is ascribed to titanium and is most pronounced in Kashan pale rubies which have raised titanium levels.

Rubies from Thailand and many from Sri Lanka are found to be markedly different chemically from synthetic stones, but the trace element composition of Burmese, Kenya, Chatham, Verneuil, Kashan and Knischka stones (the lighter reds) causes their colours to be indistinguishable from synthetic stones (their excitation spectra are also undiagnostic). However absorption minima in the UV allow effective separation of natural from synthetic specimens.

5 Colour

5.1 THE PHYSICS OF COLOUR

Before examining the various ways in which colour can be produced in gemstones we should understand that the colour of any object resides in the white light in which we view it. When all the wavelengths which combine to give white light are present the object will appear colourless; when by the operation of colour-causing mechanisms some of the wavelengths making up white light are absorbed, the unabsorbed wavelengths combine to give the object a residual colour. What the colour actually is depends on which wavelengths have been absorbed. This refers to body-colour and not to play of colour, interference or dispersion, all of which give areas of all the spectrum colours. A number of optical mechanisms can give rise to the colour of a gemstone, including dispersion, interference, diffraction and scattering (which may give other optical phenomena too).

Dispersion

Dispersion is the breaking-up of white light into its component colours: red, orange, yellow, green, blue, indigo and violet. White light can be dispersed by passage through a prism when the red component is refracted least and the violet most, so that a spectrum results. Stones vary in their ability to separate (disperse) the different colours, depending on the refractive index of the mineral (Fig. 5.1); generally the higher the RI the greater the dispersive power (although diamond is not as highly dispersive as might be predicted by this criteria it is still the most highly dispersive of the commoner naturally occurring colourless gemstones). Many coloured stones are also highly dispersive, although their own colour usually masks the effect to some extent; sphene, demantoid garnet and benitoite are in this group. Of the various artificial materials made to simulate diamond, rutile and strontium titanate are full of fire; the latest successful simulant, cubic zirconia, has a higher dispersion factor than diamond but the value is not as high as that of rutile, which almost resembles opal in its display of colour.

Dispersion is best displayed when the stone is cut as a brilliant. When a

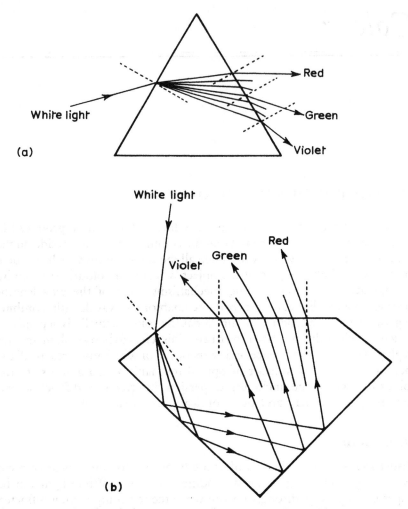

Figure 5.1 (a) White light entering a glass prism is dispersed into its spectral colours; (b) Dispersion and total internal reflection of white light entering a gemstone.

well-cut diamond is examined under many points of tungsten light the flashes of red, orange, yellow, green and blue, with an occasional purple or violet, combined with the sharp reflection from the surface, give the stone an appearance which cannot be matched even by the best of simulants. Stones with a low dispersion factor are hardly worth cutting as brilliants and other shapes can be selected; in coloured stones there are fewer colours remaining to be dispersed after absorption has taken place, so that the property is less important than body colour.

Dispersion cannot be measured by the standard refractometer because it is calibrated for use only in sodium light; where it is necessary to measure dispersion (as for a new material) the RI is measured twice, in light of wavelengths as close as possible to the red and violet ends of the spectrum, respectively, usually with a goniometer. The dispersion value is found by simply subtracting the lower RI (red) from the higher (violet). Normally the goniometer is used for this purpose. Some values are: diamond 0.044, zircon 0.038, benitoite 0.047, sphene 0.051. Among the synthetic stones are rutile 0.280, strontium titanate 0.190 and cubic zirconia 0.058 to 0.066, the variation depending on the stabilizer employed. Low figures are shown by quartz at 0.013, emerald 0.014 and corundum 0.018.

Interference

Interference of light at thin films or cracks, and the resulting bands of different colours, arise from the same effect as the colours in oil slicks or soap bubbles (Fig. 5.2). When a parallel beam of monochromatic light falls on a thin glass wedge an individual ray is partly reflected by the front surface and partly by the back surface, the latter having travelled a greater distance by the time it rejoins its companion and leaves the glass altogether. Thus the two waves will be different in phase, and will show interference effects. If the trough of a one wave coincides with the peak of another, the 'destructive' interference causes 'extinction' of the light, since the waves

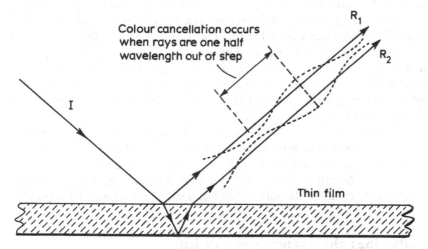

Figure 5.2 Representation of the production of colour in a thin film by interference between reflected rays. The wavelength or colour related to the extra distance travelled by R_2 is either cancelled where this is equal to half a wavelength (as shown) or reinforced when it is a whole wavelength (i.e. when the two rays are in phase).

cancel each other out. If, however, the phases of the two waves are such that they reinforce each other, the reflected light is brighter than the incident light. These effects, and those intermediate, can be seen together in one crystal, depending on the number of wavelengths across the wedge at different places. Where the light is monochromatic this operation will produce a series of light and dark bands. If white light is used, different colours will give bands at different positions. At the thinnest end of the wedge the order of the combined colours is: black, grey, blue–grey, white, yellow, orange and red – these are known as first-order colours. The second order which comes next along the wedge consists of violet, blue, blue–green, yellow–green, orange and red. Next come violet, blue, green–blue, etc. This sequence of colours is of great use in determining the refractive index of minerals in thin sections.

Interference colours in gemstones usually arise from cleavage cracks within the stone; they can often be seen in topaz and in other stones with an easy cleavage. The colours seen are produced by a combination of refractive index and the thickness of the film. In quartz (rock–crystal) that has been heated and suddenly cooled by plunging it into cold water, cracks which can cause interference are developed inside.

Diffraction: play of colour

The play of complete spectrum colours produced by diffraction is best seen in opal. A 'diffraction grating' is produced by the spheres of cristobalite silica containing water and the voids between the spheres which contain more silica and water with a different composition. Depending on the opal, diffraction takes place from the spheres or from the voids; it is important that there should not be too great a difference in RI between the two or light will be randomly scattered, making the opal milky white. If the spheres are irregularly arranged or are not all of the same size there will be no play of colour.

The thin layers of which iris agate is composed also give rise to diffraction which is seen by transmitted light. Labradorite feldspar usually gives green, yellow, red or blue colours on a cleavage surface viewed in the appropriate direction. This is due to repeated twinning; here the colours appear to be closer to the surface than in opal. The background is greyish.

Scattering: chatoyancy and asterism

The light scattering mechanisms which give rise to chatoyancy (the cat's-eye effect) and asterism (the star effect) may also cause colour. Chatoyancy is caused by light reflected from needles or tubes arranged in parallel (the band of light runs at right angles to their length) and is best seen in cat's-eye chrysoberyl and (as a less sharp eye) in quartz. Tourmaline and other gemstones occasionally provide chatoyant material.

Reflections from sets of needles (usually though not always rutile) give a star effect when they are arranged at 120° to each other in the basal plane of corundum and some other species. The stone must be cut as a cabochon for the effect to be seen, preferably under a single light source. While most star stones show their effect best by reflected light (epiasterism) star quartz should be viewed by transmitted light (diasterism). Some garnets show four-rayed stars and various crystallographic operations occasionally give three-rayed or 12-rayed stars.

Scattering gives the lustre or orient of pearl from the overlapping plates of aragonite, and other reflections from platy inclusions can give the special kind of lustre known as aventurescence and schiller. In sunstone (oligoclase feldspar) the inclusions are hematite and in green aventurine quartz they are chrome green fuchsite mica.

The moonstone effect (adularescence) arises most probably from a combination of scattering of light from very fine particles at the blue end of the spectrum and from a layered structure of intergrown orthoclase and albite. The synthetic spinel sometimes offered as a moonstone imitation gets its colour from similar fine precipitated particles scattering light at the blue end of the spectrum.

5.2   THE CHEMISTRY OF COLOUR

There are thought to be at least 15 causes of colour, although not all are applicable to gemstones. All involve the interaction of electrons with light and many of the best-known stones contain impurity elements which cause the colour. Stones in which the colour is attributable to a particular colouring element which forms part of its normal chemical composition are known as idiochromatic; allochromatic stones take their colour from an element which is foreign to the stone – an impurity. Both idiochromatic and allochromatic stones are covered by the first and in many ways the most important of the theories of colour causation which apply to minerals and gemstones. This is crystal field theory. All crystal field colour can be attributed to the behaviour of unpaired electrons.

Crystal field theory

Ruby is such a good example of crystal field theory in operation that it is normally chosen to illustrate the theory. Ruby is mainly corundum, Al_2O_3 which in itself is colourless; the addition of about 1% chromic oxide (Cr_2O_3) gives the familiar red. Trivalent chromium ions take the place of trivalent aluminium ions. In trivalent chromium compounds three of the six unpaired electrons in the outer shell are used in forming chemical bonds, leaving the other three to undergo the transitions which give the colour. Each ion of chromium is surrounded by six ions of oxygen in a distorted octahedral shape; this shape and the distances and positions involved give an electric field (crystal field or ligand field) at the chromium ion.

In Fig. 5.3 A is the ground state (the state of lowest energy). When light is illuminating the stone the electron is excited from the ground state (A) to level C or D (selection rules forbid the transition from A to B). The A to C transition corresponds to the energy of green light and the A to D transition to the energy of violet light, so that these two colours are absorbed. The remaining, unabsorbed light has energies corresponding to red light, and thus the stone appears red. In returning to the ground state, selection rules again come into play; the transition from C and D to B is preferred to that from C and D direct to A. The subsequent transition from B to A is accompanied by the emission of red fluorescence. This can be observed by illuminating the stone with green or violet light or with radiation of higher energy such as UV or X-rays. The UV in daylight is sufficient for an enhanced red glow to accompany that already given to ruby by absorption.

In emerald the symmetry shown by the chromium and aluminium ions is the same as for ruby, but the strength of the crystal field is less. This causes

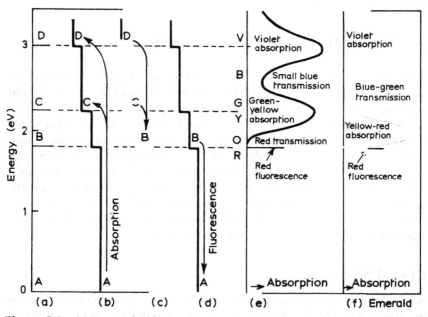

Figure 5.3 (a) Energy levels A to D of chromium in ruby; (b) to (d) absorption and fluorescence; (e) the resulting absorption spectrum in ruby; (f) the equivalent spectrum of chromium in emerald. (Reproduced with permission from Nassau, 1980.)

a lowering of the C energy level from about 2.25 eV to 2.0 eV. As a result almost all the red light is absorbed and the blue–green is transmitted. Alexandrite, which is red in tungsten light or candle-light and a bluish-green in daylight, occupies a position intermediate between ruby and emerald.

When the environment surrounding the ion of the colouring element is not spherically symmetrical the colour is not either; instead it varies, usually in intensity only, with direction, showing the phenomenon of 'pleochroism'.

Chromium is one of the transition metals (elements 22–29 of the periodic table). All have unpaired outer shell electrons which operate to cause colour by the crystal field theory. They are:

- *TITANIUM*, responsible, with iron, for the blue in sapphire, blue zoisite and probably benitoite.
- *VANADIUM*, giving green in some beryl varieties, in some korneru-pine, green (tsavolite) grossular garnet and in a synthetic emerald which contains no chromium. Vanadium also gives a green to red colour in a corundum variety of the alexandrite type.
- *CHROMIUM*, which gives the red of ruby and red spinel and green in emerald and green jadeite. It is also responsible for pink topaz, green (hiddenite) spodumene, grossular garnet and both red and green alexandrite.
- *MANGANESE*, which gives the pink to morganite beryl, pink tourmaline and kunzite spodumene, the red to beryl and the green and yellow to andalusite.
- *IRON*, which gives blue aquamarine, blue and green tourmaline, chrysoberyl and green quartz, and some of the green, yellow and brown jade.
- *NICKEL*, which gives a green to some opal and to chrysoprase.
- *COBALT*, which gives a pink colour to virtually all natural minerals but when added to man-made gemstones it gives a dark blue.

The above examples are all examples of allochromatic colouration; idiochromatic examples are:

- *CHROMIUM*, in uvarovite garnet (green) and crocoite (orange).
- *MANGANESE*, in rhodochrosite and rhodonite (orange-pink) and spessartine garnet (orange).
- *IRON*, in peridot (green), almandine garnet (red) and lazulite (blue).
- *COPPER*, which gives the blue in turquoise, azurite and chrysocolla, the red in cuprite, and the green in malachite.

The presence of unpaired electrons is vital to the operation of crystal field theory but variable valence, a characteristic of the transition metals, can also be important since it can give rise to different colours. Divalent copper,

Cu^{2+}, with two electrons used in bonding, has one unpaired electron to give coloured compounds in this cupric state. Both ferrous (Fe^{2+}) and ferric (Fe^{3+}) iron can cause colour by the action of their unpaired electrons.

Colour centres Colour centres are defects in the crystal, caused either by an extra (unpaired) electron trapped at a location where no electron is usually present (electron colour centre) or a hole where an electron is absent from a location normally occupied by a pair of electrons, leaving one unpaired (hole colour centre). These rogue electrons behave, when excited, in the same way as those on transition metal ions and can thus cause colour (Fig. 5.4).

Fluorite, most often a purple colour, often has a trapped–electron type of colour centre, and was in fact one of the first minerals to be investigated in this context. The most important of the several colour centres which can operate in this mineral is known as the F-centre (the F standing for the German word for colour, *Farbe*). Fluorite has one Ca^{2+} ion with two F^- ions, the two fluorine negative ions to one calcium positive ion giving electrical balance. If one fluorine ion is missing an electron has to occupy its vacated space to preserve electrical neutrality. It is thus trapped and may be stimulated to higher energy levels in the same way as unpaired electrons on transition metal ions. This electron colour centre can give colour and fluorescence under appropriate conditions. Smoky quartz has the other type of colour centre, the hole colour centre. Here Al^{3+} ions replace Si^{4+} ions; the maintenance of electrical neutrality is carried out by Na^+ or H^+ in the vicinity. Irradiation of quartz contaminated with aluminium may cause one of a pair of electrons to be ejected from its position on an oxygen ion close to one of aluminium, leaving an electron unpaired. This gives a hole colour centre with accompanying energy levels and absorption of light. Heating a quartz crystal in which the displaced electron is trapped in

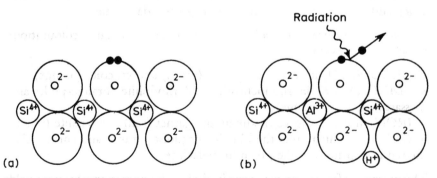

Figure 5.4 (a) The schematic structure of quartz SiO_2. (b) The smoky colour centre produced on irradiation. Each black disc represents an electron with a unit negative charge.

another site may return the electron to its original place, so eliminating the smoky colour; a temperature of 400°C would do this. Further irradiation (X-rays, gamma rays, neutrons) re-instates the smoky colour once more.

Iron in quartz as an impurity may give a green or yellow colour and some of these stones when irradiated will turn to an amethyst colour which can be returned to yellow or green on heating. Some gemstones whose colour is caused by this type of operation may lose their colour simply by exposure to light or heat; some irradiated topaz (coloured yellow or brown), some yellow irradiated sapphire and green irradiated kunzite fade rapidly. The colours of other stones in this class are stable. The case of the Maxixe (natural) and Maxixe-type (artificially coloured) blue beryl is particularly interesting; the colour obtained by treatment (see below) is both spectacular and fugitive.

The phenomenon of phosphorescence (emission of light after irradiation stops) takes place when the previously excited electron (operating in any kind of colour creating mechanism) gives off light when returning from the C to the A level. For phosphorescence to occur the thermal energy at room temperature must be sufficient to allow the movement to take place slowly enough for light to be emitted. Fluorite and diamond may show phosphorescence (not all specimens by any means) and some calcite and fluorite may display thermoluminescence (when heating is needed to help the electron escape from the BC trap).

Molecular orbital theory

In the crystal field mechanism electrons involved in the production of colour belong to a single ion or defect. It is also possible for electrons to orbit around two or more ions where the structure of the material is, in the main, ionically bonded. Electrons can also orbit around two or more atoms where the structure is mainly covalently bonded. The theory embracing both of these conditions when they cause colour is called molecular orbital theory. Energy levels and absorption of light are again involved.

One type of colour-causing operation classifiable under this theory is charge transfer. This can take place between ions of two different transition metals or between two ions of the same substance with different valencies. A good case in point is the blue colour in sapphire which is caused by the following operation:

$$Fe^{2+} + Ti^{4+} \rightarrow Fe^{3+} + Ti^{3+}$$

Electrical balance is maintained but at the same time energy changes involve absorption of light, thus giving the colour; other blue gemstones (benitoite, blue zoisite) may be coloured in this way. Iolite may get its colour from transfers between two different valency states of iron, thus:

$$Fe^{2+} + Fe^{3+} \rightarrow Fe^{3+} + Fe^{2+}$$

In the case of lazurite, the principal (blue) constituent of lapis–lazuli, groups of sulphur ions are held together by electrons which orbit about the whole group. The electrons have excited states and cause the blue colour.

In some organic materials such as coral, amber and pearl (conchiolin) the colour is caused by organic compounds which act as dyes in which the colour arises from transitions which affect the electrons in organic molecules, still part of molecular orbital theory. Artificially dyed gemstones such as agate and turquoise also come into this class.

Band theory

In band theory the electrons involved in colour production belong to the whole crystal rather than to ions, defects or groups of atoms. In metals, electrons can move from one atom to another, since each atom lends its electrons to a common stock; this 'sweep' of electrons explains the electrical conductivity of metals and also their high metallic lustre. In these conditions energy levels are very close together so that a continuous energy spectrum is approximated – this is known as a band of energies (Fig. 5.5). Starting with that of lowest energy, each energy level is filled with electrons in turn until all the electrons are used up. The number of electrons that can be fitted in at each energy level is different. The outer filled band is known as the Fermi surface; beyond this there are many unoccupied energy levels which provide the levels into which excited electrons are elevated. The difference in the number of electrons accommodated by the different energy levels gives a distinctive shape to the band for each material and causes differences in reflectivity at different energies. This can cause

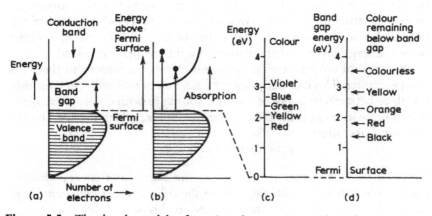

Figure 5.5 The band model of semiconductors (a), with light absorbing transitions (b) leading via the colour scale (c) to the band-gap colour sequence (d). (Reproduced with permission from Nassau, 1980.)

differences in colour, as in gold where there is more absorption at the blue end of the spectrum, giving the metal its yellow colour.

In some materials there is a band-gap between the Fermi surface and a conduction band at energies higher than that of the surface; these substances are covalently bonded in the main so that electrons are shared. Diamond and sphalerite, both colourless, are wide-gap semiconductors while some grey to black minerals are narrow band-gap semiconductors. Some orange-yellow minerals such as greenockite, CdS, are intermediate (only orange, yellow or red colours can be found in this context).

Looking closely at these substances we see that while the energies below the Fermi surface are filled with electrons (this area is the valence band) there are no electrons in the gap nor in the conduction band. When the substance is irradiated (by light, for example) the light with energies higher than that of the band-gap is absorbed but not that with lower energies. Where the energy of the band-gap is small we have a narrow band-gap semiconductor and absorption is possible at all visible energies. The material will be black or grey and opaque. It will also conduct electricity at room temperature. Where the energy of the band-gap is large absorption can only take place in the UV; this gives a transparent colourless substance with electrical insulating properties when pure, diamond for example. Approximate energies for the wide band-gap would be over 3 eV and for the narrow band-gap about 1.5 eV. At around 2.5 eV only blue, violet and higher energies are absorbed so that the complementary colour (yellow) is produced (e.g. greenockite). Where the band-gap is 2 eV all visible light except red is absorbed and this is how proustite and cinnabar get their colour. Impurity levels in semiconductors may also give rise to fluorescence and phosphorescence.

The band-gap energy in diamond is about 5.5 eV and normally the stone is colourless as we have seen. Diamond is made of carbon, which has four outer electrons which are in the valence band. When an impurity such as nitrogen is added (this has five available electrons) it is known as a donor because it gives up an extra electron when it replaces an atom of carbon. The donors (either small groups of atoms or individual atoms) form an energy level in the band-gap, 3 eV below the conduction band. In this case light of 4 eV energy or higher can be absorbed, the electrons being raised to the conduction band. This is how the yellow colour of natural diamonds is caused. Where there is much more nitrogen the energy level alters and the colour is an intense green. Where larger quantities still are involved the nitrogen atoms merge into clusters which do not act as donors, and the stones are colourless.

Boron has one less electron than carbon, so that when boron substitutes for carbon an acceptor level in the band-gap is created. This can take an electron from the valence band and light may be absorbed by the transition. The blue colour in diamonds containing boron arises in this way.

5.3 COLOUR FILTERS

Colour filters were devised originally for the testing of emerald at a time when virtually all synthetic emeralds appeared bright red through them; glass emerald imitations appear green through the filter. The best known of the colour filters is the Chelsea filter developed at the Chelsea Polytechnic (and at the Precious Stone Laboratory of the London Chamber of Commerce) by B. W. Anderson and C. J. Payne. The filter (really two) allows two narrow bands of colour to be transmitted, one in the deep red (about 700 nm), the other in the yellow–green (a narrow band around 570 nm). All stones examined through it are either one of these two colours or a rather indeterminate colour which is quite clearly neither red nor green. Stones coloured by chromium transmit in the deep red, whatever their body colour. In ruby the fluorescence shown by many specimens enhances the red observed and can be used to distinguish them from red glass (admittedly a fairly experienced eye is needed). The finest emeralds show a decided red through the filter whereas green glass and soudé composites remain green. On the other hand emeralds from India and South Africa show little change. Other important green stones such as demantoid garnet, green fluorite and green zircon show pink through the filter.

Some blue stones coloured by cobalt also show red through the filter; these include cobalt glass and cobalt-coloured blue synthetic spinel. In this way the paler stones at least can be distinguished from aquamarine or pale blue sapphire which appear green to greyish-green. Stained green chalcedony, showing pink when chromium is added, can be distinguished from chrysoprase which remains green. Some care should be taken with blue spinel as it seems that there are some natural cobalt-bearing gem-quality specimens about, which appear blue. Natural green jadeite remains green but dyed jadeite tends to show pinkish through the filter.

A variation on the Chelsea filter is known as the Sterek. This again is a two-colour filter and transmits red at about 670 nm upwards and also a broad band of colour in the blue–green, centred at about 520 nm. Aquamarine shows an intense blue–green through the Sterek filter which can easily be contrasted with the pink of the synthetic spinel imitation. Chromium-bearing stones such as natural and synthetic rubies and alexandrites show a brighter red than they do through the Chelsea filter. The Sterek can distinguish most natural emeralds from synthetic stones.

The emerald filter developed by the Nohon Hoseki-Kizai Company of Japan operates in the same way as the Chelsea filter and is useful in the detection of dyed jades. The GemMaster ruby discriminator made by R. Lary Kuehn Inc. is a green filter which identifies chromium-bearing stones. When a stone is held under a strong white light and viewed through the filter the colours seen range from bright blue for ruby, dark greyish-blue for red spinel and dark purple for red glass. Pyrope garnet shows a dark brown. There is no means of distinguishing between natural and synthetic stones.

If natural and synthetic emeralds are placed in blue light and the group examined with an infra-red filter, the synthetic stones will show a brilliant red while the natural ones will show scarcely any reaction.

5.4   THE SPECTROSCOPE

The spectroscope is a very quick and easy way to test a surprisingly large number of gemstones with no ancillary apparatus other than a suitable light source. Both rough and cut stones can be tested and there is no need to remove set stones from their settings (Fig. 5.6).

Spectroscopes analyse light reflected from the surface of, or transmitted through, a stone. Which method is adopted depends on the stone, even opaque ones often having a slightly translucent area near the edge which can be used to transmit light. Rays of the different colours which make up white light are refracted to different extents when passed through a glass prism or its equivalent. A ribbon-like spectrum can be produced by the prism or by passing light through a diffraction grating – a transparent flat glass plate upon which up to 15 000–30 000 parallel lines to the inch are ruled (in practice the lines are printed photographically on the glass, having been reduced from a larger master negative). The spectrum thus produced is a product of both diffraction and interference and is less bright than the prism spectrum, as multiple spectra are formed on either side of the main one, so diluting the effect. On the other hand, the grating spectrum shows the wavelengths evenly spread whereas in the prism spectrum the wavelengths at the red end are bunched together due to dispersion.

The spectroscope, whether using a series of prisms or a diffraction grating, is a short metal or plastic tube; at one end is a focusing eyepiece and at the other a small slit, adjustable on the better-quality models, to allow light into the instrument. In the prism type the spectrum is produced by arranging three components of crown-flint-crown glass (five are sometimes used) in series so that sufficient dispersion can be obtained (around 10° is a common value). It is generally arranged that one of the rays of the beam emerging from the instrument is in line with the incident ray (usually the ray at 589.3 nm) (Fig. 5.7).

The slit is the most important part of the spectroscope as the spectrum seen is a series of images of the slit. If the slit is opened wide the spectrum will not be pure since the slit images will overlap. The spectroscope is focused by pushing the draw-tube home for observations at the blue and violet end of the spectrum and drawing it out about a quarter of an inch for

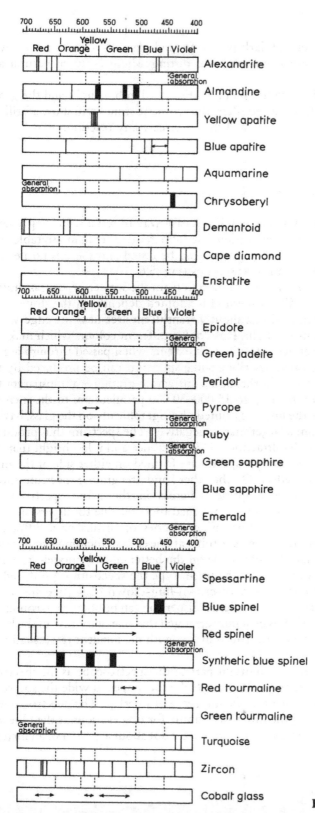

Figure 5.6

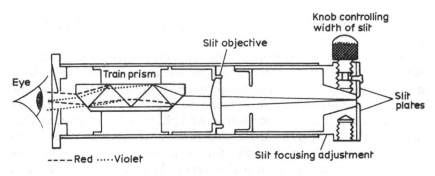

Figure 5.7 A direct vision prism spectroscope.

observations at the red end. These aids should be borne in mind by the novice user since they make a great difference to ease of observation.

A series of fine dark lines can be seen when a spectroscope is pointed toward the sun or merely out of the window on a fairly bright day. These cross the spectrum from one end to the other and appear as vertical almost hair-like lines. These dark lines are called the Fraunhofer lines after their discoverer and arise from the absorption of some wavelengths of the bright continuous spectrum emitted by the sun's glowing core by metallic atoms in the cooler gaseous atmosphere surrounding it. The Fraunhofer lines are very fine compared with those seen in solids (including gemstones) because the atoms in the solids absorb in a much more restricted way than gaseous atoms. The vertical Fraunhofer lines should not be confused with dark horizontal lines which are the result of dust on the slit and which can be eliminated by opening the slit a small amount. The stone can be placed on the stage of a microscope for transmitted light, ensuring that only light which has passed through the stone reaches the observer. A number of stands are available for obtaining reflected light. If a high-powered source of light is used (a 500-watt projection bulb in an asbestos-lined box is often recommended) care should be taken to prevent the specimen overheating. This is most easily achieved by allowing the incident light to pass through a water-filled flask. The shape of the flask allows it to function as a condensing lens as well as a filter. A similar flask filled with copper sulphate helps observations at the blue–violet end of the spectrum and also shows up emission lines in the red much more clearly since the background red is absorbed by the blue filter.

Ideally observations should be carried out in a darkened room. Avoid dazzle at all costs since the eye takes up to 15 minutes to regain its effectiveness. If bands do not appear which experience suggests should be there, try a different lighting technique or a different instrument. It is always fatal to hurry spectroscope work. The slit should be constantly adjusted where possible and the head moved to get the various parts of the spectrum in view. It should be remembered that many lines and bands show as little more than dark stains on the spectrum and that the drawings in books are of necessity idealized.

There are many spectroscopes on the market. It is possible to obtain an instrument which will give wavelength measurements but in practice this is an unnecessary refinement in gemstone testing since the pattern of lines and bands is sufficiently familiar or ascertainable without actual measurement being necessary. An adequate source of light is more important than gadgetry and the many quartz-iodine bulbs now available give a very intense beam of light most suitable for spectroscope work (they must be kept cool, however). Some manufacturers make up a complete assembly which includes spectroscope, wavelength scale, illumination for both, provision for filters (coloured and polarizing) and employ fibre-optic guides to bring the light to the specimen. These are most useful and the equipment made by Eickhorst of Hamburg is particularly good.

Some metallic elements absorb some of the wavelengths from white light when they are present in a gemstone (or a solution or glass) and they absorb in such a way that they colour the materials in which they are contained. These are the transition metals, described earlier (p. 75). These elements (or most of them) give a characteristic pattern of lines and bands which enable the experienced observer to recognize them at once and so ascertain at least some of the chemical composition of the unknown.

Before beginning a discussion of the characteristic absorption spectra of the colouring elements we should remember that not all coloured gemstones owe their colour to these elements so not all coloured stones will show the characteristic patterns associated with them. Some gemstones which show colour owe it to the operation of other optical mechanisms and show no absorption in the visible – diamond is a good example. Conversely, some colourless stones will show fine line absorption spectra which have no great influence on colour, as in the case of zircon. These fine line spectra are usually due to the presence of uranium (as in zircon) or to one or other of the rare earths (as in many synthetic stones, although in the case of garnet-type oxides such as YAG and GGG the stones are coloured).

Chromium is easily the most important of the colouring elements and shows an absorption spectrum which should be quite simple to recognize. There is a doublet at the red end of the spectrum formed of two lines very close together, accompanied by two or more lines on the orange side and by a broad absorption band in the yellow or green (the exact position of this band has a considerable effect on the colour of the stone). There are often narrow lines in the blue and the violet is strongly absorbed, to such an extent that the spectrum appears to be shorter than expected, an effect known as a cut-off.

Spectra due to iron are characterized by less well-defined lines and bands; both ferrous (Fe^{2+}) and ferric (Fe^{3+}) iron give characteristic spectra and the colours of the stones are red, green, yellow or blue. Spectra due to manganese show two bands in the blue and stronger (sometimes very intense) bands in the violet; this absorption region may extend into the UV. Stones are orange to pink in colour. Although cobalt–bearing stones 'ought' to be blue, natural cobalt–bearing minerals are pink; thus it should be assumed that a blue stone whose absorption spectrum proves the presence of cobalt are synthetic. However, this may not always be the case as some natural blue cobalt-bearing spinels have been reported. The characteristic spectrum consists of three strong and broad bands in the yellow, green and blue–green.

Copper as seen in turquoise gives two narrow absorption bands in the violet which are often very hard to see. Vanadium as seen in synthetic corundum intended to imitate alexandrite gives a strong sharp line at 475 nm in the blue. There is no characteristic absorption spectrum for titanium nor for nickel. Bands due to uranium are found in zircon and when seen at their strongest occur over the whole extent of the spectrum. Those due to rare earths also occur as fine lines, sometimes over the whole spectrum and sometimes in well-defined groups. We usually speak of lines in this case rather than bands (though the tendency is to use both terms rather loosely). Any rare earth lines appearing strongly always indicate a synthetic product or a glass. Before examining the spectra of many of the important stones readers should be aware that although the British and European convention is to print the spectrum with the red end at the left, the reverse is the case in the United States. Please check to see where your textbook is published!

The absorption spectra of gemstones

The absorption spectra of gemstones differ not only because different elements are involved but also in the strengths of the lines and bands relative to each other. In the following descriptions notably strong bands will be marked 's', weak bands 'w'.

1. *ALEXANDRITE* shows well–defined lines and a central band. The latter's position shifts to the red, on changing the viewing light from short to long wavelength, causing a colour–change. It is centred at 580 nm, about half-way between its position in emerald and in ruby. The doublet in the red is at 680.3 and 678.5 nm and may be seen in some circumstances as an emission (bright) line. Other lines are at 665, 655 and 645 nm (w) and in the blue at 473 and 468 nm.

2. *ALMANDINE GARNET* which owes its colour solely to iron shows a classic spectrum (with zircon, one of the first to be described in 1866 by Church). Three broad bands are located in the yellow at 576 nm, in the green at 527 nm and in the blue–green at 505 nm (s). Others are found at 617 and 462 nm (w). The 505 nm band turns up in pyrope and spessartine as well as in almandine, due to isomorphous replacement.

3. *APATITE* in blue colours does not show the didymium bands but has bands at 631 and 622 nm in the orange, 511 nm (s) in the green, a band at 490 nm (s) and a broad band at 464 nm (w). These are seen in the yellowish ordinary ray.

4. *AQUAMARINE* in its green form (i.e. before being heated to give the steely blue more commonly seen today) shows an iron absorption spectrum with a narrow line at 537 nm in the green (in the extraordinary ray only). This is best seen by reflected light. Blue aquamarine shows two rather vague bands – in the blue at 456 nm (w) and in the violet at 427 nm (s in larger stones). The fine dark blue beryls from the Maxixe mine and also the irradiated Maxixe-type beryl seen in recent years (before it fades) show an unusual absorption spectrum with bands in the red at 697 and 657 nm and a weaker band in the orange at 628 nm.

5. *BLUE SAPPHIRE* shows the iron spectrum described for yellow and green sapphire but there are some important variations. Stones from Australia show all three bands but those from Sri Lanka may show only traces of the strongest band at 450 nm. This is best seen in the ordinary ray and should be sought in the direction of the optic axis, in this case the vertical axis of the crystal. In synthetic blue sapphires three vague bands can sometimes be seen, the central one of which is close to the 450 nm position as in the natural stone; however, 450 nm band is better defined and narrower in the natural stone.

6. *BLUE SPINEL* shows an iron spectrum that frequently confuses the student by its apparent complexity. The main band is centred in the blue at about 459 nm and there is a narrow band at 480 nm (both are similar in strength). There are also rather faint bands in the green at 550 nm, and in the yellow and orange, at 592 and 632 nm, respectively.

7. *BLUE SYNTHETIC QUARTZ* is also coloured by cobalt with a very similar spectrum to that shown by cobalt glass.

8. *BLUE TOURMALINE* has the band at 497 nm also shown by green tourmaline.

9. *BLUE ZIRCON* normally displays the strongest lines characteristic of the zircon spectrum (at 653.5 and 659 nm). They are fine and narrow.

10. *BLUE ZOISITE (TANZANITE)* has a strong broad band in the orange at 595 nm and weaker broad bands in the green at 528 nm and in the blue at 455 nm.

11. *CHROME-BEARING CHALCEDONY (MTOROLITE)* has a strong clear doublet in the deep red but not much else. Stained green chalcedony shows an indistinct band in the red at the doublet position with weaker bands, each of the latter being accompanied by a high transparency region like those seen in emerald.

12. *CHROME DIOPSIDE* contains both iron and chromium; the chromium shows narrow lines in the red with the doublet centred at 690 nm. Two lines in the green–blue are at 508 and 505 nm.

13. *CHRYSOPRASE* owes its colour to nickel and shows a faint line in the red at 632 nm.

14. *COBALT GLASS* shows similar bands to synthetic blue spinel with the central band being the narrowest of the three. The presence of a band at 505 nm will indicate the almandine section of a garnet-topped doublet, a frequent combination used as a blue sapphire imitation.

15. *DEMANTOID GARNET* has a ferric iron absorption spectrum with some chromium to give lines in the red. These are at 701 nm (the single line usually seen of the doublet) with other lines at 640 and 621 nm (w). The band due to iron is very strong and is centred at 443 nm, giving a cut-off effect. It may be seen separately in paler stones.

16. *DIAMOND*'s absorption spectrum is important in diagnosis and particular care needs to be taken with yellow stones. The majority of the colourless stones show no absorption spectrum but the off-coloured 'Cape' stones with a hint of yellow show absorption bands in the blue and violet. The most prominent of these is at 415.5 nm and this is seen best when light is transmitted through the specimen. A blue filter is recommended. Other bands can be seen at 478, 465, 451, 435 and 423 nm. Where a stone has been irradiated to change its colour the 415.5 nm band may still appear and show the stone's history (remember that only indifferently coloured stones will be worth treating).

Diamonds which have been irradiated by neutrons to give a green colour throughout (altering to yellow on heating) show absorption bands at 592, 504 and 498 nm. These bands are much easier to see if the stone is placed near ice or otherwise cooled. Some books give the first band as 594 nm. If a yellow stone shows these bands in addition to the band at 415.5 nm it is clear that it has been treated. Natural brown diamonds show bands at 504 and 497 nm, but the former is the stronger, whereas in treated stones the reverse is true. Green treated stones should show bands at 504 and 498 nm but some are reported to show no absorption in the visible. In recent years there have been reports of yellow diamonds, known to have been treated, which do not show the expected band at 592 nm. This is thought to be the result of annealing the stone (up to 1000°C). The absence of this band does not therefore prove that a yellow diamond is natural, for which cryogenic techniques must be used.

Various interesting absorption spectra, bearing out the dictum that diamond is unique, are reported in the literature. Some brownish-pink stones known to have been irradiated, showed bands at 637, 620 and 610 nm (the last two forming a pair) together with the 592 nm band. A bright jonquil-coloured diamond (yellow) showed bands at 493, 451 and 430 nm and was proved to be natural on the basis of its inclusions. A deep orange irradiated stone gave a series of bands at 592, 568, 504, 498, 500 nm and displayed general absorption from 620–590 nm. A blue treated diamond was reported to show an emission line near 504 nm and many pink diamonds with an orange to yellow–orange fluorescence show a band near 575 nm. These stones were probably a faint pink before treatment. A reddish-brown stone known to have been treated showed a band at 640 nm but no band at 592 nm. The only way to be sure about this kind of example is to use cryogenic techniques with a spectrophotometer to record bands too faint to be seen with the eye.

Of the diamond simulants only synthetic rutile shows an absorption spectrum. This is a strong absorption of the deep violet so that the spectrum appears to have been cut off. Imitations of coloured diamonds provided by YAG and its analogues and by cubic zirconia CZ will usually show rare-earth spectra. If they do not, a reflectivity meter test will identify them.

17. *EMERALD* has a strong absorption doublet in the deep red at 683 and 680 nm; the lines are about twice as far apart as those in ruby and are accompanied by two weaker lines at 662 and 646 nm; these are somewhat diffuse. Beside each of them is a 'window' (an area of high transparency). Another line at 637 nm is at its strongest in the ordinary ray. There is a broad but weakish absorption band in the yellow and in chromium-rich stones a line at 477.4 nm may be seen in the ordinary ray spectrum.

18. *ENSTATITE* can also be coloured by chromium in addition to iron. The main band is at 506 nm (s) and there are also bands at 548, 483, 450 nm (w). In chromium-rich stones, such as the bright green enstatites found with diamond at Kimberley, a doublet occurs at 687 nm with some weaker bands.

19. *GREEN GROSSULAR GARNET* can be coloured by chromium and shows, in the hydrogrossular variety, a broad band in the orange centred at 630 nm.

20. *GREEN JADEITE* shows characteristic chromium lines in the red, the lines being more diffuse than those in emerald. There is general absorption of the violet but more important is a strong band at 437 nm, due to iron. The slit of the instrument has to be opened a little wider to allow this important band to be seen. It is most prominent in the paler greens and in the lavender and even white jadeite; there is another band at 432 nm (w). Green dyed jadeite shows a broad and diffuse band in the red from the dyestuff.

21. *GREEN SAPPHIRE* shows the classic iron spectrum with bands at 471, 460 and 450 nm. The two stronger bands virtually coalesce but the 471 nm band is sufficiently separated to make identification simple.

22. *GREEN TOURMALINE* is coloured by iron and all the red end of the spectrum is absorbed down to about 640 nm. There is a narrow band at 497 nm.

23. *GREEN ZIRCON* shows, especially in the dark green 'high' stones from Burma, a beautiful and extensive absorption spectrum in which up to forty bands can be seen. Most green stones from Sri Lanka have fewer bands than this, sometimes showing ten to twelve in all. The strongest band

in either case is that in the red at 653.5 nm. It should be noted that only in high stones are the bands really sharp. In low zircons there is only a broad and woolly band to be seen near 653 nm though some low stones show a narrow band at 520 nm. Low stones heated to about 800°C show an anomalous spectrum with bands at 687, 668.5, 652.5, 589, 574, 560.5, 473.5 and 451 nm.

24. *IOLITE* shows an iron spectrum which varies according to direction of view; in blue light there are bands at 492, 456 and 437 nm and in yellow light there are bands at 593 and 585 nm, forming a distinctive double band. A narrow band at 535 nm may also be seen.

25. *PINK TOPAZ* owes its colour to the heat-treatment of some brownish material from Ouro Preto, Brazil; it contains sufficient chromium to show a red fluorescence between crossed filters and gives a weak emission line in strong illumination at 682 nm which in deep-coloured and large stones may show as an absorption line. The fluorescence doublet in pink sapphire is much stronger.

26. *PERIDOT* shows the iron spectrum of bands at 493, 473 and 453 nm, the first two having a darker centre with diffuse edges. In the spectrum of the β ray an additional band at 529 nm (w) can be seen.

27. *PYROPE GARNET* can owe its colour to chromium or iron or both. Those stones of a bright red colour (Bohemian garnets) may come from Kimberley, Arizona or Czechoslovakia and show a broad band in the yellow–green centred near 575 nm; the almandine band at 505 nm can usually be seen where the green and blue meet. Lines in the red are not often seen.

28. *RARE-EARTH SPECTRA* are described for yellow natural apatite. Where neodymium is used alone as a dopant the stone will be lilac and the lines will be grouped in the yellow and green. Where praseodymium is used alone the stone will be green and the lines grouped in the blue and violet. These elements are found in some yttrium aluminium oxide (*YAG*) where the lines are sharper than those seen in apatite. A similar effect is seen in doped scheelite. Rare-earth ions sometimes give a line fluorescence; this has helped to distinguish synthetic from natural scheelite, something which is always difficult, as both may show rare-earth spectra. However neodymium-doped material shows no fluorescence under crossed filters while natural material with didymium traces shows a line fluorescence with lines at 650, 620 and 558 nm.

29. *RED SPINEL* Chrome-rich Burmese stones show a group of emission (bright) lines in the red; there may be up to five, the two central ones being the brightest, and that at 686 nm the most prominent. Again a blue filter assists observation and a powerful light source is needed. Verneuil red spinels (rarely seen) show only a single emission line although some of the zinc-rich flux-grown red crystals show the group very prominently. Broad absorption in the green is centred at 540 nm but there

are no lines in the blue, the best way to distinguish red spinel from ruby.
30. *RED TOURMALINE* has an absorption spectrum due to manganese; it shows two narrow bands in the blue at 458 and 450 nm. There is also a narrow band at 537 nm appearing within the broad absorption band which covers the green (this is seen in some brownish-red stones).
31. *RED ZIRCON* shows an absorption spectrum which is characterized by a strong narrow band in the red at 653.5 nm, one at 659 nm (w) and some ten others ranging through the whole spectrum. These bands are ascribed to uranium and appear in most zircons, even colourless ones. The red stones from Expailly, France, do not show absorption bands, however.
32. *RUBY* shows a strong doublet in the deep red at 694.2 and 692.8 nm, usually seen as a single bright fluorescence line at approximately 693 nm. This is a very sensitive test for chromium in corundum. As little as 1 part in 10 000 Al_2O_3 will cause the line to be seen and even in apparently colourless sapphires or blue sapphires it is sometimes observed (especially in Sri Lanka stones). Other lines include 668 nm and 659.5 nm, followed by a broad band centred near 550 nm covering most of the yellow and green. This is followed by lines in the blue at 476.5, 475 and 468.5 nm, and the violet is generally absorbed.

In stones with a trace of iron (especially 'Siam' stones) the fluorescence line is less strong. Observers are recommended to use a blue filter for better viewing of both the fluorescence lines and those in the blue. In the ordinary ray the broad absorption band is stronger and broader than in the extraordinary ray.

When deep-coloured natural or synthetic rubies are examined by transmitted light in a direct beam the lines in the red will show as (dark) absorption lines.
33. *SINHALITE* shows an iron spectrum, similar to that of peridot but with an extra band at 463 nm. The full spectrum is therefore 493, 473, 463 and 450 nm with a weaker band at 527 nm and perhaps another at 436 nm.
34. *SPESSARTINE GARNET* when virtually pure is orange, but inclines to red as almandine molecules increase their presence. The spessartine bands, due to manganese, are at 495, 485 (w), 462 and 432 nm (s). Occasionally further bands at 424 and 412 nm (s) can be seen.
35. *SYNTHETIC BLUE SPINEL* is coloured by cobalt and shows a cobalt absorption spectrum which has three broad bands in the orange, yellow and green at 635, 580 and 540 nm. In spinel the central band is the widest (in cobalt glass it is the narrowest). If the stone is a very deep blue the bands may merge. Red is strongly transmitted.
36. *TURQUOISE* has a copper spectrum which is notoriously difficult to see. It consists of bands in the violet at 432 and 420 nm, both about equal in strength and width. Light should be passed through a translucent edge or the specimen viewed by reflected light. There is also a faint band in the blue at 460 nm. If you have unexpected ease in seeing the turquoise absorption

spectrum the stone may have been impregnated with plastic. Use of a blue filter assists observation.

37. *YELLOW APATITE* shows the characteristic rare-earth spectrum of didymium (comprising neodymium and praseodymium). This shows a group of fine lines in the yellow (the strongest of which are at 584 and 578 nm) and a group of fine lines in the green with a centre near 538 nm.

38. *YELLOW CHRYSOBERYL* shows an iron spectrum characterized by a strong band where the violet begins at 444 nm. As the colour of the stone deepens the band strengthens and in deep-coloured stones two weaker bands at 505 and 485 nm can be seen. The presence of the 444 nm band in cat's eye chrysoberyl distinguishes it from quartz cat's-eye.

39. *YELLOW ORTHOCLASE* shows a ferric iron (Fe^{3+}) spectrum with bands at 448 (s) and 420 nm (w) in the blue and violet.

40. *YELLOW SAPPHIRE* shows the three bands in the blue which are best seen in green sapphire. The strongest is at 450 nm and can be seen in stones with a low iron content such as those from Sri Lanka. This is accompanied by bands at 471 and 460 nm. The band at 471 nm is a little distant from the other bands and this helps to distinguish yellow sapphire from chrysoberyl where a solid block of absorption is closer to the blue.

41. *YELLOW SPODUMENE* shows another ferric iron (Fe^{3+}) spectrum with bands in the violet at 438 (s) and 432.5 nm (w).

42. *YELLOW ZIRCON* shows the spectrum of uranium much more fully than red zircon. The strongest bands are at 691, 662.5, 659, 653.5 (s), 589.5, 562.5, 537.5, 515, 484 and 432.5 nm. The golden yellow stones made by heat-treating brown crystals show a less strong spectrum, although the bands at 653.5 and 659 nm can always be seen. It is suggested that reflected light may be found useful if bands are too faint to be seen by transmitted light.

Still the best account of absorption spectra (from which much of the above is drawn) is a series of articles by B. W. Anderson and C. J. Payne in *The Gemnologist* from September 1953 to December 1956. Unfortunately sets of this periodical are now extremely rare.

5.5 ALTERATION OF COLOUR

Heat treatment

The ability to alter the colour of gemstones by heating has been known at least since classical times. Amber can be darkened; the aquamarine variety of beryl is routinely altered from green to a bright steely blue; yellow beryl can lose its colour and orange beryl can be turned to pink. Blue sapphire can be darkened or lightened; brownish or purplish ruby can be changed to give a better red; pink corundum can be altered to orange; asterism can be induced or removed and colour added to pale or colourless stones. The colour of irradiated diamonds can be altered, ivory can be darkened, amethyst can be made colourless, yellow, brown, green or milky. Smoky quartz can be turned to a greenish-yellow or can have the smokiness removed. Kunzite can be turned purple or violet from very pale original material. Yellow, brown or blue topaz can be turned colourless; blue or blue-green tourmaline can turn to green and the colour of red stones lightened. Brown zircon is heated to give reddish, colourless or blue material; green zircon can be altered to blue or yellow. Zoisite can be heated to give the deep purple-blue tanzanite.

There is a wide variety of techniques and apparatus used in heating stones; none are very complicated and in general no very high temperatures are required (stones would crack, in any case). Critical factors are the maximum temperature reached, the time for which that temperature is sustained, the rate of heating and cooling, the chemical nature and pressure of atmosphere and the nature of any material in contact with the stone. Methods of heat treatment are notoriously hard to come by since, for commercial reasons, operators are unwilling to disclose their secrets.

The nature of the heated stones varies so widely, even within species, that predictions as to how a particular specimen will behave are unreliable. Nassau (1984) in his book *Gemstone enhancement*, from which this chapter draws freely, lists the various processes by which heat affects gem materials. The darkening of amber results from charring and/or oxidation; removing the colour from blue or brown topaz or zircon, the 'pinking' of brown topaz, the change of amethyst to pale yellow or green and the change in smoky quartz to greenish-yellow or colourless are the results of the destruction of colour centres. A change in hydration or aggregation by heat is responsible for the alteration of carnelian to orange, red or brown, and sapphire to deep yellow or orange. Metamict states can be reversed as in zircon and oxidation states can be changed, usually with oxygen diffusion, as in changing aquamarine from green to blue, amethyst to citrine, sapphire to various colours, and brown and purple ruby to red. Asterism in corundum can be induced to form when heat causes rutile crystals to precipitate out at right angles to the c-axis of the crystal. Re-heating can remove stars already formed. Colour addition by impurity diffusion causes diffused colour and asterism in sapphire. A rapid change in temperature causes cracking which gives fingerprints in sapphire and makes 'crackled' quartz. Reconstruction and clarification gives flow under heat and pressure, examples being reconstructed and clarified amber and reconstructed tortoiseshell.

Oxidation and reduction of minerals during heating can be critical, especially for blue sapphire. Varying amounts of oxygen can affect the valence state of an impurity; when, for example, iron is present in Al_2O_3 in either the ferrous (Fe^{2+}) or ferric (Fe^{3+}) state. The two states can be altered from one to the other with a change from oxidizing to reducing conditions, as will be seen later when corundum is discussed.

Diffusion Heat treatment can cause impurities to diffuse into a stone to alter the colour or to produce asterism. Chromium can be diffused into a corundum to turn it into a ruby; this takes a long time and higher temperatures are required than those needed for other diffusions because colouration needs to proceed right through the stone. Diffusion methods involving burying the stone in a powder to give stars reach only a few fractions of a millimetre (typically 0.010–0.025 mm) inside the stone.

Cracks into which dye can penetrate by diffusion can be caused by heating a stone and dropping it into water. Even without a dyestuff in the water the cracks will themselves produce attractive colours through interference. This is typically done with rock crystal. Imitation inclusions to give the impression of natural corundum can be produced if fine cracks can be made by uneven heating as with a blowtorch. If the stone is re-heated the cracks may be partially healed to give characteristic 'fingerprint' inclusions typical of corundum.

Amber can be clarified by placing it in rape-seed oil and heating it for many hours. Cooling must be effected at the same slow rate. It is possible that the air or water filling the bubbles that cause the cloudiness escapes by diffusion during the heating process.

Irradiation

The colour of gemstones can be altered by exposing them to a variety of radiations, including visible light (with an energy of 1 to 3 eV), or ultraviolet (UV) (with an energy of about 5 eV). More alterations are produced by exposure to X-rays or gamma rays (10 000 and 1 000 000 eV). A typical unit to produce gamma rays from cobalt-60 gives an irradiation intensity of one million rads or 1 megarad. Colourless quartz will turn smoky in less than 30 minutes under these conditions.

Energetic helium nuclei (alpha rays) and energetic electrons (beta rays) can also be used. Electrons of high energy are produced most commonly in linear accelerators. Although energies up to one billion eV can be obtained, very high energies may cause radioactivity in the stone and in any case electron bombardment can only give surface colouration as the electrons are strongly absorbed. Protons (positively charged hydrogen nuclei), deuterons (the nuclei of the heavy isotope of hydrogen) and alpha particles have also been used in irradiation, although radium, which leaves the specimen radioactive through decay products, is not used today. Neutrons (uncharged particles of unit mass) will pass right through stones, thus giving the possibility of uniform colouration (their lack of charge means that they are not repelled by atoms and thus interact forcibly when striking a nucleus. Some radioactivity may be induced by neutron irradiation.

For most irradiation purposes energies of from 5 to 10 eV are needed. Gamma rays are, however, quite often used since they give uniform colouration, consume no power and do not induce localized heating or radioactivity. Nassau (1984) gives a useful table of the major changes in gemstones after irradiation: beryl and aquamarine turn from colourless to yellow, blue to green, pale colour to deep blue Maxixe type; corundum turns from colourless to yellow or from pink to the orange-pink padparadschah colour; colourless and pale coloured diamonds may turn black, blue, green, yellow, pink, brown or red. Pearls darken to grey, brown, blue or black (not a true black but a very dark blue–grey). Colourless, pale green or yellow quartz will turn smoky, amethyst or citrine colour combined with amethyst as a parti-colour. Spodumene and kunzite turn yellow or green, topaz will turn from colourless to yellow, orange, brown or blue (some specimens may fade). Colourless and pale-coloured tourmalines will turn to yellow, brown, pink, red, bicolour green and red, or blue to purple; some stones may fade after treatment. Colourless zircon may turn brown to reddish.

Many stones have been irradiated naturally at some time or times during their life; any colour change so caused can be subsequently lost or altered before mining by heat. This may explain why some stones vary in colour (e.g. not all topaz is blue or brown).

Irradiation and colour centres The alteration of gemstone colour by irradiation is often an effect on colour centres which, as discussed earlier, are responsible for the colour of a number of species. Old colourless glass exposed for a long time to bright sunlight may turn purple, that colour being driven off by heating and restored to greater than its previous intensity by gamma-irradiation. This shows a colour centre in operation.

Colour centres in a substance involve either the loss of one electron from a place where one would normally be found (hole colour centre) or the presence of an extra electron (electron colour centre) – see pp. 76–7. Irradiation can create both types of colour centre at the same time by transferring an electron from its usual paired state to a higher energy level, and trapping it there. If the trapped electron is freed by light or heat the original state is restored and the colour lost by fading. When the trap is shallow, even room temperature in the dark may be enough to release the electron (this is about 0.1 eV of energy). Less shallow traps need more energy to release the electron but it could still be released by bright sunlight, for example. Some irradiated yellow sapphire, brown topaz and the deep blue Maxixe beryl fade for this reason. Yellow sapphire, with some other gemstones, comprises both fading and non-fading colour centre forms. Only a fade test, involving exposure to light or other radiation, can distinguish which is which. Still deeper centres will normally give stable colour as in smoky quartz, amethyst, blue topaz and stable brown topaz. These stones may need temperatures up to 500°C to bleach.

Chemical treatment

One of the commonest chemical treatments is bleaching; it is used for pearls, both dark specimens and those tinged with green; coral, too, is bleached. In most cases the bleaching agent is hydrogen peroxide or bright sunlight. Brown tiger's-eye may be bleached to give a paler colour.

Impregnations with wax (colourless or coloured) to hide unsightly cracks have long been carried out. Materials like turquoise are stabilized by this treatment to consolidate a powdery nature and also to darken the colour. Light scattering is reduced as the surface is smoother after the impregnation. There are rare cases of a coating being applied to a stone in order to make them harder (synthetic sapphire film over synthetic rutile). Emerald and other stones are frequently oiled or even impregnated with plastic which does not seep out of the cracks when the stone is warmed.

Materials like chalcedony are frequently dyed and this practice, of considerable antiquity, is commonly accepted by the trade. Concentrated hydrochloric acid dissolves the iron in the stone which is then disseminated throughout the stone to give a yellow–red–brown colour as a thin film on internal surfaces after treating and warming. Chalcedony can be made deep yellow, brown or black by using sugar in solution followed by sulphuric acid treatment.

Surface alterations include wax or oil film treatment, inking, making mirrors on the back of the stone, setting it in a coloured closed mount, and synthetic overgrowth. Purple can be applied to the back facets of a yellowish diamond to improve its colour, purple being the complementary colour to yellow. Varnish and lacquer coatings have been used and a large number of different materials placed behind stones to give interesting colours; a salmon-pink is a characteristic colour for foiled glass. Even the wings of insects have been used. Some backings have star-like rays engraved on them, the star being visible at the top of the stone.

Specific treatments

Amber Some amber darkens on its own, perhaps from air–oxidation. Bubbles in amber (less than 1 μm in diameter) can be removed by heating in oils, the process being slow.

Andalusite may turn pink from olive-green when heated; brown faces become colourless above 800°C.

Benitoite may owe its colour to iron–titanium charge transfer. On heating for 19 h at 600°C the blue colour gradually disappears; some colour may be restored by irradiation with cobalt-60.

Beryl (for further details refer to Sinkankas, 1981). Pink stones (coloured by manganese) may fade with heating to 500°C. Iron-coloured beryl (yellow, green, blue) may contain either Fe^{2+} or Fe^{3+} or both; green stones contain Fe in two different sites: Fe^{2+} in an interstitial site between the SiO_6 units gives a blue which heat cannot alter. Many specimens also contain some Fe^{3+} substituting for Al^{3+} which on its own gives yellow or with the Fe^{2+} (blue) gives green. When the beryl is heated the Fe^{3+} in the channel gains an electron from a trap to form Fe^{2+}, which does not give any colour from absorption in the visible. Heat therefore bleaches yellow beryl to colourless and turns green beryl into aquamarine. Irradiation can reverse this process and further heating restores the colourless or blue state. Beryl containing both iron and manganese changes from orange into morganite if stones are buried in sand or charcoal and temperatures from 250 to 500°C are used. All heat-induced changes can be reversed with gamma rays but are stable to light. Some Californian morganite is reported to fade from peach to a stable pink in daylight.

The dark blue Maxixe or Maxixe-type beryl which fades on exposure to light is most attractive and it is sad that the colour can never be stabilized. The original stones were found at the Maxixe mine in Brazil in 1917 and in about 1971 stones of a similar colour and with similar behaviour appeared on the market; they are said to be the result of a process involving a pink beryl from Barra de Salinas, Minas Gerais, Brazil.

The blue colour is caused by a colour centre which can be produced in appropriate material with the right precursors. This appears to be a nitrate impurity in the original natural material and a carbonate impurity in the Maxixe-type stones. These two ions have four atoms with 24 electrons in their outer shells. Both can lose one electron on irradiation and this forms a 23 electron hole centre:

$$NO_3^- \rightarrow NO_3 + e^-$$

$$CO_3^{2-} \rightarrow CO_3^- + e^-$$

The free electron is trapped, perhaps at a hydrogen ion, to give a hydrogen atom:

$$H^+ + e^- \rightarrow H$$

Colourless or pink materials give a deep blue, yellowish ones give bluish-green or green. The electron centre is unstable so that slight radiation, even daylight, restores the starting colours by releasing the electron, typical fading being loss of half the colour depth in one week of exposure to bright sunlight. Heat can also bleach the irradiated stones. The blue material on the market in recent years has been given the trade name Halbanita and this has gained its deep blue from exposure to UV for one to a few weeks. Anomalous dichroism and absorption spectra characterize the stones; the deeper colour is shown by the ordinary ray, whereas in aquamarine it is carried by the extraordinary ray, and there are absorption bands at 697 and 655 nm with weaker bands toward the yellow. Emerald overgrowth on a colourless seed is described by Sinkankas (1981) in the chapter on synthetic stones.

Oiling and waxing of emerald are common; some stones take their colour from the evaporation of a solvent rather than from coating, the colour concentrating in cracks. Cleaning followed by impregnation under vacuum conditions is practised in Colombian emerald markets. Stones are cleaned by boiling in methyl alcohol or ethyl alcohol and then placed in aqua regia (a mixture of concentrated nitric acid with 3 to 4 volumes of concentrated hydrochloric acid; very dangerous to handle). After drying, oil is applied. If the oil is coloured, a stone can be recognized as having received this treatment on immersion, the colour being concentrated in particular areas. The dye may give its own absorption spectrum. Colourless oil is harder to spot, although the outline of the cracks may still be seen and gentle warming may give the smell or feel of oil; those commonly used include light machine oils, cedarwood oil, Canada balsam and clove oil. Solvents may remove some of the oil. An emerald which was found to have been coated gave a very low RI of 1.48. On removal of the coating the RI was found to be 1.579, 1.588, characteristic of Zambian emerald.

Chalcedony This is so frequently dyed that the practice arouses no adverse comment and has no commercial significance. Heating turns pale colours to brown and red or to milky white, both types being stable. Organic dyes may fade while inorganic colourants remain stable. Dreher's *Das Farben des Achates,* published locally in Idar-Oberstein in 1913, gives a useful summary of the dyeing processes used in Germany at that time. All colouring agents were inorganic. After cleaning with soda solution and then with acid to eliminate grease and traces of iron the agates are washed. To give red, 0.25 kg of iron nails were used; green was obtained by two separate saturations, of chromic acid and ammonium carbonate. Nickel was also used on occasion. Black was obtained with a sugar solution followed by steeping in concentrated sulphuric acid. For blue, potassium ferricyanide was used followed by immersion in a warm saturated solution of ferrous sulphate with some drops of sulphuric and nitric acids. Many of these processes are dangerous. Many dyed chalcedonies substitute for other gem materials and should be tested by conventional gemmological methods.

Coral is frequently dyed to give an ox-blood red or angel-skin pink (this has no orange tinge). Colours may be fugitive or stable depending on the dye used. Black coral can be bleached to give a fine golden colour. Cavities in coral have been filled with epoxy resins. Blue opaque coral beads with a plastic coating have been reported. Gas bubbles can be seen in the coating which is fairly soft, yielding to the point of a pin. The plastic nature of the coating gives an acrid odour with the hotpoint.

Corundum This material can be altered in a wide variety of ways; since the mineral occurs in so many different colours there is always the impetus towards improvement as well as outright alteration. Inclusions can be removed or induced and asterism produced or eliminated. The best summary of the colour changes of corundum so far encountered is that by Nassau (1984).

Pale green, pale yellow or nearly colourless corundum can be heated to give a yellow, golden or golden-brown. This is due to Fe^{3+} (it may or may not be substitutional) and is often a fine dark stable gold; temperatures needed range from below 1500°C to about 1900°C and an open crucible is used so that oxygen is available for the conversion of Fe^{2+} to the desired Fe^{3+}. Fading yellow stones are much more likely to have been irradiated than heated. Chromium-bearing pink sapphire with some iron content may be heated to give the stable padparadschah orange–pink colour.

Irradiation produces two types of colour centre, one stable, the other not. Both can occur at the same time in yellow sapphire. Nassau calls the two centres yellow stable colour centre and yellow fading colour centre (YSCC and YFCC). If yellow colours are found with chromium the padparadschah orange–pink is formed on irradiation; blue sapphire can be turned green. Some natural yellow sapphires contain Fe^{3+} yellow and others the YSCC. As the YFCC fades so rapidly in light (a few hours in sunlight) it is unlikely to be seen in cut stones. Some synthetic corundum may be irradiated to give a yellow or brown but these colours fade. Both colour centres are removed by heating which converts yellow to a paler yellow or to colourless, padparadschah to pink and green to blue. Further irradiation restores the colours as long as they originate from a colour centre in the first place.

Heating for some time between 1600 and 1900°C causes titanium dioxide to dissolve in the alumina and thus eliminate asterism. Heating for an extended period at 1300°C will bring the star back, the rutile crystals growing in thickness with increase of time.

The combination of iron and titanium oxides can give colourless, yellow, green, blue–green, blue and black in corundum. The actual colour depends upon how the mineral has been heated and the colours may be modified by small amounts of such elements as chromium.

Much information on the alteration of colour in corundum is jealously guarded by practitioners in the art but some definite features have been established. Whitish or pale blue corundum of poor colour and often containing rutile crystals in interlocking masses ('silk') has been called Geuda material. This can be heated in air up to about 1200°C to remove any blue present. A deep blue, on the other hand, is developed by heating up to perhaps 1900°C at most; this will also remove the 'silk'. The higher temperatures can be held for a shorter time and thus save heating costs. It is believed that the stones are heated in an alumina crucible after being coated with a borax-based solution. They are surrounded by charcoal in the crucible to give a reducing atmosphere; some additives are employed to reduce cracking and for other purposes. Borax appears to be used most often. The heating may take a few hours or days. It may be repeated where necessary. The treatment damages the surface and this is a useful clue for the gemmologist.

Very dark blue sapphires can be lightened and one report states that a few minutes at 1200°C will be enough (Australian sapphires have not been metamorphosed like those from Sri Lanka which are the ones most often treated). Some near-colourless corundum may turn to a bicolour blue–yellow. Dark brownish or purplish rubies long known as Siam (or Thai) rubies may be heated to remove the blue or green component which darkens the more desirable red. (Darker rubies appear less frequently on the market nowadays, so this report may be substantiated.)

Corundum is often oiled or waxed just as beryl is and the techniques are similar. A coloured oil is reported to be used in Thailand. Foiling and coating are rare in corundum gems although dyeing the back was known to Cellini in 1568.

A recent method of improving the colour of ruby is to fill surface cavities with glassy substances which makes the stone appear clearer. The fillings are prone to fracture and any stone with a table facet showing signs of damage should be regarded with suspicion. Differences in surface lustre will also be apparent and stones with a weight which will appear incorrect for the size of the stone. A stone examined by the Gemological Institute of America measured 12.78 mm in length and contained a glass-filled cavity 7.40 mm long. Immersion, as so often, is a useful test and the occasional RI (one stone gave 1.56) will betray the presence of glass. Gas trapped between stone and filling will show up as uneven voids. Some sapphires have been found to have a glass coating on the surface and this is thought to result from the preparation of the stone for heat treatment (a coating is applied before the treatment begins). A colour-enhanced corundum bead was found to have dye lining the drill hole. This was sufficient to give the whole stone a purplish-pink colour. Sometimes a red string is used for the same purpose.

Oiling can be identified as in the beryl gems, but identification of those stones treated by other methods may not be so easy. The method of diffusion gives colour in a thin surface layer (in orange and blue faceted stones and cabochons) of a few tens of micrometres. The colour variation can easily be seen when the stone is immersed and the layer can of course be polished away. Asterism can also be induced (see above). Bleeding of colour around cracks and pits helps to identify diffusion.

Multiple girdles and pock-marks identify heat-treated corundum; they arise from the need to repolish the stones after treatment since they are marked by the heat or by the borax coating. Stress cracks around inclusions are common. The 450 nm absorption line will not usually be seen in colour-enhanced blue sapphires nor in yellow stones which may show less fluorescence than their colour suggests would be likely. Melting of rutile needles during heating can cause impurity diffusion from molten drops and these may show up as diffuse orange–yellow haloes. Blue colour-enhanced sapphires may show abnormal dichroism of violet and greenish-blue or greyish-blue and there may be a chalky green fluorescence. Absence of silk should also be grounds for suspicion. Colour-enhancement of ruby takes place at lower temperatures so some of these signs may not be so clear. Interestingly it has been noted that heat-treated corundum tends to 'plink' on a hard surface rather than 'plonk'.

Carbon dioxide as an inclusion in corundum can occur inside negative crystals. On heating the gas expands and may even explode out of the stone and damage it; as much corundum is now heated as a matter of course to see if the colour improves, the incidence of CO_2 inclusions has declined. The presence of such an inclusion can be taken as proof that heat treatment has not taken place.

Diamond On irradiation, diamond may give blue, green, brown, orange, very dark green and yellow from stones which were originally off-white. There are two processes, one using electrons and the other using radium salts; the latter causes radioactivity in the stones and thus is never used other than experimentally. In addition to radium, neutrons from a nuclear reactor, protons, alpha or heavy particles from a cyclotron or linear accelerator have been used. Sir William Crookes buried diamonds in radium salts in about 1904; the treatment made the stones dark green and the colour resulted from fast-moving nuclei recoiling from their disintegration becoming implanted into the surface. Heating does not remove the radioactivity. Virtually all processes give a similar result in that an absorption band GR 1 is formed extending from the GR 1 line in the infra-red at 741 nm to well into the yellow–green. This gives a blue–green, dark green or black. Heating of the stone during the irradiation can lead to other colours being formed. Most radiation, apart from gamma rays and neutrons, colour only a little way into the stone and an umbrella-like effect can be seen surrounding the culet when cyclotron irradiation is carried out from the side of the stone. Localization of colour can be seen in other positions and should always be sought in testing a diamond. High-energy electrons penetrate deeper than heavy particles and give better colouration. The stones become a blue or blue–green and the main GR 1 band is created. All neutron and electron-produced colours used today are stable. Odd examples of colour change include a greenish-yellow fluorescing diamond which turned after irradiation to an orange–red with a bright orange–red fluorescence (perhaps a stone known as Type 1b, see page 176).

Colours produced by irradiation can be modified by subsequent heating. Usually the stones change from blue to green to brown to yellow and back to the original colour. Heating destroys the GR 1 band starting at about 400°C and lines at 595 (594, 592), 504 and 497 nm are formed, these being at their strongest at about 800°C, giving (from the blue) a brown, green, yellow or orange. By 1000°C the 595 nm line disappears although the colour of the stone does not change significantly so that its absence does not mean that a yellow diamond is naturally coloured. In natural yellows the 504 nm line is stronger than the 497 nm line but this situation may be reversed after irradiation. If the line at 415.5 nm does not disappear with irradiation, it shows that the original stone was a member of the Cape series. Those absorption lines characteristic of the brown series can be altered by heat. Stones known as Type 1a can be changed to a bright yellow by heating under high pressure to near 2000°C for short periods only. Type 1b stones have been changed to 1a with lightening of their yellow colour.

The so-called chameleon diamonds are greyish-green to yellowish-green naturally but change to a fancy yellow in the dark or on gentle heating. Exposure to light or to UV for a minute or two restores the original colour. This presumably is the work of a very shallow colour centre. Diamond overgrowth on diamond has been reported with the presumed intention of weight increase. The drilling out of inclusions with laser beams followed by acid etching is widely practised; so long as the hole is perpendicular to a facet near the inclusion the optics of the stone will not be badly affected.

Coating and foiling, usually bluish to make a yellowish stone look a better colour, are still practised; even a touch of blue under the prongs on the girdle can improve the look of the stone. Coated diamonds have been re-appearing in the trade; the coating is very thin and gives a greyish appearance. Sometimes it is in the form of a single band on one side of the girdle; in other cases the coating is on both crown and pavilion sides of the girdle. A 9.58 ct fancy pink diamond was replaced at an auction sale by a diamond weighing 10.88 ct which had been painted all over with pink nail polish; the true colour of the stone was light yellow. Had the sale gone forward, the price would have dropped about $500 000 to between $12 000 and $15 000.

For testing treated diamonds a spectroscope is required with facilities for cooling the stone in a vacuum, to make some absorption lines easier to see (especially those at 595 and 504 nm). Woods and Collins (*Journal of gemmology* 20, 2, 1986) found that it was not possible to colour a diamond yellow or brown artificially by radiation damage and heat treatment without producing at least one of the absorption lines at 595 nm, 1936 nm or 2024 nm. Where all three are absent it is safe to say that the diamond is naturally coloured. Liquid nitrogen is used to record lines in the visible – it is not needed for the observation of lines in the near infra-red.

A report on a treated diamond shows how careful the gemmologist now has to be: the stone, greenish-yellow and weighing 1.23 ct, showed no trace of treatment and was labelled as natural after two weak lines above 500 nm were seen and interpreted as the 504 line; there was also an unusual line followed by general absorption below 500 nm. No fluorescence was detected under UV. When the stone was tested at 20°C with a spectrophotometer, traces were seen of the two lines in the green, the strong absorption in the blue and traces of lines between 750 and 700 nm. At the temperature of liquid nitrogen (−196°C), a doublet at 737 and 732 nm was seen instead of the usual lines at 741 and 723 nm. There was strong development of the 594 line and a group of lines at 516, 506, 512 and 503 nm were ascribed to radiation damage. Other lines were recorded at 489, 478, 475, 468, 464 and 462 nm with a strong line at 445 nm. The lines at 489 and 445 are due to irradiation. Running a test in the infra-red did not show the line at 690 nm which had previously been claimed to persist throughout irradiation. It appears that this line may not develop in diamonds with a high nitrogen content.

Ivory can be heated gently to give an aged appearance with darkening and dark specimens can be lightened with hydrogen peroxide. Walrus ivory is known to have been stained green and copper salts will give ivory or bone a blue colour on heating. It is not usually possible to detect darkened or bleached ivory.

Jade can have its green colour lightened by heating but many stones of indifferent quality are dyed. The most difficult process to detect, however, is the hollowing out of a fine quality white translucent jadeite and placing a second piece into the hollow, the colour coming from a green gelatinous material inside the hollow; the back is sealed by a third piece of jadeite. The absorption spectrum shows woolly bands in the red from the dye. Some dyed jades show strong fluorescence (orange) under long-wave UV.

Lapis lazuli This mineral can be marred by white calcite inclusions which can be dyed; the colour concentrates in cracks and can be removed with acetone (or nail polish remover) on a swab. Some lapis lazuli is waxed rather than dyed to improve the colour. The wax does not permeate the parts of the lapis where pyrite and calcite are present and the coating thus appears uneven.

Opal If opal is heated, all its water is driven off, so destroying its colour. Water can be re-introduced, however, in a vacuum of at least 700 mm Hg pressure and the play of colour is then restored. Black dyed opal is made by immersing the stone in a sugar solution for several days. It is then cooled and rinsed before being placed for one or two days in concentrated sulphuric acid which is heated to 100°C. The stones are then washed, rinsed in bicarbonate of soda and washed again. They can then be polished and in this way an apparently black opal is produced. The carbon giving the colour can be seen as dots in the stone. Smoking opal to produce an apparent black is carried out with stones from Jalisco, Mexico (the very porous hydrophane opal). The stones are preformed then wrapped in brown paper, or wrapped after soaking in old sump oil or wrapped in layers of newspaper; the wrapping, whatever its nature, is then heated to carbonize it. Carbonaceous material thus comes from the wrapping.

A report by Mitchell (1982b) shows that a number of opals appearing on the market have been oiled to reduce the prominence of cracks. Naturally anyone offering such a stone for sale should state that it has been oiled. If the stone is examined with light passing across the crack from one side the crack will be easier to see than if it passes down the length of the crack. A smell of oil or a tackiness are also signs which should be taken seriously. Oil is chosen to have a refractive index as close as possible to that of the opal.

Pearl Pearls can be darkened with gamma irradiation to give a black (really a bluish-grey). Freshwater Lake Biwa pearls turn silver–grey when irradiated by neutrons. All types of pearl can be bleached, usually with hydrogen peroxide. Pearls which have been worn for a long time can be affected by acids from the skin of the wearer as well as by cosmetics and perfumes. Very rarely the top surface can be peeled off to expose a brighter one below but this technique is difficult.

Many pearls are dyed after first bleaching, the tints being pale and not always needing to be disclosed under trade regulations. Various natural organic and inorganic dyes can be used. Mother-of-pearl beads are also sometimes dyed. Most dyestuffs will concentrate in cracks or in the drill-hole. Cultured black pearls are now made, natural blacks coloured by silver nitrate dissolved in a dilute ammonia solution. X-rays are strongly absorbed by the silver; X-ray fluorescence also reveals silver. There is no red fluorescence under long-wave UV as with natural black pearls.

Quartz Iron in the ferric (Fe^{3+}) state can give quartz a golden-yellow or reddish-brown colour (citrine). Iron in the ferrous (Fe^{2+}) state gives the green prasiolite. Pink rose quartz owes its colour to a combination of titanium with iron. Smoky quartz and a greenish-yellow quartz owe their colour to a colour centre involving an aluminium impurity. All quartz contains some aluminium, usually about one Al^{3+} for every 10 000 Si, as well as similar small amounts of hydrogen and alkali metal ions. (When Al^{3+} replaces Si^{4+} one positive charge is missing so that the electrical balance has to be restored by one H^+ or Na^+ for each Al^{3+} present; they are known as interstitial ions because they occupy any available unoccupied space in the crystal.) If one of a pair of electrons is ejected by irradiation it may become trapped at the H^+:

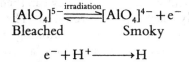

$$[AlO_4]^{5-} \underset{}{\overset{\text{irradiation}}{\rightleftharpoons}} [AlO_4]^{4-} + e^-$$
Bleached   Smoky

$$e^- + H^+ \longrightarrow H$$

The $[AlO_4]^{5-}$ is the hole colour centre which gives the dark colour to smoky quartz; bleaching is brought about by the reverse of the second reaction on heating, which releases the trapped electron and allows the first reaction to be reversed. The amethyst colour centre operates in a similar way, with Fe^{3+} replacing Al^{3+} and in this case it is the $[FeO_4]^{5-}$ hole colour centre which causes the colour;

$$[FeO_4]^{5-} \rightleftharpoons [FeO_4]^{4-} + e^-$$

When the smoky stone is heated, the reverse reaction produces $[FeO_4]^{5-}$ which gives pale yellow citrine which can be irradiated to give amethyst. Pale yellow synthetic quartz grown with iron under oxidizing conditions, to give Fe^{3+} in place of Si^{4+}, will give green quartz when heated under reducing conditions with an Fe^{2+} content in place of Si^{4+}. On irradiation the Fe^{2+} is first converted into Fe^{3+}, and this can then form the amethyst colour centre when further irradiated. Greened amethyst is the result of heating to reverse the process.

The smoky colour produced in quartz by irradiation derives from what is called the A_3 broad absorption band close to 440 nm which is accompanied by bands at 690 and 550 nm. The A_3 band originates from a colour centre which consists of a hole on one of the four oxygens next to an aluminium ion that has substituted for silicon, with the electron trapped at a neighbouring hydrogen or alkali ion. The A_2 centre derives from the absence of an electron from two of the four oxygens next to the aluminium and the A_1 from an electron missing from three of the four oxygens. Virtually all quartz contains some aluminium so that a smoky colour can arise on irradiation, this is stable to light, although it can be lightened by heat. Heat will turn natural or irradiated smoky quartz through greenish-yellow (honey quartz) to colourless; this colour is not due to iron and its origin is at present uncertain.

Much citrine in commerce comes from heating amethyst at about

450°C. If the temperature is raised to about 500–575°C the colour passes to orange–red and higher temperatures still will give a milky appearance. Some pink quartz may be intensified by irradiation and heat-bleached stones may have their colour restored. Much blue quartz gets its colour from colloidal-size rutile crystals or from cracks which cause Rayleigh scattering (see pp. 72–3). Heating from 300° to 1000°C can give a violet colour or eliminate all colour.

Rainbow quartz, made by heating rock crystal and plunging it into a cold dye-filled liquid, has already been mentioned. Much quartz is foiled and brown tiger's-eye can be lightened by removing the iron in the tubes by bleaching agents containing chlorine. Surface coatings to give deep yellow have been seen on faceted colourless quartz. The coatings act as interference filters and may be applied to the lower central area of the pavilion. (The pavilion is the lower part of a faceted stone, i.e. below the girdle. The upper part is called the crown.)

The amethyst–citrine variety of quartz can be obtained by heating sectored amethyst from Minas Gerais, Brazil, either by a medium heat (the change being reversible by irradiation) or by a more intense heat followed by irradiation and a second more gentle heating. Synthetic quartz may be treated in the same way. The second process cannot be reversed to bring back the all-amethyst appearance. The temperatures used are in the region of 450°C.

Spodumene The variety of spodumene known as kunzite can be irradiated to give a deep green which quickly fades. This is due to the change from Mn^{3+} to Mn^{4+} rather than to a colour centre. A true colour centre operates in the irradiation of kunzite from the Malagasy Republic which turns brown, fading quickly to green which itself quickly fades. An intense orange spodumene varying to greenish-yellow recently found on the market advertised as citrine, was radioactive from treatment in a nuclear reactor. The colour was stable to short exposures to sunlight. Brown irradiated spodumene when heated to about 80°C turns to green which will bleach to give the original pink at about 200°C. At temperatures of about 500°C kunzite fades but the colour can be restored by irradiation and heating. A clear pink can be obtained from a purple stone by heating at temperatures in the 100–250°C range.

Topaz Topaz, with its colours of yellow–red–brown, blue, pink to violet and colourless, has long been a subject for colour alteration. All colours are the result of colour centres. Most topaz will turn brown on irradiation and this forms at two different rates, the brown stable colour centre (BSCC) fades either very slowly or not at all while the brown fading colour centre (BFCC) fades quickly. Not all irradiated colours fade; whether a particular one does or not appears to depend upon composition, but is not well understood. If a natural brown topaz is heated to drive off the colour which is then brought back by irradiation, the original fading stone will still fade.

Some topaz turns from brown to green on irradiation and following heat or exposure to light the yellow or brown component will be driven off, leaving a blue colour which is stable to light. A very intense deep blue is formed when topaz is strongly irradiated, as in a nuclear reactor; some blue stones have been made radioactive by this process; it is said that only pure topaz can safely be treated in this way since the persisting radioactivity is brought about by the activation of impurities. Some brown topaz containing chromium will turn pink on heating – gentle heating in a test tube over an alcohol flame is sufficient. This treatment gives a stone with an emission line at 682 nm. The original colour can be brought back by irradiation.

Identification is not always easy. Natural pink topaz is said to have a much lower pleochroism than the heated kind; irradiated brown stones have a lower refractive index (1.61–1.62) than natural browns (1.63–1.64) with a specific gravity of 3.53 compared to 3.56–3.57. Blue irradiated stones show a large thermoluminescence at about 360°C which is not seen in natural blue stones.

Tourmaline Tourmaline has a great variety of compositions and colours; irradiation can intensify an existing red or yellow colour or create it. Very pale pink stones deepen their pink colour or turn red; blue and green stones may turn purple; yellows may change to an orange or peach colour; medium green stones can go grey. Some greens may change to a red–green bicolour. Irradiation can intensify yellow and turn pink stones orange or colourless. Blues and greens may turn bluish-green or yellow. Blue–green dark stones can be lightened by heating to about 650°C; some stones may turn an emerald green but as heating drives off essential water (at about 725°C) care should be taken. There seems to be no way to identify treated tourmalines.

Turquoise Turquoise often has its pores filled with oil or wax or is stabilized by plastics. Paraffin wax is a favourite, the stone when dry is soaked in warm melted wax for a few days. Dyeing is not usually used since the other processes improve the colour on their own. Coating is commoner with surface impregnation being protected by a colourless layer of lacquer. Most treatments can be seen under the microscope. The hotpoint will show plastic coatings and wax impregnations; local concentrations of dye are of course suspicious.

Zircon Zircon is often heated to bring low (metamict) stones back to the high, fully crystalline state. Brown zircons are turned blue or colourless by heating them while surrounded by charcoal at about 1000°C for a few hours. This gives some blue, some colourless and some off-colour stones. The off-colours are heated to about 900°C in air which leaves some colourless and some yellow, orange or red. Very slow cooling is thought to be important, especially after the oxidation heating. Sometimes zircon can be discoloured by heating (notably blue and colourless stones) the colour being restored by further heating or by irradiation.

Zoisite The tanzanite variety of zoisite begins as brownish crystals with a strong pleochroism of violet–red/deep blue/yellow–green. Heating for about 2 h at 370°C alters the yellow–green component into a deep blue with the violet–red to deep blue pleochroism. Stones are cut first; some deep blue crystals are found but these are exceptions. Heating above 900°C will drive off water and turn the stones yellow.

6 Fashioning

It has not always been appreciated that a gemstone contains interior beauty and its surface appearance, attractive though this may be, is only part of the total power of attraction: such crystals as the octahedra of diamond and spinel were left virtually untouched in early times. The fashioning of stones began with the grinding down of natural crystal faces.

6.1 FASHIONING OF DIAMOND

The polishing of diamond began with the removal of the 'skin' or coating which frequently covers the crystals and it is not certain when or how faceting as we understand it began. Tavernier in 1665 found that some kind of diamond polishing was carried out in India; they polished the natural crystal faces and if the stone was clear did no more than that. If the stone was included small extra facets were added in an attempt to cover the inclusion but there was no attempt to improve the shape or to arrange for light to be totally internally reflected from the back facets to the crown.

In the early days of diamond polishing the regular forms were the diamond point and the diamond table. The point was formed from the natural faces ground to a regular shape where possible; many small diamonds accompanying larger coloured stones in rings were fashioned in this way. The diamond table was derived from the regular octahedron by grinding down one corner until the table facet was half the square cross-section in width. The opposite corner was ground down to give the small culet parallel to the table. This was wasteful and was accomplished only very slowly. The earliest diamond tables had their angles between the crown and pavilion (the upper and lower portions of the stone) formed by the slope of the natural octahedron faces (54° 44'); later on the angle was reduced to 45° which makes the facets of the crown and the pavilion form a right angle with each other across the girdle.

The rose cut (Fig. 6.1), which in its most symmetrical shape consists of 24 regular triangular facets with a flat base, is still used for small stones and was probably developed in India. There are several varieties of this cut, one of

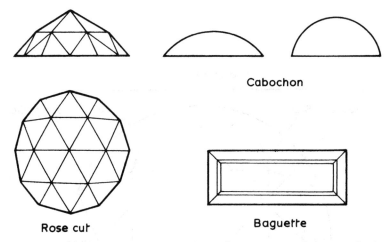

Cabochon

Rose cut   Baguette

Figure 6.1 Gemstone cuts showing the early rose cut and the simple domed cabochon form (which is used for opaque, cat's eye and star stones). The baguette is often used for small diamonds.

the most attractive being the briolette, a modified double rose (in French the term is used to denote a pear-shaped diamond).

The brilliant cut

The brilliant cut probably originated from the table cut during the seventeenth or eighteenth centuries during which time extra facets were added to the original simple cut. The present-day brilliant (Fig. 6.2) when it appeared, with 33 facets in the crown and 25 in the pavilion, together with the culet where present, is so superior to all others in the way in which it allows light to pass through and back out of the stone that it was quickly adopted even though more of the crystal was used up than in previous forms of cutting. In early brilliants the girdle was rectangular or square but the modern brilliant has an octagonal shape with a rounded girdle (Fig. 6.3). The names of the facets have changed over the years and there have been small changes in their size and distribution; different polishers use different names but the simplest ones are easiest to remember:

table (1)
upper girdle facets (16)
stars (8)
crown mains (kites) (8)

pavilion mains (8)
lower girdle facets (16)
culet (1)

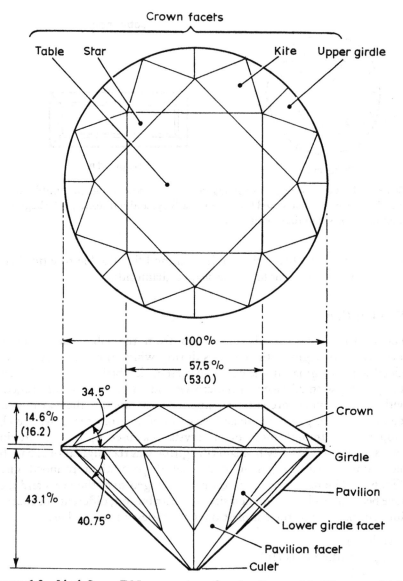

Figure 6.2 Ideal Scan. DN proportions for the diamond brilliant cut (Tolkowsky proportions are shown in brackets where these differ).

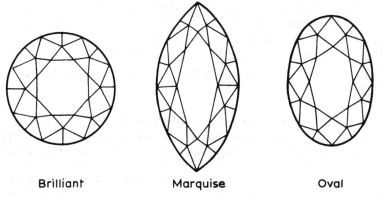

Brilliant   Marquise   Oval

Figure 6.3 Three variations of the brilliant cut. All three cuts have thirty-three crown facets, twenty-four pavilion facets and a culet.

Although the brilliant is the ideal cut for most diamond crystals, small stones are often made as eight- or Swiss-cuts with fewer facets than the standard brilliant. Large stones will have extra facets. Some diamond polishers consider that the girdle should be polished with the aim of increasing the brilliance of the stone, but this makes little difference in fact.

Ideal brilliants

In 1919 Marcel Tolkowsky proposed an angle of 40° 45′ between girdle and pavilion as the ideal angle to give the most dispersion with the most brilliance. Greater angles would increase brilliance but diminish fire. The angle between the girdle and the crown facets was calculated by Tolkowsky to be 34° 30′ for the best results. Small facets at an angle of 42° need to be added to refract rays which have been twice reflected; these are the upper girdle facets; lower girdle facets in a similar position on the edge of the pavilion are added to stop loss of light. Star facets on the crown are about 15° to the horizontal and help to prevent loss of light through the back of the stone. Although decreasing dispersion they increase the number of rays that can be dispersed. The stone proposed by Johnson and Rosch in 1926 modified the Tolkowsky design but gave lower brilliance when seen from the side and the thick crown reduced the spread for the weight. The cut proposed by Eppler in 1940 is used in Germany for high quality diamonds and the cut is sometimes known as the European cut. Here the angle between girdle and crown is 33° 10′ and between girdle and pavilion 40° 50′.

Not only are the angles between girdle, crown and pavilion of great importance if a well-cut stone is to be made; the proportions of table to total width, thickness of crown and pavilion, and depth, are all critical (see

Table 6.1). In general, too shallow a pavilion angle gives a blank effect in the centre of the stone, making it appear dark. This effect is called a fish–eye. If the pavilion angle is too great there is little dispersion. In practice the angles are not always fashioned precisely according to one or other of the published systems and there is a certain amount of latitude given to the polisher.

Table 6.1 Proportions of the Tolkowsky brilliant, the Eppler brilliant, and the stone quoted as ideal in the Scandinavian Diamond Nomenclature (Scan. DN). (Dimensions expressed as percentages of girdle diameter) (from Bruton, 1978)

	Tolkowsky	*Eppler*	*Scan. DN*
Diameter of girdle	100%	100%	100%
Diameter of table	53%	56%	57.5%
Thickness of crown	16.2%	14.4%	14.6%
Thickness of pavilion	43.1%	43.2%	43.1%
Depth of table–culet	59.3%	57.7%	57.7%
Angle of crown facets	34° 30′	33° 10′	34° 30′
Angle of pavilion facets	40° 45′	40° 50′	40° 45′

Other cuts

The step or square cut (Fig. 6.4) used for diamond (with bevelled corners the term emerald cut is used) lacks brilliance and tends to show a dark hole in the centre of the stone; to some extent the hole (sometimes known as a window) can be avoided by cutting the pavilion facets at an angle greater than the critical angle of 24° 26′. Too great a depth causes the diamond to lose brilliance and the best results with this cut are obtained when the width of the bottom pavilion facets is equal to the width of the table.

In the past few years a modification of the square or emerald cut has been developed. The aim was to improve the fire without sacrificing some of the weight which has always been higher with the square cut than with the brilliant. The combination of an emerald-cut pavilion with a modified brilliant-cut pavilion, the one being superimposed on the other, gives more scintillation (the sparkle caused by multiple flashes of coloured and white light) since when the stone is tilted the steps on the crown break up the reflections from the pavilion to give a fountain-like pattern. The dispersion is approximately equal to that of a brilliant. A dark centre is also avoided. This cut, developed by the South African polisher Basil Watermeyer, is called the Barion cut.

In the 1960s a London polisher, Arpad Nagy, introduced the Princess (now the profile) cut. This uses flat crystals to give a large surface area

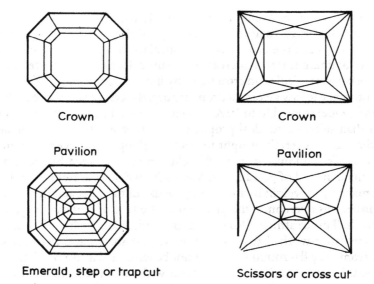

Crown   Crown

Pavilion   Pavilion

Emerald, step or trap cut   Scissors or cross cut

Figure 6.4 The more traditional cuts which are used principally for coloured gemstones.

which is cut by V-shaped grooves on the back. Dispersion is low but the stones are bright. Flat 'portrait' crystals are also used for the rose cut. In crystals with a parallel top and bottom, as in the portrait stone, light is refracted at the upper and lower surfaces so that print can be read through the stone. Where the bottom is left flat and the top cut at an angle, as in the rose cut, all light can be reflected from the back to the front if the lower angles are twice the critical angle; however, this will give no dispersion. To obtain dispersion the cut has to be altered so that light enters and leaves the stone at different angles; this somewhat diminishes the life of the stone so that in the end the rose cut can never be entirely satisfactory.

Diamond polishing

The polishing of a diamond falls into several stages: cleaving or sawing, bruting, grinding and polishing. First of all the rough crystal is subjected to the scrutiny of the works manager who will decide what kind of finished stone can be made from it. This naturally depends upon the shape and state of the crystal but market demands are equally important. It may be more desirable at some times to cut several stones from one piece of rough and at another time to go all out for the largest single stone possible. The conservation of weight must always be borne in mind. It is possible that the crystal may contain inclusions so placed that cleavage will be preferred to sawing (again with conservation of weight in mind) and the manager will decide on this before passing the stone on. He will also decide the make of the stone (its nearness to ideal proportions). With lower quality goods it is desirable to get as close as possible to an ideal make without losing weight; one way of doing this is to increase the stone's diameter by shortening the distance between girdle and crown and by increasing the width of the table. It is customary to saw an octahedron through the centre and make each half a spread (wide) stone. Up to 10% of weight is saved by spreading the stone rather than adhering to ideal proportions. Larger stones may be spread to take them over a certain weight to make a selling point, even though the ideal make is at a lower weight. Weight can also be saved by leaving the girdle thick instead of polishing it. Macles (thin diamond twin crystals) are often made up in weight by increasing the size of the culet and the width of the girdle; this overcomes the problem of the lack of material to give the full depth. This treatment also spreads the table. Cuts such as the baguette are quite easily worked from thin rough; pear shapes and marquises are made from very flat rough which cannot be fashioned into a brilliant and is still too thick to be sacrificed to other cuts without wastage. In the pear and marquise cuts the pavilion may be much shallower than in the brilliant.

The manager or designer will decide where the girdle of the finished stone will lie, designating the thinner side as the table and the thicker side as the pavilion. Inclusions should be kept to the upper part of the stone, unless they can be cleaved or sawn away; if they lie in the lower part they will be reflected a number of times and seem more profuse than they really are. The whereabouts of inclusions may be ascertained by polishing a small window on the crystal, placing it between crossed polars, and looking to see if there is an extinction pattern which indicates strain. Such strained crystals are limited in the ways in which they can be cut.

Cleaving If cleavage is decided upon, a nick is made in the crystal to be cleaved by drawing a diamond crystal with a sharp cleaved edge across it. The nick is called a kerf and will be parallel to one of the octahedral cleavage directions. The crystal to be cleaved is cemented into a stick (wooden rod) and a thin, blunt-edged steel blade is placed in the kerf. The blade is sharply tapped and the stone should split. The process is a highly specialized art and is attended by high risks.

Sawing Most diamond crystals are sawn by the rotating action of a phosphor–bronze disc charged with diamond powder which is held in place by a paste made up of the powder and olive oil. The disc, which is between 0.06 and 0.15 mm thick and 7.5 to 10 cm in diameter, revolves at 4500–6500 rpm. The disc is made to rest against the crystal by an adjustable weight. It takes about 40 minutes to saw through a quarter carat octahedron and a one carat crystal may take up to a day if sawing proves difficult. Such a crystal would take about 0.1 ct of diamond powder and lose about 3% of its weight during the sawing process.

Bruting After sawing comes bruting (or rounding). The crystal is placed in a lathe-like machine, cemented into a dop mounted in the chuck. Rounding is accomplished by holding another diamond against the first, the second being cemented into a dop at the end of a stick about 61 cm long. The stick is held steady under the bruter's arm against the first stone which is rotated until as rounded a shape as possible is obtained. Allowance has to be made for the presence of inclusions in various parts of the stone and the stone is so mounted in the machine that the fewest possible chips need to be removed to achieve roundness. The bruter may leave part of the original skin of the crystal on the girdle to show that he has got the maximum possible diameter from the crystal. The rough section is known as a natural and can be seen on the girdle of many polished stones. The stone used to round the first specimen is then bruted itself, the dops in machine and stick being interchangeable.

The facets The last stage in the polishing of a diamond is to put on the various facets. Strictly speaking, the facets are first ground and then polished but the methods and the equipment used are very similar. A horizontal wheel, called a scaife and made of cast iron, is made to revolve at about 2500 rpm. The diameter of the scaife is about 25–30 cm and the surface is scored so that it can retain diamond paste (similar to that used in sawing). The stone is held against the scaife in a dop whose angle can be adjusted and the dop itself is attached to a rod known as the tang which is held firm on the bench by stops. The surface of the scaife is usually divided into three sections; the first, prepared by dry diamond powder, is used for testing (i.e. finding that point on a facet at which the best start can be made). The running ring on which the facets are ground to size is followed by the polishing ring which gives the best possible finish. This is done by smoothing out the grooves caused by the grinding. This last ring is towards the edge of the scaife, and experts will make a scaife last for up to eight weeks before the surface needs renewal. The polisher can cope with more than one stone at a time, two to four being usual.

For each stone the polisher has to decide on a plan of campaign based on the relationship of the original crystal and its varying hardness directions to the final brilliant cut stone. The directions chosen for grinding are tried out on the testing ring of the scaife; sound is a good guide since a diamond being ground in the appropriate direction gives off a faint ring. If the sound increases in volume the direction taken by the grinding may be moving off the ideal grain direction. Such an unsatisfactory direction will also show on the surface which will show greyish.

The first eighteen facets are put on by the cross-worker (cross-cutter).

These are the larger facets on crown and pavilion and the table is put on first. After these facets have been ground the stone goes back to the bruter so that the girdle can be made perfectly round by a process called 'rondisting'. The stone returning to the cross-worker from the bruter now has the first eighteen facets polished. The curved grooves left by the grinding have to be taken off before the stone goes to the brillianteerer who puts on the remaining facets. He begins with a star facet on the crown and completes his work by putting on the culet. Although this is omitted in many stones, its presence minimizes the risk of fracture, and is therefore a desirable feature.

On completion of the polishing the stone is checked and cleaned; for the latter process an acid bath is used to remove traces of dust and oil. It is often said at this stage that the diamond will never again look so well; this is simply because so many set diamonds are never cleaned.

In recent years automatic cutting machines have largely replaced the skill of the cross-worker and brillianteerer for stones up to one carat.

6.2 OTHER STONES

Stones other than diamond are known as coloured stones (even when colourless) and are cut by a lapidary. Coloured stones are so fashioned as to display their body colour, colourless ones to display dispersion or some other attractive optical effect such as the glow in moonstone. Star stones and cat's-eyes are also the province of the lapidary. Very low-cost material may be tumbled (rotated several times in a cylinder with grinding powders of increasing fineness) but this method is usually reserved for polishing pebbles. The cabochon cut with a flat or slightly convex base and a domed top is used for opaque or translucent materials and also to show off asterism or chatoyancy. The stone to be polished is held in a dop-stick against a wheel revolving on a horizontal axis; this is the traditional way of polishing agates and very large wheels at which the operator has to take a prone position can be seen in the museums at Idar-Oberstein in West Germany. The first grinding is done on a coarse wheel and initial polishing done with emery paper. Final polishing is done on a wheel made from hard felt and with a powder such as cerium oxide.

The lapidary needs grinding equipment for faceting too, and also a lap, similar to the scaife of the diamond polisher. It may be made of aluminium or cast iron if hard grit such as corundum is to be used; alternatively it may be made from copper impregnated with diamond powder. Facets are put on with the assistance of a mechanical dop whose angles can be adjusted by reference to a scale marked on it. The pavilion facets are normally cut first after the girdle is rounded; a suitable lap might be diamond-impregnated copper with a mesh size of 360. Faceting then proceeds until the stone is ready to be reversed, at which time the crown facets are cut, followed by the table. It is interesting to note that the diamond polisher's operations are in the reverse order. There are many types of lap and a wide variety of polishing powders on the market and a textbook on gem cutting should be consulted for further details.

Cabochons

The cabochon form (Fig. 6.1) in which the stone is domed with a flat back, is the oldest known form of cutting. Sometimes the cabochon was hollowed out as in the carbuncle form of almandine garnet, the aim being to show some of the very deep red rather than allow the stone to appear black. Cabochons shaped to form beads are still widely used in oriental jewellery and the cabochon is the only way in which to display a star or cat's-eye. Some cabochons are hollowed and backed by coloured foil to enhance colour.

Selection of rough

Rough material with cracks extending deep into the interior should be avoided as should crystals with large and well-developed cleavages (these will show either as flattish surfaces with a characteristic pearly lustre or be inside the crystal and marked out by interference colours). A skilled lapidary will be able to cut difficult material such as kunzite or sillimanite but the beginner should not attempt them. Many crystals are brittle and heat-sensitive: check a faceting textbook before embarking on any faceting process.

The selection of rough material to be cut should also include a close look at colour distribution. The finished stone will be no use if it cannot show its best colour through the table and if extensive colour banding minimizes the dramatic effect that a strong colour should give. Amethyst, of the better-known stones, frequently shows colour zoning and it is also marked in Kashmir sapphire (though Kashmir rough is not readily available). Where colour occurs in one part of the rough only, it may be possible, providing the dimensions of the rough are suitable, to place the coloured spot so that the table is completely coloured, even if the stone appears colourless when examined from other angles. Sri Lanka blue sapphire used to be seen in which this effect had been skilfully handled. Stones with a high birefringence such as high zircon need to show this effect as little as possible through the table since the doubling of the back facet edges will make the stone appear fuzzy.

In most cases immersion of the rough will greatly assist the finding of unpalatable flaws inside the stone; a liquid which matches the refractive index of the rough gives the best results, although water is better than nothing.

Star stones and cat's-eyes

Choice of material for star stones and cat's-eyes is quite difficult. The rutile or other crystals which cause the star lie in only one series of parallel planes at right angles to the optic axis; the position of this axis has to be found first, then a rounded shape should be polished on the crystal at one end of the axis to enable the lapidary to judge the quality of the star and to ensure that it will be centrally placed. Some asteriated crystals show a silky sheen on the surface which helps to fix the position of the asterism. Corundum crystals show triangular surface markings on the ends so rough material which shows these but no crystal faces can still be correctly oriented. The use of a liquid such as methylene iodide helps to show up the sheen caused by the rutile crystals. Examine rough under a single overhead light.

Chatoyant rough should be chosen in much the same way; material which approximates the spherical as closely as possible should be chosen in preference to other shapes. Moonstone rough which is sold as thin plates (the shape results from the operation of two easy cleavages) will not give a satisfactory sheen to a finished stone since the adularescence only shows along thin edges. Rough which has been water-worn without producing cleavages often gives good results; if they were not hard, cleavages would have started to develop during the progression from host rock to alluvial deposit.

The lapidary buying rough will soon find that it is quite expensive and that material offered on one day may not be available the next. Commoner material commands fairly standard prices in bulk but rare stones are sold individually with individual prices. Prices for the best rough will not be far below those of the finished stone at wholesale rates, although difficult material may have a slightly lower price because the dealer did not want to risk so much money in the first place on crystals which might have broken in transit.

7 Gemstones in commerce

7.1 DIAMOND GRADING

In the past few years the grading of diamonds for colour, clarity and cut
(with carat weight these criteria make up the 'four C's' so astutely made a
part of diamond advertising campaigns) has become *de rigueur* and thus
makes up a large part of the income of gemmological laboratories. The
boom in investment stones helped to fuel the setting-up of a number of
different grading schemes which have now narrowed down to a very few,
which resemble each other to some extent, differing only in the names used
for the different criteria.

Colour grading

Colour is assessed by how much a diamond diverges from the accepted
finest white. The term blue–white, so often used in popular literature, is
now discouraged as it is too easily misused. It originated with some
exceptional diamonds from the Jagersfontein mine; the stones, which are
extremely rare, gave so strong a blue fluorescence even in daylight that the
white was tinged with blue. Other stones with a strong blue fluorescence
come from the Premier and the former Williamson mine in Tanzania. The
so-called Premiers have a yellowish colour with a blue fluorescence which
gives an overall oily bluish appearance.

Anyone embarking on the colour grading of diamonds will soon find
that there are 'many shades of white', no matter how paradoxical it may
sound. In colour grading the room light must contain no colour, the
decoration must be dead-white and the stones examined against a white
background. Fluorescence, whether in the stone under test or from parts of
the testing apparatus, must be completely eliminated. Sunlight contains
UV radiation which can stimulate a bluish fluorescence in a diamond with a
yellowish body colour; the blue and the yellow cancel each other and the
stone appears whiter than it really is. In the northern hemisphere diamonds
are graded in a north-facing light and in the southern hemisphere in

southwards-facing light; even the time of day in which grading takes place is important since at the beginning and end of the day the rays of the rising and setting sun contain too much red light. Natural light is, therefore, not often used for grading, and many light sources have been designed with diamond grading in mind.

Perhaps the simplest effective grading method (which also demonstrates how white does in fact usually 'draw' some colour) is to place a set of master stones, already graded, in the fold of a white card. The stone to be graded is placed alongside each control stone in turn until a place is found in which it fits into what comes to be seen as a sequence of colours. Yellow and brown are the tinges of colour most commonly seen. This method is surprisingly instructive, but for commercial grading and certainly with large or important stones some form of quantitative measurement is necessary. This will be one of the many available kinds of photometer, some of which are made specifically for the diamond market. In the earliest and simplest form of photometer the transmission factor of the stone is measured by passing light through a yellow filter into the table facet, the reflected and refracted light being measured by a photo-cell and meter. A blue filter serves to collect light at the 415.5 nm mark. The transmission quotient is obtained by dividing one figure into the other. This photometer is designed for white to yellowish stones but brown stones can be assessed in a similar way. An instrument which takes reflectance measurements at 412, 420 and 416 nm, feeding the readings into a calculator programmed to compute the effective area of the 415.5 nm absorption band has been used at the Diamond Grading Laboratory and later at the Independent Gemstone Testing Laboratory in London; the apparatus was devised by R. V. Huddlestone.

One of the older systems of colour nomenclature is the Scan. DN system, the most recent grades of which are: River, Top Wesselton, Wesselton, Top Crystal, Crystal, Top Cape, Cape, Light Yellow and Yellow (see Webster, 1983). More recent schemes have tried to eliminate any reference to yellow because it has come to be associated with low quality; such terms as 'slightly tinted white' and 'tinted white' have been substituted. The term Jager is still occasionally used for a stone with an exceptional blue fluorescence. Wesselton mine stones in the early days were found to have less of a yellow tinge than stones from other mines. Cape stones (from the Cape of Good Hope) referred to the generally yellower appearance of the majority of South African stones compared to those from India or Brazil. At the present time colour grading is carried out according to three main systems. These are the Scandinavian Diamond Nomenclature (Scan. DN; above), the Gemological Institute of America (GIA) system and that devised by CIBJO, the organization of European Jewellers. The GIA system uses a series of letters with D as the top grade and CIBJO uses more descriptive terms.

Clarity grading

Grading of diamonds for clarity is based on the inclusions found in the stone. These may be cleavage cracks, fracture or tension cracks, feathers which are cracks observed at right angles to their plane, all kinds of mineral inclusions, structural defects seen as cloudiness, or twinning lines which are seen as fine curved lines crossing the pavilion facets of a brilliant. These twinning lines represent the basal planes of twin crystals and are not really curved; the apparent curvature is due to the cone formed by the pavilion facets. Although these planes are usually colourless they may show a faint brownish or yellowish tint. The grading of a stone for clarity also includes features of the cutting: tiny cleavage cracks along the girdle (fringes), small percussion marks on the surface only (not penetrating into the stone), damaged culets which may have cracks associated with them, extra facets, naturals (traces of the original crystal) and polishing marks.

The terms used to describe clarity vary from one grading system to another, but in general the classification runs from internally flawless (no inclusion seen with a 10 × lens) to 3rd pique (containing large inclusions). The latter are easily seen with the naked eye. Intermediate terms are: very very slightly imperfect (VVS), very slightly imperfect (VS), slightly imperfect (SI) and 1st–2nd pique (see Table 7.1). Clarity is normally depicted in most grading systems as a diagram of the number, size and type

Table 7.1 Clarity grading

Grade	Description
Loupe clean or internally flawless	Completely transparent and inclusion-free. Tiny external faults may be ignored
VVS 1 and VVS 2	Very very small inclusions, very hard to find under 10 × magnification
VS 1/VS 2	Very small inclusions, hard to find under 10 × magnification
SI	Small inclusions, easy to find with 10 × magnification but not visible through the crown with the naked eye
P1	Inclusions immediately visible with 10 × magnification. Hard to find through the crown with the naked eye; brilliancy not impaired
P2	Large or numerous inclusions easily visible through the crown to the naked eye and slightly reducing the brilliancy
P3	Large or numerous inclusions very easily visible through the crown with the naked eye and reducing the brilliancy

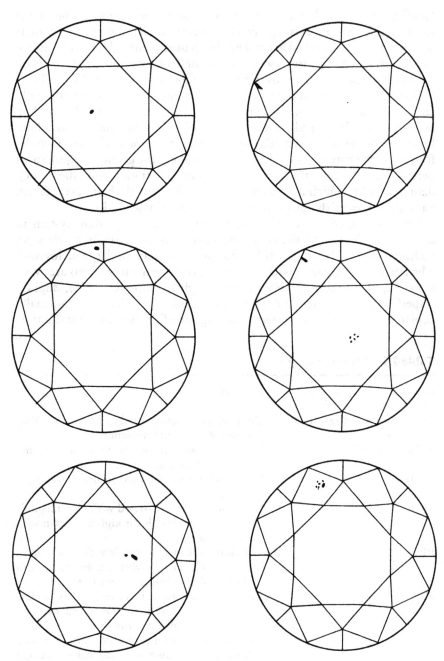

Figure 7.1 High-to-low clarity grades in diamond.

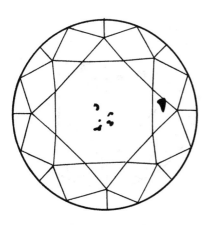
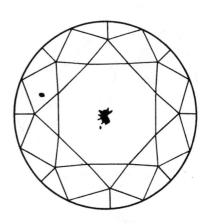

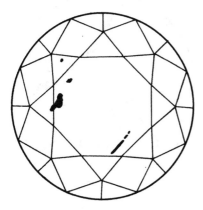
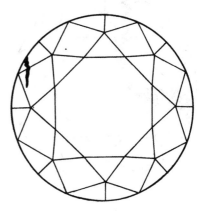

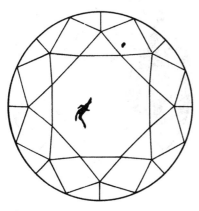
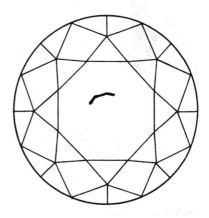

Figure 7.1 *Cont.*

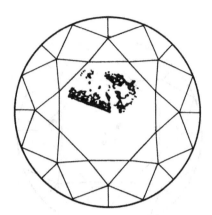

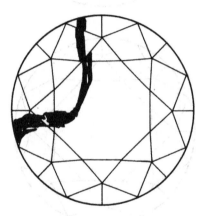

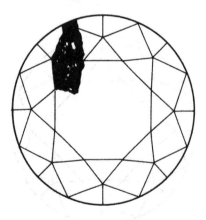
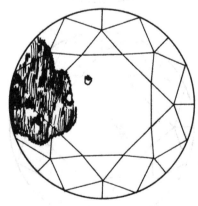

Figure 7.1 *Cont.*

of inclusions or external features. Diagrams of the various diamond cuts are provided on work sheets and the grader marks the inclusions in so that the customer can see how the final assessment of clarity has been calculated (Fig. 7.1).

Proportions

The proportions of a diamond are also taken into consideration when assigning a grade. The history of diamond proportions is reviewed in the chapter on fashioning but briefly it can be said that there has been a succession of 'ideal' brilliants, each design endeavouring to give maximum dispersion with brilliance and table size. The amount by which these three criteria can be reconciled is a measure of the success of a particular style of cutting and it should be remembered that the fashions themselves change over the years. Grading systems have been developed in which a stone can be compared with one or other of the ideal designs (this is done only for brilliants); there are 'proportionscopes' on the market which show the proportions of the stone to be graded against those of the ideal stone so it can easily be seen how great the difference is. Height of crown, angles of girdle and depth of pavilion are taken into account, as are the straightness of facet edges and the central positioning of the table. The central position of the culet is also important, as is its size. Growth lines and burn marks (the light milky spots arising through overheating during polishing) are also taken into account and there are symbols depicting all these features. The symbols are added to the work sheet which already shows the presence of inclusions.

7.2 BUYING GEMSTONES

Those wishing to start a collection of gemstones should try to get both rough and cut specimens. Many examples of the important gemstones can be obtained, providing that they are not so large and fine that they command high prices; those with numerous inclusions or with slightly damaged surfaces can provide interest and excitement every bit as much as their fellows which are destined for setting in jewellery. Included stones, in fact, are more interesting since the study of the inclusions may identify the way in which the stone was formed. Lower quality stones will also serve perfectly well as test and display material.

How does the student or collector obtain his specimens? Many people think that a jeweller will supply loose stones, but this is only true of some; the majority would have to approach a gemstone dealer to satisfy a request for loose faceted stones and a request for rough material would have to be passed on to a firm dealing with the type of stone wanted. It is quite customary for a collector to approach a dealer direct; almost all are in areas of large cities which have been devoted for centuries to the gemstone and jewellery trade; London's Hatton Garden is a good example. Some firms specialize in one stone only (diamond, emerald, pearl and opal seem to be the chief specialities); others deal with a wide range. The rare stones may be found either with the very large dealers to whom such stones often come automatically since they are so well known; or with small specialist dealers who deal in fine stones for only a part of their professional time. The twin towns of Idar-Oberstein in West Germany house hundreds of jewellery firms, with many specialists among them. Some also deal in rough gem material and visiting them (make an appointment first – a directory of all the firms is published every year) is essential for the serious gemmologist. Surprisingly often good stones turn up in the auction rooms and sometimes fetch quite reasonable prices. They may occur in mixed lots but these are sensibly put together so that it is unlikely that you will have to buy a grandfather clock to secure a single gemstone.

Most cut stones are sold by the carat (1 ct is 0.2 g) and are kept in stone papers (small parcels) which have all relevant details marked on them (sometimes the price is in code). Larger and more important stones occupy parcels on their own while small stones, especially calibré material (small stones all cut to the same size and designed for the jewellery manufacturer) may be fifty or more to the parcel. Dealers customarily have one price for the single stone and a picking/lot price for groups of the smaller ones. As customers will pick the best stones from a lot, the picking price for a stone is higher than the price that will be paid when it is bought as part of a lot.

Rough gemstones

Rough gem material is usually purchased at the mine by agents of important dealers. By the time it gets to the collector the best specimens will have been creamed off but a persistent enquirer will still be able to get good material from an appropriate dealer. Although small advertisements in the lapidary journals carry much useful information and those advertising are almost all completely honest (as is the trade in general) many will offer 'mine run' specimens which are best avoided. The low price asked should give a clue. Mine run material is frequently the sweepings left after good material has been separated from its matrix by a hammering known as cobbing.

Assessment

Those taking the plunge and buying fine quality stones will soon notice that certain colours and qualities attract higher prices than others which appear almost equally attractive. There are always examples of unusual colours turning up in the market (a blue garnet is yet unknown but in recent years a red beryl has been found and cut) and these will always be costly unless a large quantity is found. It is usually wise to buy the unusual if you intend your collection to be truly representative since flooding of the market with once-rare stones is not a common occurrence. The collector should not scorn synthetic stones of whatever origin nor should he ignore those stones whose colour has been altered artificially. All these specimens show some scientific interest and all are beautiful. Those interested in man-made crystals should make the acquaintance of people working in the field of crystal growth.

Preferred colours

The colour characteristics of better-known stones can be used to attempt to form a grading system – always a difficult undertaking. Not only do eyes see colours differently, but preferences vary from country to country as well. Some colour preferences are listed below:

● *ALEXANDRITE:* the best stones from the Ural Mountains show emerald-green in daylight and rich purple under tungsten light. This combination is extremely rare. Sri Lanka alexandrite has a less pronounced colour change and inclines to a brownish rather than an emerald-green.
● *AMETHYST:* a rich purple with a change of colour from reddish-purple under tungsten light to bluish-violet in daylight fetches the highest prices and is known as Siberian. The names Uruguay and Bahia are used for other qualities, both being somewhat paler than the Siberian stones. Some purple–pink stones have been called Rose de France (and are imitated by synthetic corundum).
● *AQUAMARINE:* although most aquamarine is mined as sea–green crystals which give their name to the variety, most aquamarine on the market is a medium blue with no hint of green. This shade of blue has been called Fortaleza blue from the Brazilian locality of that name.
● *BLUE SAPPHIRE:* a rich somewhat violet–blue is preferred, a colour which may appear violet–blue in daylight and purplish-blue under tungsten light. This colour is especially characteristic of the finest Burmese stones. Kashmir stones have a velvety appearance and despite colour zoning are highly valued. Pale violet–blue stones are characteristic of many Sri Lanka sapphires. Australia also produces fine blue stones, although they are also renowned for producing inky (virtually black) specimens.
● *CAT'S-EYE CHRYSOBERYL:* the eye should be silvery-white and very sharp, seen against a background of translucent honey yellow or brown. Some prefer a greenish brown but olive greens are less desirable.
● *CITRINE:* the best stones are a rich yellow with orange overtones and range from brown to an orange–red. Rio Grande and Madeira are terms

used for heat-treated amethyst which has been turned to reddish-yellow or reddish-brown. Palmyra stones are more orange–yellow after heating.

• *EMERALD:* the preferred colour is an intense bluish-green but it is fair to say that many also like a yellower green which is much brighter. Siberian emeralds are bright and the term is sometimes used to denote any bright green emerald irrespective of origin. The Sandawana emeralds from Zimbabwe are also a superb bright green, keeping their colour down to the smallest sizes, as do Pakistan stones. Top class emeralds keep their value whatever the size of stone.

• *OPAL:* black opals command the highest prices and should have as dark a background as possible with maximum variety and intensity of colours. The matrix should be translucent. Red is always a desirable colour. If possible the patches of colour should be evenly distributed over the stone and be even in size. Traces of milkiness detract from value. Some of the finest Mexican opal, resembling water opal, is faceted and shows a play of colour by both reflected and transmitted light. This variety has been called contra luz opal.

• *RUBY:* the preferred colour is traditionally known as 'pigeon's blood' and is a rich red with a hint of purple. Orange- and brown-tinted reds, characteristic of the rubies from Thailand, still known in the trade as Siam rubies, are next in esteem. Sri Lanka (Ceylon) ruby is inclined to pink but still most attractive. It grades into pink sapphire which has a distinct colour of its own. Star rubies rarely combine a fine colour with a central star, most stones having a reddish purple colour with little translucency. The best Burmese star rubies combine a rich crimson red with a sharp central star and considerable translucency.

• *TOPAZ:* stones with a medium reddish–orange with some brown have been called 'Imperial topaz' and are characteristic of some Brazilian deposits. Sherry-brown and pale brown stones may fade, though not all do.

8 Descriptive section: inorganic materials

INTRODUCTION

Throughout the descriptive section the chemical composition, crystal system, physical properties and other data are given in the same order for each species. Mineral families such as corundum and beryl are described as families so that ruby must be sought under corundum and emerald under beryl. In order to avoid overloading the text with figures which have no more than academic significance, I have not provided items such as dispersion for more than a few species in which this property is important. Abbreviations used are:

H = hardness
SG = specific gravity
RI = refractive index
DR = double refraction

The letters (ω) and (ϵ) signify the ordinary and extraordinary rays in uniaxial materials and the letters (α), (β) and (γ) indicate the three refracted rays in biaxial materials. GIA is the Gemological Institute of America.

Though many of the species described have never been used in jewellery they may very well be sought by the growing number of gemstone collectors and every attempt has been made to ensure completeness.

ACTINOLITE

$Ca_2(Mg, Fe)_5Si_8O_{22}(OH)_2$, forming a series with tremolite and ferro-actinolite. Monoclinic, forming elongated bladed crystals often twinned. Two directions of cleavage. H 5.5. SG 3.05. RI (Tanzanian material) (α) 1.619–1.622 (β) 1.632–1.634 (γ) 1.642–1.644. Biaxial negative. DR 0.022–0.026. Pleochroism yellow to dark green. May show faint absorption line at 503 nm.

Found in contact metamorphic limestones and dolomites, magnesium-rich limestones and ultrabasic rocks; some crystals of a dark green colour have been faceted, chiefly from the deposits in the Malagasy Republic. Chatoyant material with an SG of 3.0 and RI 1.63 is reported. The jade mineral nephrite (q.v.) consists chiefly of actinolite. Named from the Greek word for ray, referring to actinolite's frequently fibrous nature.

ADAMITE

$Zn_2(AsO_4)$ (OH). Orthorhombic or monoclinic, dimorphous with parada-mite and forming a series with olivenite. Although the usual colour is yellowish-green to yellow, colourless transparent crystals with some pink and violet specimens have been found at Mapimi, Mexico and stones have been cut from them. They have a hardness of 3.5, SG of 4.48 and an RI (α) 1.722 (β) 1.742 (γ) 1.763 with a DR of 0.041. Crystals show prism, pinacoid and dipyramid forms. Named after a French mineralogist.

AMBLYGONITE

(Li, Na)Al(PO_4) (F, OH), forming a series with montebrasite which has more (OH) than F. Both species are said to be faceted but virtually all reported faceted stones are montebrasite. Triclinic, forming equant to short prismatic crystals (i.e. same dimensions in three directions), commonly twinned, with one direction of perfect cleavage and another direction of good cleavage. H 5.5–6. SG 3.0–3.1, montebrasite 2.98. The RI of amblygonite ranges from 1.578 to 1.591 for the α index and from 1.598 to 1.612 for the γ index; equivalent values for montebrasite are 1.594 to 1.611 and from 1.616 to 1.633. The optic sign is positive for montebrasite and negative for amblygonite, RI and optic angle decreasing as Na and F content increase; the change occurs at the point where the (OH) content is around 60%. DR 0.021. The minerals show variable but unimportant luminescence.

Amblygonite and montebrasite are minerals of granite pegmatites, most faceted material being a clear lemon yellow, usually from Brazil; faceted montebrasites usually from Arassuahy, Brazil. Any stones which are optically positive may have come from the montebrasite location at Mogi das Cruces, São Paulo, Brazil. A lilac variety is reported from Karibib, Namibia. Amblygonite is named from a Greek word referring to the bluntness of the crystals; montebrasite from Montebras, France.

ANALCIME

$NaAlSi_2O_6H_2O$. Cubic, crystals commonly showing icositetrahedra. H 5–5.5. SG 2.22–2.29. RI 1.487 (mean).

One of the zeolite group of minerals, analcime is usually colourless but may show tinges of yellow or pink from impurities. Cut stones are rare and unlikely to exceed 2 ct. Analcime occurs as a secondary mineral in basic igneous rocks. The area of Mont Saint Hilaire in Quebec, Canada, provides good crystals but analcime occurs world-wide. Named from a Greek word meaning weak, referring to the weak electrical charge developed when crystals are rubbed.

ALEXANDRITE see CHRYSOBERYL

AMETHYST see QUARTZ

ANDALUSITE

Al_2SiO_5. Orthorhombic, most gem material found as nearly square prisms or pebbles. Distinct cleavage in one direction. H 6.5–7.5. SG 3.13–3.17. RI (α) 1.629–1.640 (β) 1.633–1.644 (γ) 1.638–1.650. Biaxial negative. DR 0.007–0.011. Strong pleochroism, olive green to flesh red. A manganese-bearing green (non-dichroic) variety has an absorption spectrum consisting of two areas of absorption in the yellow and the green with general absorption of the blue and violet; this gives a knife-edge at 553.5 nm with fine lines at 550.5 and 547.5 nm with fainter lines at 518, 495 and 455 nm. A colourless andalusite found in Brazil as part of a bicoloured crystal was cut into a separate stone with low RI due to low iron content or the complete absence of iron. The readings were (α) 1.630 (β) 1.634 (γ) 1.638 with DR 0.008. Other andalusites from Brazil had a Mn_2O_3 content as well as being rich in iron. Typical RI were (α) 1.642 (γ) 1.653 with DR 0.011. The colour was a pleasing yellowish-green.

Andalusite's chief claim to importance as a transparent ornamental mineral is its attractive pleochroism; most stones are cut with the olive-green colour visible through the table facet and the red occurring as flashes at the ends of the stone. Another, opaque, variety chiastolite has carbonaceous cross-like inclusions in greyish-brown crystals which are cut across their length to display them. Andalusite is found in metamorphic rocks, particularly in gneisses and slates; it also occurs in schists. Most gem-quality material comes from Brazil where pebbles are found in stream beds, near Santa Tereza, Espirito Santo. Though andalusite crystals are usually almost square orthorhombic prisms with broken ends a golden yellow crystal from Minas Gerais, Brazil, had an almost monoclinic appearance. A crystal from Sri Lanka with similar shape is recorded. Named from Andalucia, the first reported location. Chiastolite is named from the Greek letter chi whose shape resembles a cross; good material is found in the Nerchinsk area of Transbaikalia, USSR, and in Zimbabwe.

ANGLESITE

$PbSO_4$. Orthorhombic, crystals tabular, prismatic, with a cleavage in one direction. H 2.5–3. SG usually 6.38. RI (α) 1.877 (β) 1.883 (γ) 1.894. DR 0.017. Dispersion 0.044. Biaxial positive. Weak yellow fluorescence under short-wave UV.

Although pale shades of yellow or blue occur, anglesite is most commonly colourless with a high dispersion (equal to that of diamond). It is too soft for strictly ornamental use but a well-cut stone is very attractive provided that the dispersion shows to best effect. Anglesite occurs as a secondary mineral in lead deposits and arises through the oxidation of galena (PbS). Though named for the British locality of Anglesey, most gem quality material comes from Tsumeb, Namibia or Bou Azzer, Morocco.

ANHYDRITE

$CaSO_4$. Orthorhombic, forming equant crystals, although cleavage masses are more commonly used ornamentally; there is a perfect cleavage in one direction and a nearly perfect one in another direction. H 3–3.5. SG 2.9. RI (α) 1.570 (β) 1.575 (γ) 1.614. Biaxial positive. DR 0.044. Violet crystals show colourless to pale yellow, pale to rose violet pleochroism; some red material (German) gives red fluorescence in long-wave UV.

Cleavage masses of anhydrite of a violet, blue, purple, whitish-grey or brownish colour are used ornamentally and some stones have been faceted from Swiss and Canadian localities. Anhydrite is a rock-forming mineral found in gypsum beds and limestones; it can also be found in hydrothermal veins, cavities in basalts and other igneous rocks. Material from the Faraday mine, Bancroft, Ontario, Canada is purple; that from the Simplon tunnel area of Switzerland a paler purple to pink; fine blue masses are reported from Mexico and a greyish-white variety from Volpino, Italy (vulpinite) is cut into cabochons. Named from the Greek word meaning waterless (from its composition).

APATITE

$Ca_5(PO_4)_3(F, OH, Cl)_3$. The name apatite refers to a group of minerals in which considerable replacement may occur so that there are silicate and carbonate apatites as well as the more common fluor-apatite which provides most gem material. Hexagonal, crystals prismatic, pyramidally terminated with a plane of symmetry at right angles to the six-fold axis of symmetry. H 5. SG 3.10–3.35. RI (ω) 1.632–1.649 (ϵ) 1.628–1.642. Uniaxial negative. DR 0.001–0.013. RI and DR figures vary widely with compositional variation. Blue apatite shows distinct pleochroism, blue and yellow. Some varieties, particularly the yellow–green crystals, show a rare earth spectrum (didymium) consisting of two groups of fine lines; a group of seven lines near 580 nm and five lines near 520 nm. Apatite shows a wide range of luminescent effects but they are not particularly useful in identification.

The apatite group includes a wide range of colours, those best known as faceted gemstones being the yellow–green, green and purple varieties. Found in a number of different environments, gem material being commonly an inhabitant of igneous rocks, particularly pegmatites. Yellow–green material, which accounts for most crystals, comes principally from Durango, Mexico; violet crystals are found in Maine (Androscoggin County), USA and formerly at Ehrenfriesdorf, Saxony, East Germany; a green apatite which has been given the unnecessary name trilliumite comes from Canada; most blue apatite is Brazilian but blue material is also found in Burma and Sri Lanka. A golden-green variety has been reported from Kenya and chatoyant yellow and blue–green stones have been found in Sri Lanka and Burma. Green cat's-eyes are reported from Brazil. Very fine apatite cat's-eyes have been reported from Tanzania. They are yellowish-green, green, greenish-brown to red with a fine, sharp eye which is caused by fine goethite fibres. RI (ϵ) 1.632–1.636 (ω) 1.636–1.640 with a very constant birefringence of 0.004. SG is 3.22–3.35 and there is a didymium absorption spectrum. A blue–green apatite of gem quality is found at Gravelotte, Transvaal, South Africa. Fine yellowish-green apatite (often known as asparagus stone) is found in the Spanish province of Murcia. A massive sky-blue variety mixed with lapis and found in Siberia has been called lazurapatite. Named from a Greek verb meaning to deceive, as apatite was confused with other minerals in earlier times.

APOPHYLLITE

$KCa_4Si_8O_{20}(F, OH) \cdot 8H_2O$. Strictly speaking there are several members of the apophyllite family (e.g. fluorapophyllite, chlorapophyllite, hydroxyl-apophyllite). Tetragonal, forming pseudocubic crystals with a perfect cleavage in one direction. H 4.5–5. SG 2.3–2.5. RI (ω) 1.534–1.535 (ϵ) 1.535–1.537; the index varies. Uniaxial positive or negative. DR very low (0.001 or less), some specimens appearing virtually isotropic.

Apophyllite can reach so high a degree of whiteness that it may appear silvery; green Indian material may sometimes be faceted. Apophyllite occurs as a secondary mineral in basic igneous rocks, mainly in the Bombay area of India. Named from a Greek word explaining the mineral's tendency to exfoliate on heating.

AQUAMARINE see BERYL

ARAGONITE

$CaCO_3$, dimorphous with calcite. Orthorhombic, forming pseudohexagonal crystals, frequently twinned with a distinct cleavage in one direction. H 3.5–4. SG 2.93–2.95. RI (α) 1.530 (β) 1.681 (γ) 1.685. Biaxial negative. DR 0.155. Luminescence variable; some Sicilian material may phosphoresce green in long-wave UV.

Faceted colourless aragonite seldom reaches large sizes (cut calcite is commoner and larger); cut stones from Horschenz, Germany are a straw yellow and reach 10 ct. Named from Molina de Aragon, Spain, a source of the material. A yellow stalagmitic material from Karibib, Namibia, sold as aragonite, is calcite.

AUGELITE

$Al_2(PO_4)(OH)_3$. Monoclinic, forming thick tabular crystals with perfect cleavage in one direction. H 4.5–5. SG 2.69–2.75. RI (α) 1.574 (β) 1.576 (γ) 1.588. Biaxial positive. DR 0.014–0.020.

Colourless crystals of augelite have been faceted, most material coming from the Champion Mine, Mono County, California, USA where it occurred in an andalusite deposit. The augelite is now rare at this locality. Named from a Greek word referring to its glassy lustre.

AXINITE

(Ca, Mn, Fe, Mg)$_2$Al$_2$BSi$_4$O$_{15}$(OH), a group name. With magnesium, the mineral is magnesioaxinite; where iron > manganese it is ferroaxinite and the converse is manganaxinite. Triclinic, forming wedge-shaped crystals with a cleavage in one direction. H 6.5–7, varying with direction. SG 3.26–3.36 (magnesioaxinite 3.18). RI (α) 1.674–1.693 (β) 1.681–1.701 (γ)1.684–1.704; magnesioaxinite has (α) 1.656 (β) 1.660 (γ) 1.668. Biaxial negative. DR 0.010–0.012. Pleochroism intense in all coloured specimens with cinnamon-brown, violet, olive-green, yellow and colourless variations of colour. Absorption lines may be seen at 532, 512, 492, 466, 444, 415 nm (those at 492 and 466 nm may be broad; the 415 nm line may be intense). Luminescence is variable with some material from Franklin, New Jersey showing red in short-wave UV; magnesioaxinite from Tanzania shows orange–red in long-wave UV with a dull red in short-wave UV. A cinnamon-brown stone with strong pleochroism and RI (α) 1.675 (β) 1.681 (γ) 1.685, DR 0.010 and SG 3.314 has been found to be the iron endmember of the axinite group. Stones contained black feather-like structures which may be healing fissures.

Axinite is found in areas of contact metamorphism and metasomatism; most gem quality material, of an unmistakable clove-brown with purple dichroism, comes from Baja California, Mexico and fine crystals have been found at St Christophe, Bourg d'Oisans, Hautes-Alpes, France; a violet coloured material has been found at Rosebery, Montagu County, Tasmania, Australia. Magnesioaxinite of a pale blue is found in Tanzania. Named from a Greek word meaning axe, with reference to the mineral's crystal shape.

AZURITE

$Cu_3(CO_3)_2(OH)_2$. Monoclinic, forming large tabular crystals or masses with a perfect cleavage in one direction. H 3.5–4. SG 3.77. RI (α) 1.730 (β) 1.758 (γ) 1.836. Biaxial positive. DR 0.110. Pleochroism strong in different shades of blue.

Azurite is found as a secondary mineral in copper deposits and is frequently mixed with malachite (azurmalachite) in the massive form (i.e. comprised of many minute crystals) of both minerals. Facetable azurite crystals are small, the best coming from Tsumeb, Namibia. The Bisbee and Morenci areas of Arizona, USA produce fine masses and some crystals. Named from a Persian word alluding to the colour. The name 'Royal gem azurite' was given to a tough azurite mixture with malachite and other copper minerals, found at the Copper World mine, Nevada, USA.

BARITE

$BaSO_4$. Orthorhombic, forming tabular crystals and masses with a perfect cleavage in one direction. H 3–3.5. SG 4.3–4.6. RI (α) 1.636 (β) 1.637 (γ)

1.648. Biaxial positive. DR 0.012. Various types of luminescence, none of great use in testing.

Although faceted stones are known, cabochons are much more common. Material from England (Cumbria) yields large crystals and deposits in France provide yellow–brown crystals (the commonest colour). An attractive blue variety is found in the area of Sterling, Colorado, USA. The massive white variety is used as a quasi-marble. Barite is found in low-temperature hydrothermal vein deposits and in a number of other environments. It is named from the Greek word meaning heavy and the mineral was once known as heavy spar.

BAYLDONITE

$PbCu_3(AsO_4)_2(OH)_2$. Monoclinic (perhaps orthorhombic) forming fibrous concretions and masses. H 4.5. SG 5.5. RI (α) 1.95 (β) 1.97 (γ) 1.99. Biaxial positive. DR 0.040.

Yellowish-green cabochons have been cut from material found as a secondary mineral in lead–copper deposits, notably from Tsumeb, Namibia. Named after John Bayldon.

BENITOITE

$BaTiSi_3O_9$. Hexagonal, forming crystals with a trigonal axis of symmetry and a plane of symmetry at right angles to it (the mineral is the only one known with this property). H 6–6.5. SG 3.64–3.68. RI (ω) 1.757 (ϵ) 1.804. Uniaxial positive. DR 0.047. Dispersion 0.046. Strong pleochroism, colourless and blue. Intense blue fluorescence in short-wave UV only. Some near-colourless benitoite has been found to show a dull red fluorescence under long-wave UV. As the blue becomes more pronounced in the mineral this effect diminishes.

Benitoite, whose dispersion helps to distinguish it from blue sapphire, the fine blue colour of which it somewhat resembles, was discovered in 1906/07 and was first thought to be sapphire. It occurs in a massive fine-grained white natrolite in San Benito County, California, for which it is named.

BERYL

$Be_3Al_2Si_6O_{18}$ with some Fe, Mn, Cr, V and Cs. Hexagonal, forming prismatic elongated or flattened crystals often etched or striated; also found as rolled pebbles or in masses. Beryl displays a wide range of colours, ranging from green in emerald and blue to greenish-blue in aquamarine to golden yellow, pink or peach colour in morganite and an orange- to near ruby-red in the manganese-bearing beryl from Utah to which the name bixbite has (somewhat unfortunately) been given. There is also a colourless beryl, goshenite. H 7.5–8; there is an indistinct cleavage but beryl is brittle. SG: goshenite 2.6–2.9; morganite 2.71–2.90; aquamarine 2.68–2.80; emerald 2.68–2.78. RI: goshenite (ω) 1.566–1.602 (ϵ) 1.562–1.594 with DR 0.004–0.008; morganite (ω) 1.572–1.592 (ϵ) 1.578–1.600 with DR 0.008–0.009; aquamarine (ω) 1.567–1.583 (ϵ) 1.572–1.590 with DR 0.005–0.007; Maxixe beryl (containing caesium) (ω) 1.584 (ϵ) 1.592 with DR 0.008. The SG and RI of emerald vary with place of origin (see Table 8.1).

Beryl is uniaxial negative and has the low dispersion of 0.014. Pleochroism is notable in the deeper coloured specimens, aquamarine showing blue to colourless or greenish-blue, morganite showing pale pink to deep bluish-pink, golden beryl showing brownish-yellow to lemon-yellow, emerald showing yellowish-green to blue–green. In aquamarine the strongest colour belongs to the extraordinary ray and thus cannot be seen at its deepest without a dichroscope or polarizing filter.

Colour

The colour in emerald arises from the replacement of aluminium by chromium and/or vanadium. Both elements are often present. The absorption spectrum shows noticeable differences according to whether the ordinary or the extraordinary ray is examined; the ordinary ray shows two narrow bands in the red, the first being a doublet at 683 and 680 nm, the second a sharp line at 637 nm. A broad, weak absorption with ill-defined edges covers much of the yellow from 625 to 580 nm and there is a narrow line in the blue seen only in very chromium-rich stones which also show a line at 477.4 nm. In the spectrum of the extraordinary ray the doublet is stronger but the line at 637 nm is absent and replaced by two diffuse lines at 662 and 646 nm, the latter line being bordered on the short-wave side by transparency patches which are characteristic for this spectrum. The broad absorption is weaker and nearer to the red and there are no lines in the blue.

The colour of green aquamarine is due to iron in two different sites. If the iron is in an interstitial site within structural channels Fe^{2+} gives a blue colour which is unaffected by heat. Many specimens also contain some Fe^{3+} substituting for Al^{3+}; this by itself gives the colour in yellow or golden beryl or green aquamarine when combined with the blue Fe^{2+}. In aquamarine there is a broad band in the violet at 427 nm and a more feeble diffuse band at 456 nm; in the extraordinary ray there is also a narrow band at 537 nm and the other bands appear more strongly. This is not seen in the blue aquamarines which owe their colour to heat treatment of the green

Table 8.1 A recent collation of emerald constants, by Gübelin (1982b)

| Country of origin | RI | | | |
	ω	ε	DR	SG
Australia: Poona	1.578	1.572	0.005–0.007	2.693
Brazil: Bahia:				
Anage	1.584	1.576	0.008	2.80
Brumado	1.579	1.573	0.005–0.006	2.682
Carnaiba	1.588	1.583	0.006–0.007	2.72
Salininha, Pilao Arcado	1.589	1.583	0.006	2.70
Itabira	1.589	1.580	0.009	2.725
India: Ajmer	1.595	1.585	0.007	2.735
Colombia: Burbar	1.576	1.569	0.007	2.704
Chivor	1.579	1.570	0.005–0.006	2.688
Muzo	1.580	1.570	0.005–0.006	2.698
Mozambique: Morrua				
(Melela)	1.593	1.585	0.008	2.73
Norway: Eidsvold	1.583	1.590	0.007	2.759
Austria: Habachtal	1.591	1.584	0.007	2.734
or	1.582	1.576	0.006	
Pakistan: Mingora	1.596	1.588	0.007	2.777
Zambia: Miku	1.589	1.582	0.007–0.008	2.738
Kafubu	1.602	1.592	0.010	2.77
Zimbabwe: Mayfield	1.590	1.584	0.006	2.72
Sandawana	1.590	1.584	0.006	2.75
Tanzania: Lake Manyara	1.585	1.578	0.006	2.72
South Africa: Gravelotte	1.594	1.583	0.006–0.007	2.75
USSR: Urals	1.588	1.580	0.006–0.007	2.74
USA: North Carolina	1.588	1.581	0.007	2.73

Gübelin states that these figures represent the arithmetic means of the specimens checked which vary in number from 70 (Mingora, Pakistan) to single stones for the Anage, Salininha, Burbar, Kafubu (figures from Dr Hermann Bank, 1980) and North Carolina stones.

material. Yellow, green and blue beryls in general show two colour-causing components, broad absorption of the red to yellow (arising probably from charge transfers between Fe^{2+} in tetrahedral sites and Fe^{3+} in octahedral sites) and a strong absorption in the blue to violet. Some yellow beryl from Brazil contains Ti^{3+} in octahedral sites.

Pink beryl, or morganite, probably owes its colour to manganese, although other elements may have some role to play. At the time of writing it looks as though Mn^{2+} alone is involved in giving absorption in the ordinary ray at 540 and 495 nm and in the extraordinary ray at 555 and 355 nm. Some orange beryl from Brazil (salmon or apricot colour) was found to contain manganese, vanadium, iron and some chromium with absorption in the blue, and almost uniform transmission of wavelengths toward yellow and red. On bleaching, an apparently stable pink colour is obtained but so far the cause of the disappearance of the yellow component of the orange is not understood.

A very fine dark blue beryl was discovered in 1917 at the Maxixe mine, in the Piauí valley of Minas Gerais, Brazil, and in the early 1970s further stones of similar appearance came on to the market. These were called Maxixe-type beryls and were found to fade after exposure to light. The stones, which displayed anomalous pleochroism, with the deeper blue showing in the ordinary ray (the reverse effect from that seen in other blue beryls), showed absorption lines at 695 nm (narrow), 654 nm (strong) and weak lines at 628, 615, 581 and 550 nm. The spectra of the original Maxixe material and of the Maxixe-type stones are different; in the latter type intense absorption due to Fe^{3+} blocked the passage of blue light and gave better transmission to the green, giving a green colour to the ordinary ray. The extraordinary ray was pale yellow. The blue in this material is caused by a colour centre which can be produced in materials which have a suitable precursor. In the original Maxixe stones this seems to be a nitrate impurity and in the Maxixe-type stones a carbonate impurity. Both these ions have four atoms with 24 electrons in their outermost shells and both can lose one electron on irradiation to form a 23-electron hole centre:

$$NO_3^- \underset{\longleftarrow}{\overset{\text{irradiation}}{\rightleftharpoons}} NO_3 + e^-$$

$$CO_3^{2-} \underset{\longleftarrow}{\overset{\text{irradiation}}{\rightleftharpoons}} CO_3^- + e^-$$

The freed electron becomes trapped at some electron centre to form an atom of hydrogen

$$H^+ + e^- \rightarrow H$$

Colourless or pink starting materials give a deep blue while yellowish ones give bluish-green or green. The electron centre lacks stability so that on exposure to light the reverse equations are followed and the colour fades. Either light or heat can produce this effect, but the colour can be regained on further irradiation (Nassau, 1984).

Inclusions

The inclusions found in the beryl gemstones will often give a clue to their place of formation. The non-emerald beryls often show long hollow tubes which are usually liquid-filled; the tubes are parallel, running the length of the prismatic crystals and may have a brownish tinge. Gas bubbles are frequently found in the tubes, which are often bounded by crystal faces and are thus classed as negative crystals. Fragments of mineral matter in the tubes may help the observer to locate the tube itself. Many beryls have zones of clear material interspersed with zones containing a large number of tubes; in golden beryls the included area is often at or close to the crystal termination; in aquamarine the reverse is often the case. Crystals of biotite, hematite, pyrite, phlogopite and rutile are found in many aquamarines and sometimes when they occur in skeletal forms allow a 6-rayed star-stone to be cut. Irregular liquid droplets forming asteriated patterns were found in the famous Marta Rocha aquamarine.

In emerald, the inclusions vary with location: in stones from Colombia the commonest ones are primary liquid inclusions which in profusion cause the mossy appearance called 'jardin'. The liquid is found in cavities with a characteristic jagged profile and cubes of halite and gas bubbles are found with the liquid, the whole assembly forming a three-phase inclusion. The inclusions characteristic of the Chivor mine are crystals of pyrite and albite; those characteristic of the Muzo mine show yellow–brown prisms of the mineral parisite, $(Ce, La)_2Ca(CO_3)_3F_2$, and calcite rhombs. Emeralds from the Gachalá mine are characterized by parallel growth bands and needle-like growth tubes; as with Chivor and Muzo stones there are three-phase inclusions; pyrite crystals and six-sided cleavage cracks are also features of Gachalá material.

Brazilian stones are generally very heavily filled with two-phase inclusions, crystals of biotite, talc and dolomite and with liquid films. Those emeralds that are found in mica schists (as in deposition in Egypt, Zimbabwe and the Urals) contain mineral inclusions that have been overtaken by the growing crystal (protogenetic inclusions). These are mica, actinolite, quartz and black tourmaline. Emeralds from Zimbabwe also contain characteristic crystals of tremolite which is found as long curving needles. Brown mica platelets and curved molybdenite crystals are found in emerald from the Transvaal (the dark mica makes the whole stone dark). At least one example of a three-phase inclusion has been reported in an emerald from the Cobra Mine, Transvaal, South Africa.

Emeralds from the Ural Mountains of the USSR show bamboo-like actinolite crystals and mica plates; biotite, with actinolite, calcite and some three-phase inclusions is also characteristic of emerald from Australia. Indian emeralds contain rectangular cavities parallel to the long axis of the crystal and which contain gas bubbles. They also show biotite crystals parallel to the basal plane with fuchsite, two-phase inclusions and apatite crystals. Also characteristic of Indian stones are groups of negative twin crystals shaped like commas or hockey sticks. In Pakistan, emeralds contain two-phase inclusions, thin films, some liquid inclusions and some mineral crystals; these are euhedral crystals of calcite and dolomite; the two-phase inclusions are typically jagged and show prominent gas bubbles.

Mica and actinolite, together with two- and three-phase inclusions are found in Tanzanian emeralds from the Lake Manyara area; square-shaped cavities and tubes are also found. In Zambian emeralds, black crystals of biotite have been observed; they are very small and look like dots. Quartz crystals have been noted in some of the emerald from North Carolina. The recently-described emerald deposit at Santa Terezinha, Brazil, yields stones of a higher quality than those from other Brazilian locations. The crystals contain pyrite, chromite, talc and calcite inclusions, the pyrite occurring as sharp to slightly rounded cubes, either isolated or in groups. Chromite is in the form of black rounded crystals or octahedra. In some stones talc flakes are so profuse that they give a cloudy appearance. Flat crystals of hematite give a reddish appearance. Emeralds from the Habachtal in Austria are characterized by straight rods of tremolite with broad stems, biotite, rounded mica plates, tourmaline, epidote, sphene, apatite and rutile. Some years ago GIA reported and illustrated a fine emerald cat's-eye.

Luminescence

Emerald may fluoresce green in short-wave UV, sometimes a weak red or orange in long-wave UV. If a strong red fluorescence is seen it suggests a man-made origin and in any case the stone will show red through the Chelsea filter. All colourless beryl, as well as blue and sea-green stones, show a strong greenish-blue through the Chelsea colour filter. Those emeralds with an appreciable iron content will show no luminescence (specimens from the Transvaal and India). Morganite sometimes shows a weak lilac fluorescence.

Occurrence (details taken largely from Sinkankas, 1981)

Beryl is found in a variety of environments and has a world-wide occurrence. It is found in mica schists (particularly emerald), metamorphic limestones (emerald) and in hydrothermal veins. It is a common mineral of granitic rocks, especially pegmatites. The finest emeralds come from Colombia and were known to Colombian Indians as far back as AD 1000 or earlier. Stones were traded to other countries before the conquest by Spain. The Chivor mine location was discovered by Gonzalo Jiménez de Quesada who sent a party to the area in 1537; in 1538 he founded the city of Bogotá, later the capital of Colombia. The Muzo mine was discovered by the Spanish in about 1555; mining under Spanish overseers began at Chivor in 1555 and at Muzo in 1558. From about 1672 the Chivor workings lapsed until comparatively modern times; it is known that the Muzo mines operated continuously from 1766–1772.

In 1888 Francisco Restropo guessed at the whereabouts of the lost Chivor mine and in the following year petitioned for exploration rights. He discovered the workings in 1896 and a company was formed to exploit them but little was done until 1899. Restropo and an associate, Klein, reopened the Chivor mine in 1911; Klein left for Germany in 1914, Restropo dying at Chivor in the same year. Muzo was closed in 1925 until 1927 and was seized by a mob in 1930. It was re-opened in 1933 under Peter W. Rainier whose book *Green fire* is a detailed account of his work there. By 1946 mining rights to Muzo were given to the Banco de la República which worked intermittently from that time. Chivor was operated by Chivor Emerald Mines Inc. of Delaware, USA in 1947. It closed in 1952.

In 1954 emeralds were found at Las Vegas de San Juan, Gachalá, Cundinamarca Province, and many stones were stolen and illegally marketed. New government regulations for prospecting and mining were issued in 1955. Chivor emerald mine was in receivership from 1952. Willis F. Bronkie operated the mine as a trustee from 1957 and in 1957 a new mine, the Buenavista in Municipio de Ubalá, Cundinamarca Province, began working. By 1965 all mines were operating including the new Peña Blanca mine at Muzo. In 1968 the government-sponsored Ecominas (Empresa Colombiana de Minas) was established with authority to mine Muzo, buy, cut and sell stones. In 1970 Chivor was restored to its stock holders in the United States. Since 1976 onwards anarchy appears to have reigned at the mines.

At Muzo most of the beryl occurs in calcite veins as single crystals, clusters or fragments and little is found outside the vein. Most crystals are short, prismatic and less than several cm in length. The age of the emerald-bearing formations has been established as Cretaceous from fossil evidence: the geological column has been given as red sandstone with septarian nodules, compact sandstone, grey fossiliferous limestone between layers of grey shale with plant impressions, black carbonaceous shale and shaly limestone (which carries the emerald veins at Muzo), siliceous schists and conglomerates with jasper and flint, the whole compressed into long North–South trending folds with no igneous intrusions.

The Muzo mines lie in a deep ravine of the south-eastern-flowing Rio Itoco which joins the Rio Minero flowing north. Workings penetrate a series of black pyritiferous argillites or shales intercalated with thin-bedded limestones and folded into varying amplitudes as drag folds which have been extensively fractured and fissured due to compressional and torsional forces. Spaces formed as a result of fracturing were filled with calcite–dolomite with other species and in places emeralds are found. Emerald-bearing rocks bleach on exposure to yellowish-grey to brownish. Muzo emeralds are more yellowish than those from Chivor. Emerald from the Coscuez mine, close to and geologically similar to the Muzo mine, characteristically occurs in aggregates of crystals each with its own termination. Muzo rough is often colour zoned with a pale core but Coscuez emeralds are usually coloured uniformly throughout the crystal. A stone sold in 1985 was valued at US$25 000 per carat. Partially healed fractures are the most typical Coscuez inclusions. Emerald with a green fluorescence is reported from the Coscuez mine. The fluorescence is weak and is seen under both long-wave and short-wave UV, the latter giving the weaker effect. The cause of the emission is not known but it does not seem to result from oiling.

At Chivor the formations are about 1000 m of conformable sediments consisting of light grey calcareous shales with some lenses of carbonaceous matter. The top is formed of hard grey fossiliferous limestone, the lower member is hard blue thin-bedded limestone or calcareous shale. The geological column is (from the top) altered shale with veinlets of calcite but no emerald, shale breccia, iron oxides layer, a bluish-grey shale, a layer made up of iron oxides with albite and quartz, emerald-bearing shale with the principal emeraldiferous albite–quartz–apatite veins, another layer of iron oxides, soft altered shale with albite–quartz veins from which fine emerald may be recovered, shale with no emerald veins. Emerald crystals from Chivor tend to be more elongated than those from Muzo; they are often scattered over a vug or cavity lining.

The Gachalá deposit emeralds are found in mineralized ferruginous shales in slate beds and also in weathered decomposed arenaceous–argilla-ceous shales enclosed in beds of hard compact sandstone (Keller, 1981).

The emerald deposits of Zimbabwe produce very fine bright stones which keep their colour even in small sizes. Emerald and chrysoberyl were discovered in 1960 at the Novello prospect and the Twin Star mine about 17 km north-west of Fort Victoria; at the Novello deposit the country rock is gneissic granite with a large elongated band of serpentine intruded by pegmatite bodies and quartz veins. Emerald and chrysoberyl forms within phlogopite shells close to the pegmatites and in the pegmatite walls. At the Belingwe–Sandawana field about 120 km south-west of Fort Victoria emeralds were found in 1956; they occur in tremolite schists adjacent to granitic pegmatites intruding rocks of the Mweza range. At the Filabusi field about 88 km south-east of Bulawayo emerald is found in an area mined for scheelite and molybdenum.

The Miku emerald deposit of Zambia is south-south-west of Kitwe and named after the Miku river. Most of the emeralds are found as single crystals in a biotite–phlogopite schist with small tourmaline crystals. Large quantities of emerald are also reported from the Kafubu field to the south-west of Miku. The emeralds occur in schists rich in mica and adjacent to pegmatite bodies. Emerald from the Kitwe district of Zambia has been found to contain inclusions of apatite, quartz, chrysoberyl, margarite, muscovite and rutile. The stones are found in a biotite–phlogopite schist with tourmaline, close to pegmatites and other rocks. The RI of the Kitwe emerald is (ϵ) 1.580 (ω) 1.586 with a DR of 0.006; SG is 2.794. Chromium content is low compared with that of emerald from the nearby Miku area. This suggests that the Kitwe emerald has a quite different genesis, probably in sericitic schists (i.e. schists made up largely of a white K-mica with fine grains). Some Zambian emerald crystals show a colourless core with low RI with a 'skin' of deep green with higher RI. Stones cut from such crystals could be mistaken for doublets since the upper and lower sections would show differing RI. A Zambian emerald has been reported with an RI of 1.592 and 1.602, with a DR of 0.010 and an SG 2.77. The stone was thought to be vanadium tourmaline or vanadium grossular at first sight. Tourmaline is found in the schist host rock in which Zambian emeralds are found and also occurs as an inclusion in the emerald.

The emeralds of the Ural Mountains of the USSR have been known for certain since 1830 or 1831 when small crystals were picked up near a small brook named the Takovaya about 45 km north-east of Sverdlovsk. Stories from those days vary but the emerald deposits are described by Soviet mineralogist A. E. Fersman. The emeralds are found in biotite schists in the so-called Central Zone made up of Paleozoic metamorphic rocks following a granite contact and consisting of serpentine and other rocks, including talc schists and actinolite schists in which some emeralds also occur. Other authorities disagree with this report, stating that the emeralds are found in desiliconized pegmatites and tend to occur at the contact point between a phlogopite zone and plagioclase bodies and cores, or in the phlogopite zone itself. Sometimes fine emerald crystals are found in quartz separations in plagioclase cores. Verification is not always easy since access to the area round Sverdlovsk is restricted. The emerald crystals are usually long prismatic, often jointed and re-cemented with phlogopite; an outer colourless layer frequently encloses an inner coloured core. According to Fersman the finest emerald came from the Marinsky mines and very large ones, found early on, from the Stretensky mine.

Emerald is found in Egypt where deposits must have been known as long ago as 1925 BC. The deposits are round Gebel Sikait about 130 km north-north-west of the port of Berenice. The rocks are mostly schists, serpentines and granites, emerald occurring in biotite–quartz or tourmaline–biotite schists. Most crystals are small and deep colours are rare.

Beryl gemstones from Brazil are found in a series of highlands following the coastline of east Brazil; in these regions are many granite pegmatites which have intruded the metamorphic rocks of the Precambrian Shield. The most important emeralds come from the state of Bahia, with the exception of the Santa Terezinha deposit in the state of Goiás (see below). In Bahia the deposit at Carnaiba in the Campo Formoso district is about 77 km north of Jacobina. Emeralds are found in chromite-bearing ultramafic rocks (i.e. igneous rocks with a low silica content) outcropping in a number of places in the area. These bodies are contained in decomposed granitic rocks which have been mined, the emerald-bearing portions consisting of a vermiculite–biotite schist. At Bom Jesus das Meiras, emerald is found in an altered dolomitic marble. None of the emeralds from these deposits are of the highest quality; the deposit at Santa Terezinha in the state of Goiás produces better quality stones (Cassedanne and Sauer, 1984). The deposit is about 230 km north-west of Brasilia; emeralds were first discovered in surface material near the mine itself – eluvial matter consisting of a sandy yellow-brown argillaceous soil with angular fragments of milky quartz, limonite, quartzite and talc schist, with cubes of pyrite altered to limonite. Emerald crystals up to 1 cm in length are heavily stained with limonite. The actual emerald-bearing rock is a partially weathered talc schist. At the Garimpo de Baixo–Trecho Novo, the newest workings, the emeralds are found in a talc schist intersected by pegmatites. The stones have an SG 2.70 (some with a lot of included pyrite reach 3.05). RI is (ϵ) 1.580 (ω) 1.588 with a DR of 0.008. The colour through the Chelsea filter is pink; alternatively the stones remain inert. They are sometimes almost free from inclusions but pyrite and limonite pseudomorphs after pyrite are common; chromite, talc flakes and calcite are also often seen, as well as hematite which gives a reddish appearance to the stone.

Although emerald has been known in India since very early times the modern development of emerald mining began in 1943 when green crystals were found near the small village of Kaliguman in south Rajasthan. The crystals are found in schists; at Kaliguman the country rock is a hornblende schist in which the emerald crystals are irregularly distributed in vein-like bodies of soft talcose–biotite schists. Intruded granite pegmatites are close by. The emerald deposits of the Gravelotte district in the north-east Transvaal, South Africa are also of the schist type. These were discovered in 1927; again pegmatite bodies are in the vicinity.

In the United States emerald is found in the state of North Carolina around the small village of Hiddenite in Alexander County. The emeralds occur in contorted gneiss which contains veins of feldspar–quartz. The veins contain crystallized cavities in which the emerald occurs; these cavities are Alpine-type clefts and form a unique type of occurrence for emerald. Quartz inclusions are found in the emeralds.

In 1969 emerald crystals were reported from the deposit at Lake Manyara, Tanzania, about 110 km from the town of Arusha and west-south-west of it. Aggregations of emerald are found in lenses of biotite in schists and gneisses; alexandrite is an accompanying gem mineral. Some small pegmatite lenses are also present. The emerald crystals are found in the contact zones between schists and pegmatites; the earlier material was not of high quality but a new vein in about 1973 provided much better stones, some containing dust-like inclusions.

Emerald has been found in New South Wales, Australia, near Emmaville where it occurs as small crystal concentrations in a quartzose vein with topaz, cassiterite, fluorite, arsenopyrite and quartz. Better specimens are found at the Poona deposit in Western Australia where granitic pegmatites intrude into Archean greenstones surrounded and invaded by granites. The greenstones include hornblende schists which are altered where contact is made between the pegmatite and biotite schists. Small pegmatitic veins branching from the main ones pass into discontinuous strings of small lenses of feldspar, quartz or beryl accompanied by isolated beryl crystals within the flanking biotite schists. Emerald occurs in this beryl as well-developed hexagonal prisms.

The beryl deposit of the Habachtal, Austria, lies on the side of the Legbach ravine about 2100 m above sea level. The ravine drops west into the main Habachtal valley. The nearest town is Bramberg, which is about 70 km south-west of the city of Salzburg. The emerald deposit lies along the contact between the Central Gneiss of the Hohe Tauern and amphibolitic rocks; there is a zone at the contact representing reaction and injection rocks which are amphibolite, aplite veins, layers of biotite, talc and talc–biotite, tremolite–actinolite schists, chlorite schists and quartz. Emerald occurs primarily in biotite schists and the deposit is similar to those of the Urals. Crystals vary in size and are notable for a ready cleavage along basal planes.

In recent years there have been reports of emerald from other places but no new major deposit has been found from countries other than those already mentioned. Some emerald has been found in the Malagasy Republic where it occurs in a mica schist in the south-east of the island. The emeralds resemble those from Zambia in their appearance and constants. The location is Ankadilalana. Crystals reach only small sizes and contain goethite crystals, two-phase inclusions, negative crystals of isometric shape with liquid and gas bubbles.

Over the last year or two some beryl crystals have come from Nigeria; those I have seen grade into emerald from green beryl in a few cases.

Both emerald and aquamarine crystals have been found near Jos, Nigeria. The stones have a colour of variable intensity and RI of (ω) 1.570–1.574, (ϵ) 1.564–1.568. SG is 2.66–2.68. Zoning is seen parallel to the basal pinacoid, the hexagonal dipyramid and to the first order prism. Two different types of two-phase inclusion have been noted. The stones contain varying concentrations of vanadium, chromium and iron.

The Swat Valley in Pakistan produces fine quality emeralds which first came to light in 1958. Emerald-bearing rocks overlie a dark mica schist and are covered by a lighter green tremolite–chlorite schist, this in turn being overlain by amphibolites along a tectonic shear zone. Emerald is found in a dolomitic talc schist up to 50 m thick. The beryl and other minerals normally found in an acidic environment are here hydrothermally derived from granitic rocks and deposited in host schists after passing through basic rocks which provide the chromium. Mining is by open pit methods; the stones contain chromium and iron, the latter subduing fluorescence.

Plate 1 Diamond crystal in kimberlite matrix.

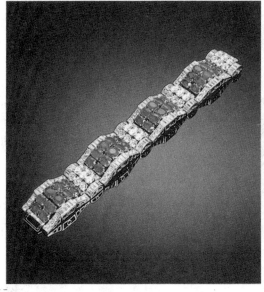

Plate 2 Ruby bracelet. Most stones are Burmese.

Plate 3 Sapphire and diamond panther clip.

Plate 4 The varieties of beryl.

Plate 5 Emerald crystal in calcite.

Plate 6 Cat's-eye chrysoberyls.

Plate 7 The varieties of garnet.

Plate 8 Spinel crystals.

Plate 9 Opals.

Vanadium appears to be absent. Mean SG is 2.77 with RI (ϵ) 1.590 and (ω) 1.595 with a DR of 0.007 – high for gem quality emerald. Stones are light red to red through the Chelsea filter but do not react to UV, due to the iron content. Wavy liquid feathers can be seen, with two-phase inclusions, healing cracks and zoning. Dolomite rhombohedra are characteristic inclusions as are jagged two-phase inclusions (Gübelin, 1982).

Deposits of small but good quality emerald crystals were found northeast of Mingora at Gujar Killi; on the occasion of my visit in 1985 the mine was producing material for the Gemstone Corporation of Pakistan but there was some danger in the wet decomposed rock in which the crystals were found. Production has now ceased.

An aquamarine crystal from the Gilgit Division pegmatites of Pakistan contained two euhedral deep blue–green translucent crystals which proved to be zircon. The 1987 report stated that the crystals showed a strong orange fluorescence under short-wave UV.

Deep blue–green coloured emerald and a light green beryl have been reported from talc–carbonate–quartz rocks near Bucha in the Mohmand Agency of Pakistan. The emerald has RIs of 1.590 and 1.600 and contains fluid inclusions as well as talc fibres and mica plates. The light green beryl has RIs of 1.575 and 1.583 with siderite, tremolite and mica inclusions.

Emerald found in the Panjshir valley of Afghanistan is a fine dark green with RI 1.578 and 1.585; the SG is 2.71 and the DR about 0.005. Two and three-phase inclusions are reported, with growth zones and fracturing.The stones occur in veins cutting through host rocks of metamorphosed limestones, calcareous slates, phyllites and micaceous schists. The veins consist mainly of quartz and albite. Pyrite is used as an indicator of the near presence of emerald, which is thought to be of hydrothermal origin.

Aquamarine

The finest-coloured aquamarine comes from Brazil where there are many important deposits. All Brazilian aquamarine seems to gain prominence in some way; in 1910 a crystal weighing 110.5 kg was found in the Marambaia region of Minas Gerais and the town of Teófilo Otoni can be regarded as the aquamarine centre of the world; almost all aquamarine production taking place within a radius of 100 km. Aquamarine is a typical mineral of pegmatites and always lies close to the quartz core in primary deposits, or is found in surrounding feldspathic material.

Any discussion of Brazilian aquamarine deposits must concentrate on one or two of the most important, since at any time there may be many small mines in operation. In northern Minas Gerais the area of the

Jequitinhonha river and its tributaries produces some important stones, in particular the 'Fortaleza' aquamarines which are found both in the main river and its streams. Best quality material is at washings (lavras) at Laranjeiras near the villages of Fortaleza and Piedra Grande; the name Pedra Azul is now used in place of Fortaleza. The famous Marta Rocha aquamarine was found in 1954 or 1955 on a farm near Teófilo Otoni; named after a Miss Brazil, the crystal was thought to be 60% facetable and the colour was a fine deep blue. Another crystal said to be deep blue to blue-green and to resemble fine blue tourmaline was found at the Pedroso alluvial mine, Rio Marambaia valley in about 1964. Fine stones are reported from the south bank of the Rio Mucuri. The finest Brazilian aquamarines are found in the deposits in the area of Teófilo Otoni. These include Marambaia, the valley of the Jequitinhonha and Aracuai rivers, and the areas of Capelinha, Maiacacheta and Governador Valadares. These districts are described by Proctor (1984). The Marta Rocha crystal was found near the village of Topazio in the first of the areas listed above. The crystal was river-rolled and weighed 34.7 kg. Some of the darkest blue aquamarine ever found was found near the hamlet of Ponto de Marambaia in 1964. All deposits are pegmatitic in origin, the peak production from the most productive mines occurring in 1981 and falling since that year. The expense of seeking and working new areas is considerable and it is unlikely that aquamarine will ever flood the market. In mid 1984 the Lavra da Invreja mine in Minas Gerais was opened and about 350 kg of medium-quality material was recovered before working ceased the same year.

Fine aquamarine is reported from a number of localities in the Malagasy Republic where the many pegmatites offer a large number of gem crystals. Although the country is more famous for pink beryl than for aquamarine, fine stones are reported from near Tsaramanga village 3 km north of Mount Itongafeno (the locality is often known as Tongafeno) where some crystals show near emerald-green cores with blue rind as well as a characteristic dark aquamarine colour; other pegmatites produce aquamarine as well as other varieties of beryl. Aquamarine occurs in a porous granite containing miarolitic pegmatite cavities in the Mourne Mountains of Northern Ireland; crystals are pale to medium-blue; some may be a deep, rich sky-blue and may be of gem quality; there are other localities in the same country, particularly at Slieve Corra. At Rossing, Namibia, aquamarine from pegmatite has a SG 2.692–2.694 and covers a range of colour from yellowish to greenish-blue.

Aquamarine is found with other colours of beryl in the Mursinska district of the Ural Mountains of the USSR. The occurrence is in pegmatites in which the beryls are found attached to the walls of cavities filled with brown clay. In 1900 over 450 kg of gem quality aquamarine was recovered from one pit at Adui near Mursinska. Fine stones were also found at Sarapulskaya in the same area. Some aquamarine has been found in Zimbabwe and there are several occurrences in the United States; the best-known are probably crystals from Connecticut where beryl has been found at Haddam Neck and in other pegmatite locations. Fine aquamarine is reported by Kazmi *et al.* (1985) in pegmatites of the Shingus–Dusso area of Gilgit, northern Pakistan. Some doubly-terminated crystals are found and some groups of a very fine blue are reported.

Morganite

Pink beryl (morganite) is found in fine qualities in the Pala district, San Diego County, southern California. Some morganite comes from the Pala Chief mine in this area and also from the White Queen mine on Hiriart Hill (from which the finest North American morganite was recovered). Stones of a rich apricot colour, probably fading to pink, come from the Mesa Grande district. In Brazil, morganite production is headed by the Corrego do Urucum mine in Galiliea, at which large crystals were found in 1973 in a very large cylindrical pocket. Kunzite was also found here. Other deposits in Minas Gerais include Sapucaia and Calisto with mines along the Jequitinhonha river. No major deposit had been discovered since 1973 (Proctor, 1984).

The Malagasy Republic is the other main producer of morganite; as with the Californian deposits, the mineral occurs in pegmatites. Some are found in the Sahatany River valley in the Mount Ibity (Mt Bity) region. Large rose-coloured, very fine beryl crystals of flattened shape are found in pegmatites emplaced along contacts of mica schists and impure marble; red tourmaline of gem quality is also found in this area. Crystals are alkali-rich and are characterized by an absence of prism faces and the predominance of the basal face with brilliant pyramid faces; some crystals are a salmon or rose colour. At a pegmatite location 8 km south of Mount Ibity fine rose- to salmon-coloured crystals are found; the name of the locality is Tsilaisina (a name adapted for describing a variety of tourmaline).

Other colours

Yellow beryl, when truly golden, is known as heliodor; this colour is found in the pegmatites at Rössing, Namibia and at a number of places in Brazil, especially Bom Jesus do Lufa, Dois de Abril, Minas Novas, Mina Urubu and Sapucaia in the state of Minas Gerais. Other significant areas include Marambaia, Ilha Alegra and Farrancho. Golden yellow beryl is found in the Mursinska district of the Ural Mountains of the USSR (see aquamarine above). Some of the crystals are a magnificent colour; one of the finest can be seen in the Leningrad Mining Institute. Golden beryls are also found at

Lewaschinagorka, east of Alabashka. A deep yellowish-red beryl is found in the area of Santa Maria do Suassui (Suaçuí), Minas Gerais, Brazil. From the Governador Valadares area of the same state a dark brown beryl with a weak asterism has been found, the effect said to be due to ilmenite inclusions. Clearer parts of the stones show that the true body colour is that of aquamarine.

A pleasing apple-green beryl, coloured by vanadium, is found in a number of places in Brazil, particularly the area of Salininha, near Pilâo Arcado, Bahia, Brazil. Vanadium probably occupies octahedral sites in the crystal normally occupied by aluminium. Raspberry red beryls are found in the rhyolites of the Thomas and Dugway Mountains, Juab County, Utah, USA. The crystals occur in gas cavities with a number of other minerals, including topaz and spessartine garnet. They have subsequently been called bixbite after the collector Maynard Bixby of Salt Lake City who discovered them. Unfortunately a completely different mineral was named bixbyite, after the same gentleman, so 'bixbite' is not encouraged. Far more attractive red beryl crystals are found in Beaver County of the same state where deposits in the extreme south-east of the Wah Wah Mountains yield crystals of a rich red colour. They are found in a whitish kaolinite or in grey rhyolite. The colour is due to manganese; specimens are usually small and the best resemble fine dark ruby. Red beryl contains a high proportion of trace elements which are low or not present in other beryls, including manganese, titanium, zinc, tin, lithium, niobium, scandium, zirconium, gallium, caesium, rubidium, boron and lead. The RI falls within the range 1.564–1.574 and the SG between 2.66–2.70. The mineral is thought to have crystallized along fractures, in cavities or within the rhyolite host from a high-temperature gas or vapour phase released during the late stages of cooling and crystallization of the rhyolite magma (Shigley and Foord, 1984).

Some beryl yields cat's-eyes and star stones; these are usually found in green beryl or aquamarine. Some black star beryls are reported from Alto Ligonha, Mozambique.

An unusual variety of emerald so far found only in Colombia (Muzo and Chivor) has been named 'trapiche' from the resemblance of the crystals to cog-wheels used to crush sugar-cane. The crystals have a central hexagonal core from which narrow bands of whitish or translucent inclusions radiate. The core is emerald which also occupies the spaces between the spokes. A report found that all trapiche crystals were untwinned and contained white feathery inclusions of albite and beryl. Muzo crystals had carbonaceous inclusions.

Beryl comes from a Greek word which was used for more than one kind of gemstone and whose meaning is uncertain. Aquamarine and heliodor describe the colour; morganite was named for the banker J. P. Morgan. The colourless variety goshenite is named for Goshen, Massachusetts, USA where it occurred at the Lily Pond mine in a pegmatite 22 km north-west of Northampton. It also occurs in Brazil, the USSR and elsewhere in the USA. Emerald comes from the Greek word for green.

BERYLLONITE

$NaBePO_4$. Monoclinic, forming tabular or prismatic crystals, frequently etched and commonly twinned. They closely resemble crystals of the orthorhombic system. There are two directions of cleavage, one of them perfect. H 5.5–6. SG 2.80–2.85. RI (α) 1.552 (β) 1.558 (γ) 1.561. Biaxial negative. DR 0.009. Dispersion 0.010. Dark blue fluorescence with phosphorescence under X-rays.

Beryllonite is a rare colourless mineral of granite pegmatites, found in gem sizes and clarity only at Stoneham, Maine, USA. Named from the composition.

BISMUTOTANTALITE

(Bi, Sb) (Ta, Nb)O_4. Orthorhombic, sometimes forming large crystals with a perfect cleavage in one direction. H 5. SG 8.84 (Brazilian crystals). RI (α) 2.388 (β) 2.403 (γ) 2.428. Biaxial positive. DR 0.040.

Found in pegmatites, bismutotantalite is rare and small as a faceted gemstone, most material of this quality coming from Acari, Brazil. The colour is light brown to black and it is named from its composition.

BOLEITE

$Pb_{26}Ag_{10}Cu_{24}Cl_{62}(OH)_{48} \cdot 3H_2O$. Cubic, forming cubes with some modifying faces. The mineral has been classed by some authorities in the tetragonal system, making the crystals pseudocubic. There are two directions of cleavage, one perfect. H 3–3.5. SG 5.05. RI (ω) 2.05 (ϵ) 2.03 (if mineral is tetragonal). Uniaxial negative. DR 0.020.

Found as a secondary mineral in copper and lead deposits, boleite, as deep blue crystals, takes its name from Boleo, Baja California, Mexico from where the only (very small) facetable crystals come. The colour is especially fine.

BORACITE

$Mg_3B_7O_{13}Cl$. Orthorhombic, forming pseudotetragonal crystals. H 7–7.5. SG 2.95. RI (α) 1.658–1.662 (β) 1.662–1.667 (γ) 1.668–1.673. Biaxial positive. DR 0.011. May show weak green fluorescence under short-wave UV.

Rare cut stones are a pleasant pale green or blue and the mineral is recovered from salt deposits, the only facetable material coming from Stassfurt, Germany. Named from the borax in its composition.

BOWENITE see SERPENTINE

BRAZILIANITE

$NaAl_3(PO_4)_2(OH)_4$. Monoclinic, forming equant and prismatic crystals, often spear-shaped and striated with a perfect pinacoidal cleavage. H 5.5. SG 2.980–2.995. RI (α) 1.602 (β) 1.609 (γ) 1.621–1.623. Biaxial positive. DR 0.019–0.021. Weakly pleochroic (light and slightly darker yellowish-green).

Brazilianite, discovered in a pegmatite in that country, was first described by F. H. Pough in 1945. Crystals, often large, are an attractive yellowish-green. The location is Conselheiro Pena, Minas Gerais; some material (although not quite of gem quality) has been found at the Palermo mine, Grafton, New Hampshire, USA. Named from the main locality.

BROOKITE

TiO_2, polymorphous with rutile and anatase. Orthorhombic, forming tabular or prismatic striated crystals with a high lustre. H 5.5–6. SG 3.87–4.14. RI (α) 2.583 (β) 2.584 (γ) 2.700–2.740. Biaxial positive. DR 0.122–0.157. Strong dispersion 0.131. Strong pleochroism, yellowish-brown, reddish-brown and orange.

Brookite is only transparent in small fragments which are rather dark and are characteristic of Alpine deposits where it may occur in schists and gneisses as well as in igneous rocks. The brown stones can be quite attractive. Named after the English mineralogist J. H. Brooke.

CALCITE

$CaCO_3$, dimorphous with aragonite. A cobalt-bearing variety is known as sphaerocobaltite. Trigonal, with characteristically profuse forms and a perfect rhombohedral cleavage. H 3. SG 2.71–2.94. RI (ϵ) 1.486–1.550 (ω) 1.658–1.740. Uniaxial negative. DR 0.172–0.190. Wide variety of luminescent effects.

Calcite occurs in almost every type of geological environment. Gem quality material may be large; Missouri, USA produces colourless crystals that have been cut and brown crystals are found in Baja California, Mexico. Sphaerocobaltite is found in Spain. Iceland spar is the name given to large clear colourless masses; alabaster, a name often and incorrectly given to a variety of calcite, should be reserved as a varietal name for gypsum (q.v.). Calcite as marble is much better known than the rare faceted stones; the name onyx is often used in this connection. Named from the Latin word for lime.

CANCRINITE

$Na_6Ca_2Al_6Si_6O_{24}(CO_3)_2$. Hexagonal, forming prismatic crystals but more commonly masses. There is one direction of perfect cleavage. H 5–6. SG 2.42–2.51. RI (ω) 1.507–1.528 (ϵ) 1.495–1.503. Uniaxial negative. DR 0.022.

Occurring in alkali-rich rocks and as an alteration product of nepheline, cancrinite is sometimes cut into yellow cabochons which may glow orange–yellow under X-rays. The best gem quality material comes from Bancroft, Ontario, Canada. Named from Count Cancrin, a Russian minister.

CASSITERITE

SnO_2. Tetragonal, forming prismatic crystals, frequently twinned in the geniculate habit. H 6–7. SG 6.95. RI (ω) 2.006 (ϵ) 2.097–2.101. Uniaxial positive. DR 0.098. Dispersion 0.071. Weak pleochroism.

With its high dispersion and almost adamantine lustre, cassiterite may arouse occasional thoughts of diamond but most stones are too dark to cut, the most attractive specimens being yellowish to light reddish-brown. As the principal ore of tin, cassiterite is found in medium to high temperature veins, metasomatic deposits and pegmatites. Many specimens are recovered from alluvial deposits. Most gem quality material comes from the Araca mine, Bolivia and some from the Erongo tin mines in Namibia. Named from the Greek word for tin.

CELESTINE

$SrSO_4$. Orthorhombic, crystals usually tabular. Two directions of cleavage, one perfect. H 3–3.5. SG 3.97–4.00. RI (α) 1.622–1.625 (β) 1.624

(γ) 1.631–1.635. Biaxial positive. DR 0.009–0.012. Bluish-white glow under short-wave UV, sometimes showing a similar phosphorescence.

Occurring in sedimentary rocks and hydrothermal vein deposits (commonly in limestones) celestine is, at its best, an attractive pale blue. Gem quality material has been found at Tsumeb, Namibia and in the Malagasy Republic (stones have lower RI). The Lampasas area of Texas, USA supplies some facetable stones. Orange crystals are reported from Ontario, Canada. Named from the Latin word meaning celestial, with reference to the blue colour of the stone.

CERULEITE

$Cu_2Al_7(AsO_4)_4(OH)_{13} \cdot 12H_2O$. Triclinic, occurring as compact masses with a sky-blue colour. H 6.5. SG 2.7–2.8. RI about 1.60.

Ceruleite, named from its colour, occurs as a sedimentary material with copper deposits and in this and its appearance is similar to turquoise. Fine cabochon material is found in Bolivia. A stabilized form is known as a turquoise imitation and has an SG of 2.58. Infra-red spectroscopy shows a band at 1725 cm^{-1} and bands characteristic of untreated material. The band at 1725 cm^{-1} is also found in plastic-impregnated turquoise.

CERUSSITE

$PbCO_3$. Orthorhombic, crystals commonly twinned and striated, very brittle and with a distinct cleavage in one direction. H 3–3.5. SG 6.55. RI (α) 1.804 (β) 2.076 (γ) 2.079. Biaxial negative. DR 0.274. Dispersion 0.055. Pale blue or green fluorescence under short-wave UV in some specimens.

With a dispersion greater than that of diamond, colourless cerussite can look spectacular when faceted, despite its softness, since it has a lustre closely approaching the adamantine. Since it is brittle and heat sensitive as well as soft and easily cleaved, cutting is difficult, although specimens suitable for cutting are not hard to find among the output from Tsumeb, Namibia, where cerussite occurs as a secondary mineral in the oxidized zone of lead deposits. Named from the Latin cerussa, given to a man-made lead carbonate.

CHALCEDONY see QUARTZ

CHAMBERSITE

$Mn_3B_7O_{13}Cl$. Orthorhombic, crystals tetrahedron-shaped. H 7. SG 3.49. RI (α) 1.732 (β) 1.737 (γ) 1.744. Biaxial positive. DR 0.012.

The brownish-purple crystals of chambersite can be recovered only by divers working in the brine storage well at Barber's Hill salt dome, Chambers County, Texas, USA. This very rare but interesting mineral is obviously named from its location.

CHAROITE

(K, Na)$_5$(Ca, Ba, Sn)$_8$(Si$_6$O$_{15}$)$_2$(Si$_6$O$_{16}$)(OH, F) · nH$_2$O. Monoclinic, forming rocks which also contain aegirine–augite, microcline and another rare mineral tinaksite which mingles an orange colour with the purple charoite and black accompanying minerals. SG 2.68. RI near 1.55.

Unlike any other known ornamental mineral charoite is easy to identify. It is named from its location on the Charo river in north-west Alden, USSR.

CHILDRENITE

(Fe, Mn)AlPO$_4$(OH)$_2$ · H$_2$O. Forms a series with eosphorite in which manganese exceeds iron. Orthorhombic, forming pyramidal crystals, often doubly terminated. H 5. SG 3.2 for the pure iron end-member. RI (α) 1.63–1.645 (β) 1.65–1.68 (γ) 1.66–1.685. Biaxial negative. DR 0.030–0.040. Distinct pleochroism, yellow, pink, colourless to pale pink.

Childrenite is found in granite pegmatites and hydrothermal vein deposits and is brown and opaque (eosphorite is orange to pink). Facetable material comes from Minas Gerais, Brazil. Named after J. G. Children, an English mineralogist.

CHONDRODITE

(Mg, Fe^{2+})$_5$(SiO$_4$)$_2$(F, OH)$_2$. A member of the humite group. Monoclinic, forming twinned crystals. H 6.5. SG 3.16–3.26. RI (α) 1.592–1.615 (β) 1.602–1.627 (γ) 1.621–1.646. Biaxial positive. DR 0.028–0.034. Pleochroism pale yellow to brownish; colourless to yellowish-green; colourless to pale green. Some golden to orange–brown fluorescence in long-wave UV.

The only known facetable chondrodite is a rich garnet red and comes from the Tilly Foster mine, Brewster, New York, USA. The mineral occurs in contact zones in limestone or dolomite. It is named from the Greek word meaning a grain since it occurs frequently in that form. A mineral which has in the last three years reached the gem markets, though only sporadically, also belongs to the humite group. This is the bright yellow clinohumite, reportedly from the USSR.

A faceted stone of 0.39 ct was examined by GIA in 1986; they report that it had an RI of 1.681, 1.642 and 1.668 with an SG of about 3.18. No absorption lines and bands were seen though the stone fluoresced a moderate to strong chalky-orange-yellow under short-wave UV.

CHRYSOBERYL

$BeAl_2O_4$ with some iron and titanium. Orthorhombic, forming tabular or prismatic crystals; trillings common. Found also in masses and as water-worn pebbles. H 8.5. SG 3.71–3.72. RI (α) 1.744–1.759 (β) 1.747–1.764 (γ) 1.753–1.770. DR 0.008–0.010. Biaxial positive. Indices vary with iron content. Distinct pleochroism with alexandrite showing deep red, orange–yellow and green; yellow–green–brown stones show shades of yellow and brown. These stones also show a broad absorption band at 444 nm, due to iron and other bands may be seen at 504 and 486 nm. In alexandrite there is a chromium-doublet at 680.3 and 678.5 nm (in some conditions this may show as an emission doublet) with other lines at 665, 655 and 645 nm. Narrow lines in the blue at 473 and 468 nm can also be seen. In the yellow–brown stones the band at 444 nm separates chrysoberyl cat's-eye from those made from chatoyant quartz. Alexandrite may fluoresce a dull red in short-wave and long-wave UV. The commonest inclusions are liquid-filled cavities with two-phase inclusions; stepped twin planes may be seen and the chatoyancy in the cat's-eye (or cymophane) variety is due to short needles and tubes parallel to the long axis of the crystal.

Chrysoberyl is found in pegmatites, gneisses, mica schists and dolomitic marbles; also as water-worn pebbles. The yellow–green stones are bright and lively; cat's-eye should at its best show a sharp blue–white 'eye' against a background of a honey brown. The eye in cat's-eye is the sharpest shown by any chatoyant gem and the name cat's-eye is enough on its own to indicate in commercial circles that the chrysoberyl variety is meant. Some cat's-eyes show a rich brown on one side of the eye while a yellowish–white is seen on the other side, according to the way in which light strikes the stones. This is the 'milk and honey' effect. Chatoyancy in some chrysoberyl cat's-eye from pegmatites in the Trivandrum area of southern India may be due to acicular crystals of sillimanite occurring along micro-fractures. In alexandrite the optimum colours are purple–red in incandescent light and intense blue–green to green in daylight. Stones over 5 ct are very rare and will command quite exceptional prices. In general the stones from the USSR are bluish-green in daylight and those from Sri Lanka a deep olive-green with less pronounced a colour change. Stones from Zimbabwe are a bright emerald-green but are heavily included and small. Quite a good colour change is shown by Brazilian and Tanzanian alexandrite. The finest cat's-eye comes from Sri Lanka; this material also comes from Brazil which produces most of the yellow–green or golden material used commercially. The finest alexandrite comes from the Takowaja emerald mines district north-east of Sverdlovsk. Alexandrite was discovered there by Nordens-kiold who named it in 1842 in honour of the Czar Alexander II since it was discovered on his birthday in 1830. Crystals occur most often as star-shaped trillings, averaging three-quarters of an inch (~ 2 cm) across. The enclosing rock is a mica schist. Fine yellow chrysoberyl is found in the Sanarka district of the southern Urals. Chrysoberyl from Sri Lanka is found in the gem gravels and does not show the crystal forms characteristic of the Uralian material. Alexandrite is found in mica schists in both Zimbabwe and Tanzania (in the Somabula Forest area and Lake Manyara area respectively). In Brazil chrysoberyl comes from various places in the state of Minas Gerais and especially from Jacuda, Bahia. A colourless variety has been found in Burma where a few alexandrites have been reported from the Mogok Stone Tract. A high reading chrysoberyl was found in a pegmatite with beryl in the Harts Range of Central Australia. RI were (α) 1.765 (β) 1.772 (γ) 1.777. There was an intense absorption band in the violet centred near 447 nm with another near 508 nm. Chrysoberyl from Anakie, Queensland, Australia has RI of (α) 1.756 (β) 1.761 (γ) 1.768 with a DR of 0.012. This is high for chrysoberyl. The SG is 3.74 and there is a strong broad absorption band covering the spectrum from 458 to 425 nm. The colour is brown. Two coloured chrysoberyls examined in Germany contained no gallium and were thus classed as synthetic since virtually all natural colourless chrysoberyl (e.g. from Burma) contains this element. Some authorities, however, maintain that this is not a certain method of diagnosis.

Chrysoberyl takes its name from a Greek word alluding to its golden colour; cymophane also has a Greek origin, the original word meaning looking like a wave (an allusion to the opalescence of some crystals).

CHRYSOCOLLA

$(CuAl)_2H_2Si_2O_5(OH)_4 \cdot nH_2O$. Monoclinic, forming aggregates and crypto-crystalline masses. H 2–4, although may reach 7 with high silica content. SG 2.00–2.45. RI (α) 1.575–1.585 (β) 1.597 (γ) 1.598–1.635. Where composition is largely quartz, RI may be that of quartz. Biaxial negative.

Chrysocolla occurs in the oxidized zone of copper deposits and is frequently mixed with silica and other compounds such as limonite. The colour depends on the amount of impurities but at its best is a fine blue. The variety Eilat Stone is frequently cut into cabochons and is a mixture of chrysocolla with turquoise and pseudomalachite although constituents vary widely; it has an SG 2.8–3.2. A touch of hydrochloric acid will stain the mineral yellow. Eilat Stone comes from Eilat, Gulf of Aqaba in the Red Sea; other deposits are in the USSR and Zaire. Named from two Greek words meaning golden and glue, with reference to a material used by the Greeks in soldering.

CINNABAR

HgS. Hexagonal, usually massive but some crystals displaying twinning are also known. Perfect cleavage in one direction. H 2–2.5. SG 8.09. RI (ω) 2.905 (ϵ) 3.256. Uniaxial positive. DR 0.351.

Cinnabar is a very fine scarlet red but the softness, brittleness and cleavage combine to restrict the number of faceted stones. The colour alters gradually on exposure to light. The finest crystals come from China where cinnabar is found in low temperature ore deposits. The name is believed to be of Indian origin.

CITRINE see QUARTZ

CLINOHUMITE see CHONDRODITE

CLINOZOISITE see EPIDOTE

COLEMANITE

$Ca_2B_6O_{11} \cdot 5H_2O$. Monoclinic, forming equant or prismatic crystals with a perfect cleavage in one direction. H 4.5. SG 2.42. RI (α) 1.586 (β) 1.592 (γ) 1.614. Biaxial positive. DR 0.028. May show yellowish or greenish–white fluorescence and phosphorescence in short–wave UV.

Colemanite is normally colourless and faceted transparent stones have low dispersion. Material comes from a number of Californian localities in particular the saline lake deposits at Boron and Death Valley. Named for William T. Coleman who owned the mine where the mineral was discovered.

CORDIERITE

$(Mg, Fe)_2Al_4Si_5O_{18}$. Orthorhombic, forming prismatic crystals with a rectangular cross section and a distinct cleavage in one direction. H 7–7.5. SG 2.57–2.61 (higher values with increasing iron content). RI (α) 1.522–1.558 (β) 1.524–1.574 (γ) 1.527–1.578. Biaxial positive. DR 0.005–0.018. Intense diagnostic pleochroism: pale yellow, pale blue, violet blue. An absorption spectrum varying with crystal direction has weak bands at 645, 593, 585, 535, 492, 456, 436 and 426 nm. An almost colourless cordierite from Sri Lanka has a very low RI of 1.530–1.540 and an iron content of only 1.14%. Cordierite with a very low RI of 1.527, 1.532 and 1.536 with a DR of 0.009 and almost colourless has been reported from Sri Lanka. The SG was 2.55 and some material was chatoyant. A colourless cordierite and specimens showing chatoyancy are reported from Sri Lanka. These specimens have a very low iron content and show RI of 1.527 and 1.532 for an iron content of 0.1%.

Cordierite (alternative name iolite, less commonly dichroite) occurs in altered aluminous rocks and in some igneous rocks. It is celebrated and fashioned for its pleochroism of which the dark violet–blue colour is the most attractive. A variety with profuse inclusions of hematite plates is reddish and known as bloodshot iolite. Much gem quality cordierite comes from the Madras area of India and more from the gem gravels of Sri Lanka and Burma. In Tanzania cordierite is found at Babati. Four-rayed star stones are found occasionally. The name iolite is from Greek words meaning a violet stone; cordierite for the French geologist P. L. A. Cordier.

CORUNDUM

Al_2O_3 with some iron, titanium and chromium. Trigonal, forming barrel-shaped prisms with flat ends; hexagonal bipyramids, tabular crystals or masses. Much ruby is found as tabular crystals with large faces at top and bottom marked triangularly and combined with the six faces of the prism and six, three above and three below, of the primitive rhombohedron. Sapphire prefers the bipyramidal habit. There is no cleavage but since the crystals show lamellar twinning there is a zone of parting parallel to the basal face and another, less commonly seen, parallel to the faces of the primary rhombohedron. H 9. SG usually near 4.00 but in the range 3.99–4.1. RI (ω) 1.757–1.768 (ϵ) 1.765–1.776; usually 1.760, 1.768. Uniaxial negative. DR 0.008–0.009, dispersion 0.018. Very pronounced pleochroism, in ruby a strong purple–red to orange–red; in blue sapphire violet–blue to blue–green; in green sapphire an intense green to a yellowish-green; in orange sapphire yellow–brown or orange to colourless; purple sapphire shows violet and orange. A colour-change corundum

from Sri Lanka showed somewhat high refractive indices of 1.764 and
1.773 with DR 0.009. The colour was red in tungsten light and greenish-
blue in daylight. Some corundum shows the occasional 12-pointed star
which is due to rutile crystals arranged at 30-degree spacings from the rays
of the more prominent six-rayed star crystals and parallel to the faces of the
second order prism. Most stars are six-rayed in both ruby and sapphire and
are formed by exsolved rutile.

Burmese rubies show a strong red fluorescence in long-wave and short-
wave UV and X-rays. Ruby from Thailand shows a less intense red due to
iron in its composition. Ruby from Sri Lanka gives strong orange–red in
long-wave UV, less strong in short-wave UV. Blue stones show no
luminescence except for some blue Thai stones which give a weak
greenish-white in short-wave UV. Sri Lanka stones may glow red to
orange in long-wave UV. Green sapphires (with a high iron content) are
inert. Yellow sapphire from Sri Lanka shows a distinctive apricot colour in
long-wave UV and X-rays, a weaker yellow–orange in short-wave UV.
Pink sapphire fluoresces strong orange–red in long-wave UV, weak light
red in short-wave UV; some colourless sapphire gives moderate light red
to orange, and orange stones show a strong orange–red, in long-wave UV.
Some stones from Montana, USA, Sri Lanka and Kashmir show dull red to
orange in X-rays.

Ruby shows a distinctive absorption spectrum in which a strong doublet
in the deep red, at 694.2 and 692.8 nm can be seen as a single bright (red)
line under normal conditions. Two fainter lines at 668 and 659.5 nm may
also appear as fluorescence lines; there is a strong and broad absorption
band centred near 550 nm covering most of the yellow and green. Three
narrow bands are in the blue, two of them close together and the other
separated, at 476.5, 475 and 468.5 nm. There is a strong general absorption
of the violet. This spectrum, which is that of chromium, is the same for
both natural and synthetic ruby. The fluorescent doublet is a sensitive test
for chromium as it may be seen when as little as 1 part of Cr_2O_3 in 10 000
parts Al_2O_3 is present. It can even be seen in colourless sapphires and in blue
Sri Lanka sapphires which contain a trace of chromium. In direct
transmitted light the doublet may appear as an absorption line in
chromium–rich or synthetic rubies. The spectrum varies with the ordinary
and extraordinary rays, the broad absorption band being broader and more
intense in the ordinary ray which is for this reason a purer red than the
yellowish red of the extraordinary ray. Some Burmese orange sapphires
show an absorption line at 475 nm, ascribed to vanadium.

A blue sapphire cat's-eye from Burma gained its chatoyancy from its
structure. It was not a single crystal but consisted of two sectors slightly
inclined against each other at an angle of 11°. In both sectors were growth
planes parallel to the hexagonal dipyramid as well as glide planes parallel to
the basal pinacoid. Chatoyancy was caused by total internal reflection at

irregular patch-like fissures on basal glide planes rather than by the usual rod or needle-like inclusions.

In blue sapphires there are three bands due to ferric iron; they are seen at their full strength in green sapphires and are also found in yellow sapphires from some localities. They are positioned at 477, 460 and 450 nm, the strongest being at 450 nm; in yellow sapphires the other two tend to merge when at full strength. In Australian blue stones the bands can be seen in full, while in sapphires from Sri Lanka (poor in iron) only the 450 nm band can be seen and this is often very faint. This band is best seen in the ordinary ray and can be seen at its maximum strength along the optic axis.

Corundum is found in crystalline limestones and dolomites which have been metamorphosed; it is also found in other types of metamorphic rocks such as schists and gneisses. It occurs in igneous rocks such as granites and nepheline syenites and also occurs in placer deposits. Corundum as a mineral occurs world-wide but gem quality stones are much rarer; fine quality rubies are far harder to find than diamonds and size-for-size the best specimens will command higher prices.

Burma

The world's finest rubies come from the Mogok Stone Tract in Upper Burma; production of ruby from this area was first recorded in 1597. In that year the King of Burma acquired the mines from the Shan ruler and they were worked intermittently until 1885. In 1886 Upper Burma was annexed by the British and the mines leased to a British company. This involved the floating of Burma Ruby Mines Ltd which went into liquidation in 1925. One of the reasons for the failure of the company was the influx of synthetic ruby on to the world's markets; another was the difficulty of coping with a difficult terrain (in 1929 an exceptionally heavy rainfall caused flooding which wrecked a lot of the machinery). The company continued to work the mine for the six remaining years of its lease after going into liquidation. In the 1930s and up to the start of the war in the area, mining by traditional native methods went on sporadically and after the war this type of mining was commercially successful until the take-over of all mines by the government in 1963. All private gem mining and commerce was forbidden from that time and the government established a series of gem auctions to which foreign buyers were (and are) invited. In general the stones offered at the auctions have not been of the highest quality, the better stones passing through illegal channels to Thailand. At the time of writing the Mogok area is inaccessible to almost everyone, certainly to foreigners, the mines being supervised by troops.

The geology of Mogok is complicated and includes high-grade metamorphic schists and gneisses, granite intrusives, including gem-bearing pegmatites, ultramafic rocks containing peridot and metamorphosed limestone (marble) in which the ruby and spinel were formed. The marble is in contact with, or interbedded with highly folded gneisses, particularly the Mogok gneiss, the chief rock of the region. It consists of a number of metamorphic rocks, including scapolite- and garnet-rich biotite gneisses. Rounded fragments of the Mogok gneiss turn up in the gem gravels. The most important of the granite intrusives is the Kabaing granite which is one of the largest rock units in the region. It contains bands of marble and in all probability is responsible for the contact metamorphism by which the rubies and spinels were formed.

Burma rubies are characterized by their 'pigeon-blood' colour; the colour is roiled and unevenly distributed in many stones; inclusions are varied and tend to be mineral rather than liquid, probably indicating that these rubies grew from a melt deficient in liquid. Short stubby rutile crystals form nests or may be seen as individuals; there are banded crystallites and rhombs of calcite, short prisms of apatite, crystals of olivine and deep yellow sphalerite crystals. Sphene is found as light yellow tablets and spinel octahedra are also seen. Negative crystals are quite common. There are few named rubies in the major world collections though there are records of outstanding stones of the past. The Edwardes ruby crystal of 167 ct was given to the British Museum (Natural History) by John Ruskin in 1887. The Los Angeles County Museum of Natural History displays the 196.1 ct Hixon ruby, an etched crystal of very fine colour. There are also several star rubies in major collections. The most useful reports on Burma ruby are by Iyer (1953) and Keller (1983).

Mining methods The *twinlon* is a small circular pit up to 12 m deep which is dug down to the gem gravel or *byon*. Mining then takes place laterally to remove gem gravel from a 10–12 m radius. More elaborate is the *hmyadwin (hmyaw)*. This is more like a quarry which is dug into a hillside for a depth of up to 15 m. Channels bring in water to wash the soil and gravels recovered. Weathering in the limestone region gives characteristic karst topography with caverns which are worked for their rich gem gravels. Mining is very dangerous in the caverns.

Thailand

About 70% of the world's fine rubies come from Thailand and have gained in importance since the virtual cessation of commercial activity in the Burma ruby market. Admittedly the Thai stones do not quite reach the magnificent colour of those from Burma – Thai stones are a darker red, some inclining to a brownish-red due to an appreciable iron content. Up to 85 or 90% of Thai rubies come from the Chanthaburi-Trat gem field near the town and in the province of the same name. The deposit is split into two; one part is to the west of the town of Chanthaburi, and includes the famous areas of Khao Ploi Waen and Bang Kha Cha, which produce blue, blue–green and yellow sapphires as well as black star sapphires. The other area, about 45 km east of Chanthaburi, in Trat province, contains the active Bo Rai/Bo Waen mining district known for ruby production. Bo Rai/Bo Waen, with the Pailin area 27 km to the north-east in Kampuchea, are today the most important ruby-producing areas in the world. Some green and some colour-change sapphires are also found in this area.

The Chanthaburi-Trat area is in the south-east of Thailand, about 330 km east-south-east of Bangkok. Gem mining was reported from the area as long ago as 1850; in 1904 Bauer noted that there were two main groups of mines in the district, Bo Nawang and Bo Channa, the two being about 50 km apart. The pits at Bo Nawang were small and about 1 m deep, sunk in a yellow–brown sand overlying clay. Rubies were found at the base of the sand in a layer 15–25 cm thick. These mines have been worked since about 1975.

An English company, Sapphires and Rubies of Siam Ltd, was formed in 1895 to mine in the country. This company was linked with Burma Ruby Mines Ltd but was not successful in the Chanthaburi district. Since 1919 the mining of gemstones in Thailand has been restricted to Thai nationals through the operation of the Siam Mining Act of that date.

The corundum deposits are exclusively in alluvial, eluvial or residual lateritic soil derived from basaltic lava flows. The basalts outcrop rarely, although blue, green, yellow and black star sapphires have been found in a presumed volcanic plug at Khao Ploi Waen (rubies are not usually found here).

The basalts are fine-grained olivine-bearing alkali basalts and locally contain spinel-rich lherzolite nodules which may have given rise to the corundum. These nodules are thought to form in the upper mantle, about 50 km deep; they may not be related to the magmas that brought them to the surface. The basalts contain augite, pyrope, calcium-plagioclase, zircon, spinel and magnetite. Spinel is locally abundant but not of gem quality. Gem deposits vary in thickness depending upon the bedrock and local topography. Sapphires are found on the surface at Khao Ploi Waen and at depths of about 3–8 m at Bang Kha Cha. In the flat Bo Rai area the gem gravels are at least 4–10 m deep and up to 1 m thick. At Bo Rai ruby is associated with black spinel, olivine and some blue, green or colour-change sapphire. Mining methods vary from primitive to sophisticated though large-scale mechanized mining was forbidden in 1980 with a view to protecting the soil at the request of farmers.

Rubies from the basalt areas of south-east Thailand can be recognized by colour and inclusions from those found in the metamorphosed Burmese limestones or in the granite gneisses of East Africa. Gübelin found that the commonest inclusions were subhexagonal to rounded opaque metallic grains of pyrrhotite, the primary sulphide in basaltic rocks, yellowish platelets of apatite and reddish-brown almandine crystals. The inclusions were surrounded by circular feathers and there were characteristic polysynthetic twinning planes. Some of the pyrrhotite was altered to goethite (since Thai rubies are rich in iron, pyrrhotite is a not unexpected inclusion).

Sapphires from the Khao Ploi Waen and Bang Kha Cha areas are different in morphology from the Bo Rai rubies which show no form. The sapphires are found as well-rounded hexagonal prisms and are usually green, blue, yellow or black, the latter sometimes showing asterism. Average size is about 3–6 mm. Large hexagonal blue–green sapphires up to 1720 ct are reported from the Bang Kha Cha area but these are not of gem quality (Keller, 1982). The asterism in a black star sapphire from south-east Thailand was found to be caused by perfectly oriented intergrowths of elongated hematite lamellae, the periodic contrast between corundum and hematite giving the effect.

Pakistan

Corundum, including ruby of gem quality, is found in the Hunza valley of Pakistan and is reported by Gübelin (1982). The stones are found in a marble that outcrops along the flanks of the Karakoram Range in the far north of the country. The marble outcrops occur below the Mutschual and Shispar glaciers near the villages of Altiabad and Karimabad. Hunza is an earlier name for the former. The corundum-bearing marble forms intercalations within garnetiferous mica schists and biotite-plagioclase gneisses. The schists are cut by pegmatites and aplite dikes. The coarse-grained marbles probably consisted originally of dolomitic calcite sediments metamorphosed by numerous intrusions of granite, aplite and pegmatite in the Eocene. The deposits are very similar geologically to the ruby deposits in Burma, the gem-bearing marble being composed of small to large calcite crystals and snow-white, greyish or yellowish in colour. The gemstones are thought to be the result of a particular metamorphic concentration process that took place at temperatures of about 600°C. The corundum occurs as fairly well-formed to well-formed crystals, pale to deep blue or pink to fine ruby red. Euhedrons of spinel are found with the corundum but there are fewer here than in the Burmese deposits. A greenish scaly material which often occurs between the ruby crystals and the enclosing marble has been found to be muscovite, containing traces of chromium and vanadium.

Ruby is the more common of the corundum varieties found at this location. Most of the red crystals are opaque to translucent, are penetrated by many fractures and show large patches of calcite (also seen at Mogok). Chromium content varies from 0.15 to 0.81%. RI have been measured at 1.762 and 1.770 with a birefringence of 0.008. The absorption spectrum is the usual one for ruby and the luminescence is also unexceptional. The specific gravity is 3.995. Most of the rubies are turbid from numerous cracks, parting planes, polysynthetic twin-planes and irregular swirl-marks as seen in Burma stones. Calcite is the commonest inclusion and forms euhedral crystals with rhombohedral twinning in some rubies. Resorbed crystals of dolomite also occur and phlogopite is common, forming red-brown flakes. Magnetite and chlorite are rarer; spinel, rutile, apatite and pyrrhotite are found; some pyrite may also be present and is often altered to goethite. None of the rutile crystals formed the silk-like appearance so common in other rubies.

Ruby from the southern portion of the Sorobi district of Afghanistan (Jegdalek ruby) ranges from a light purple-red to a deep pigeon's-blood red and the best colours equal the best Burma material. RI are 1.762 and 1.770, SG about 4.00. Moderate to strong fluorescence and purplish to orange–red pleochroism. Small needles, probably boehmite, were included.

Sri Lanka

The island of Sri Lanka is underlain almost entirely by Precambrian rocks, which can be divided into three types. The oldest unit is the Highland group which contains rocks of the metamorphic granulite facies including hypersthene gneisses, sillimanite–garnet gneisses, biotite–garnet gneisses and marbles containing forsterite and spinel. Most of the gem deposits are found in this group. The Vijayan complex is characterized by rocks of the almandine–amphibolite facies such as biotite–hornblende gneiss. The overlying South-West group of rocks belongs to the cordierite–granulite facies and consists of cordierite gneisses and khondalites. Most of these rocks contain some gem material, as do the many pegmatites in Sri Lanka, but very few specimens found in situ are of gem quality, this standard of material being found only in alluvial deposits. The host rocks of the alluvial gemstones have still not been traced, although they cannot be far away from the places in which the stones are found. The only known source of blue sapphire in situ is in a pegmatite near Kolonne in the Ratnapura district; it is found with diopside. Only moonstone and some almandine garnet are found in situ apart from this one occurrence of blue sapphire; the moonstone occurs near Meetiyagoda and the garnet is fairly widely scattered over the island (Zwaan, 1982).

The main gem-bearing alluvial deposit is the Ratnapura district in Sabaragamuwa province about 97 km south-east of Colombo. The stones are found concentrated into approximately horizontal layers at different depths in the alluvium which is mostly sand and gravel. The main locations are Ratnapura itself, Pelmadulla, Balangoda and Rakwana. There are also gem deposits near Elahera in the Central province and especially in the valley of the Kaluganga river where the gems are not concentrated into particular levels but are fairly uniformly distributed in a laterite-rich deposit of sandy gravel and clay. Near the town of Okkampitiya in the Uva province there are large quantities of hessonite garnet, the deposit being similar to those in the Ratnapura area. Gem deposits have recently been found in the Tissamaharama area (sometimes called Kataragama) in Southern province. The main localities are Amarawewa and Kochipadana. Gems are found in alluvial deposits at a depth of about 1 m. Corundum is found at most of these places; well-developed crystals are rare. Working is primitive and many stones are cut virtually where they are found.

Blue sapphire at its finest for Sri Lanka comes from the Rakwana area in the Ratnapura district; the more recently discovered Tissamaharama area is also producing good quality blue sapphire. A milky-white sapphire, locally known as geuda, can be heated to give a fine blue colour and also comes from these areas (Zwaan, 1982).

Sri Lanka ruby inclines to pink rather than to the deep red of the Burmese or the darker red of Thai stones. There is a possible confusion over what is ruby and what is pink sapphire (depending to some extent on whether buying or selling). Pink sapphire from Sri Lanka is in fact a beautiful gemstone in its own right. Sri Lanka rubies are characterized by their included rutile needles being sparser and longer than those seen in Burma stones; zircon crystals with characteristic haloes are also seen.

Yellow sapphire is also found in Sri Lanka, the stones being quite a light and bright colour; they show an apricot-coloured fluorescence which indicates that they have a lower iron content than stones from Thailand or Australia which are virtually inert. On the other hand, Sri Lanka yellow sapphires show only a weak absorption spectrum with a faint band at 450 nm. Star stones, both ruby and sapphire, are found in the country with bluish-grey, pale violet and milky white being the commonest colours. Fine coloured pieces are rare; one example is the 138.7 ct Rosser Reeves star ruby in the National Museum of Natural History, Smithsonian Institution, Washington DC, USA.

The orange sapphire to which the name padparadschah has been given, is found in Sri Lanka and has been the subject of a good deal of comment, even controversy. The name appears to have a certain cachet and, like alexandrite, possession of such a stone is regarded as highly desirable. The name is taken from the Oriental lotus (*Nelumbo nucifera*); when a lotus flower is about to open the colour is a rose-red. Early descriptions do not mention orange and the name as we know it today has come from German origins via a word *padmaragaya*, referring to the yellow–pink oriental lotus. Exactly what colour the name has meant over the years is unclear, and for this and other reasons the continued use of the name for orange–coloured synthetic corundum (Crowningshield, 1983) should be discouraged.

There are several fine examples of orange sapphires in major collections.

The Morgan collection at the American Museum of Natural History in New York has a fine orange 100 ct sapphire and a fine pink to orange crystal was recovered in the Ratnapura area of Sri Lanka quite recently – a 1126 ct, bipyramid stone. Orange to pinkish-orange corundum has also been found in East Africa in recent years.

USA

The sapphire deposits of Montana, USA produced a large quantity of industrial grade material and this alone made them economically important in the early years of this century. Gem quality material was also found and Sinkankas (1959, 1976) has estimated that about $25 million had been recovered from the Yogo Gulch mines between 1865 (their discovery) and 1929. Crystals of sapphire were found in the gravels of the Missouri River north-east of the town of Helena, Lewis and Clark County. The beds are layers of gravel lying on terraces up to 200 ft above river level. The gravel rests upon argillaceous rocks of the Belt series and contains debris ranging from sand to boulders several feet in diameter. Gold is found with the sapphires and gold mining operations have produced gemstones as a byproduct. No deposits of sapphire have been found *in situ* but they may have come from rocks eroded by the action of the river; it is believed that the gravels are all that is left of the host rocks. One sapphire-bearing dike was found in 1889 penetrating argillaceous rocks; the dike is a very fine-grained rock and contains feldspar, biotite, augite and a glassy material; sapphire is not very plentiful in it.

A number of minerals are found with sapphire in the river gravels, including cassiterite, garnet, topaz, kyanite and chalcedony. Some of the garnet is of gem quality and bright red, orange or reddish-pink. The sapphire crystals are usually small, about one quarter to half an inch in diameter and are generally some shade of pale blue or blue–green. Much of the blue is a greyish-blue, strong shades being rare. Many other colours are known, including light pink, light green, pale red, purplish-blue, purple, strong green, yellow and orange. Some of the pink and reddish-purple stones show tints of brown through the sides of the crystals. One dark blue–green stone with an alexandrite-like colour change is recorded. Crystals are frequently zoned with the palest shade in the centre; but the reverse is also encountered. Many have a silky sheen on the basal plane from platelets of hexagonal negative crystals resembling snowflakes in pattern. The roughness of the crystals presumably arises from etching while in their host rocks rather than from wear by water.

Missouri River sapphire crystals were used in jewellery soon after 1865 when a stone was sent for cutting to Tiffany & Co. Sapphires of a desirable cornflower blue were discovered in 1895 by gold prospectors at Yogo

Gulch, Judith Basin County, about 15 miles south-west of Utica, Montana; some of the stones were sent to Tiffany's. The sapphires are found in a dike outcropping at about 5000 ft above sea level and which is intruded into Madison limestones of Mississippian age and also in one place into shales. Sapphire crystals are uniformly distributed through the dike and were mined from it by removing all dike rock. Most Yogo Gulch crystals are flattened wafers of roughly circular shape, often with tapered edges, looking like small pancakes. Only some larger pieces show crystal faces. Few gems cut from the crystals are more than 1 ct in weight although a stone of 5 ct was cut from a 10 ct crystal of very fine colour and sold in 1919 for $1600 in Hatton Garden. Since most of the crystals are small the mines were hard pressed to run successfully and closed in 1929 after a succession of troubles, not all man-made. Sapphire crystals from this location are very clear and often flawless; SG varies with colour, reddish-purple stones giving readings of 3.980 while bright blue ones give 4.001–4.003. Colours vary from dark blue to light blue; some are amethystine and a few are ruby-red.

In 1947 some investors began to work the mines once more although production was slow to start. Ownership has changed hands several times but Sinkankas reported that production was running at over 250 000 carats a year (including all grades) in 1974. The crystals are 90% blue, a small ruby being found very occasionally. The crystals are sent to Bangkok for cutting, half of them being marketed there and the remainder returned to the United States for sale there and in Europe. No rough is sold. On the whole the finest Yogo sapphires are equal in colour to the best Kashmir stones and unlike the latter, are clear and bright.

Corundum is found in North Carolina in peridotitic rocks which outcrop frequently in a belt extending from Canada down to Alabama. The rocks cut ancient crystalline rocks of the eastern part of the continent and seem to be of igneous origin. The only material of commercial value comes from Jackson, Macon and Clay counties. The most important location is Cowee Creek in Macon County. Here ruby crystals are found in alluvial gravels filling the valley floor. The gravels in which the rubies are found are covered by topsoil and loam to a maximum of 5 ft so that recovery is by no means difficult and in several places digging areas are leased for short periods to visitors. The ruby crystals are associated with rhodolite garnet crystals of irregular form. They themselves show a variety of forms, some being flat and wafer-like with a hexagonal outline and edges bevelled by rhombohedral faces; others are hexagonal prisms with simple basal planes as terminations. Many crystals have triangular markings and hillocks on the basal planes and rutile inclusions occur in some specimens; others have inclusions of black ilmenite. Some star stones can be cut from rutile-bearing crystals. On the whole crystals are quite small, although there have been some very large exceptions (Sinkankas, 1959, 1976).

Kashmir

Blue sapphire from Kashmir has always been very highly regarded on account of its superb cornflower blue colour. The deposits were located in 1881 and in the next year government geologists reported to the Geological Survey of India that sapphires had been found in a glacial cirque above the village of Sumjam on the south-west slopes of the Zanskar Range. A pocket of gem quality crystals was exposed in the same year and its contents sent eventually to the commercial centres of Kulu and Simla where their blue colour aroused excitement, so that by the end of 1882 considerable sums were already being paid for crystals so that mining must have proceeded with some speed. The Maharajah of Kashmir took over control of the mines in 1883 and in the period of 1883–1887 most of the finest stones were recovered. A detailed survey of the deposit was made by T. D. La Touche who found that the first pocket was exhausted and so examined the placer deposits lying below the scree and talus on the valley floor. This exploration turned up the last of the large crystals to be found; this weighed 933 ct and was parti-coloured.

Sumjam lies in the Paddar valley; the mineralogy of the deposit is fairly simple but is not echoed by other sapphire occurrences. The steep ridge with the mines at the top is made up of metamorphic strata of coarse marble locally interbedded with biotite gneisses and schists. The gneisses are alternatively graphite- and hornblende-rich; in other places garnet is found in them. Local modification in some parts has given rise to lens-shaped bodies of actinolite and tremolite. In some places the stratified beds are intruded by feldspar pegmatites, sapphires occurring where these intrusions are free from quartz and surrounded by beds of actinolite and tremolite. The sapphire crystals are found in disc-shaped pockets of plagioclase feldspar which has been kaolinized. The pockets are usually small and they occur in some profusion. Some corundum crystals occur in contact with the surrounding graphite–biotite gneisses; it is here that pink sapphire crystals have formed, though rarely.

Almost all Kashmir sapphires show zoning and this is one of the ways in which they can be identified. The zoning is associated with very small layers of liquid inclusions which give the velvety or sleepy appearance which is also characteristic of Kashmir stones. A stone of 6.70 ct from the Hixon collection in the Los Angeles County Museum of Natural History had SG 4.03 and RI 1.762–1.770. There was weak fluorescence under X-rays and a weak absorption line at 450 nm from iron and in the red from chromium. Kashmir sapphires contain a variety of inclusions, the most common being flat films containing a brownish or yellowish liquid after the layers of liquid inclusions already mentioned. Zircon crystals with haloes and fingerprint-like structures like those in corundum from Sri Lanka are seen.

The Kashmir mines were not operating, at least officially, in 1983 and any crystals recovered would have been found by the indigenous population who are forbidden to engage in mining or in gem trading (but human nature often prevails; Atkinson, 1983).

Australia

Gem-quality sapphire is found in many parts of Queensland, Australia but the main deposit is at Anakie, 192 miles west of Rockhampton, where the sapphires occur in alluvial deposits discovered in 1870. The gem dirt may reach a depth of 50 ft and is clay-like in consistency. The sapphires occur in a number of colours, including blue, green, yellow, purple and parti-coloured stones. The Anakie area contains basalt hills above a granite basement. It is possible that the sapphire was originally in the basalt, although it could have originated in deep underlying igneous or metamorphic rocks. The best of the blue sapphires are of high quality, but it is the yellow material that is particularly notable at Anakie. Some dark opaque fragments have a bronze colour and display chatoyancy.

Sapphire is found in the Glen Innes area of New South Wales where commercial mining began in 1960. Most of the sapphire is found in the gravels of streams that dissect the basalt which has been laterized; colours found include deep blue, green, opaque black or bronze. Most of the crystals are small. The Inverell district of New South Wales yields sapphires in a wash dirt consisting principally of pebbles of basalt, felsite and fragments of sedimentary rocks. The dirt is overlain by up to 8 ft of black soil of a clay consistency. Most stones are blue, green or yellow.

A recent study of inclusions in Australian sapphire has identified uranium pyrochlore, plagioclase feldspar and magnetite, the latter occurring as opaque octahedra. These inclusions are similar to those found in sapphire from Thailand and the Khmer Republic; all three deposits are associated with Tertiary alkali basalts.

Kampuchea

Fine ruby and sapphire is found in the Pailin gem fields of Kampuchea. The fields were discovered by Burmese traders in 1874; gems come from small basaltic bodies, contrasting with those from Burmese metamorphosed limestones (ruby) and syenites (sapphire). Pailin lies on a fault zone separating a northern region of low relief underlain by the Precambrian Pailin Crystalline Complex and some Devonian and Carboniferous rocks from a southern region composed mostly of Triassic sandstones and grey-wackes. During the Himalayan orogeny in the Tertiary the southern area was uplifted by block faulting and small basaltic bodies containing gems were intruded near the fault zone with subsequent extrusion of lava. Later erosion in the Tertiary by northward-flowing rivers in particular led to the formation of boulder-strewn sands and clays of the ancient alluvium which has been reworked into the present river alluvium and fan-shaped outwash deposits. Rivers which drained from the small basaltic bodies carried the gem materials, including ruby, sapphire, zircon and garnet. Some gems have been found *in situ* in weathered basalt, others in soil profiles overlying the basement, thus proving their origin (Jobbins and Berrangé, 1981).

The Pailin fields are related to those in the neighbouring country of Thailand from which they are not too far distant. Basaltic bodies in the Indo-China area as a whole seem to fall into two groups: the smaller group of the Pailin-Chanthaburi type carrying ruby, sapphire, zircon and garnet; the other larger group is more barren. The smaller group is the older. There are blue, green and yellow stones in the westernmost deposits, and red to blue material (but no yellow sapphires) at Pailin. There are no colourless or green stones at Pailin but the rubies vary in colour from pink to brownish-red in the Phnum Ko Ngoap and Phnum O Tang bodies and the blue sapphires range from very pale to deep blue at Phnum Yat. Some colour-change stones have been found. Fluorescence is weak in general; constants are normal for corundum. One interesting inclusion reported is the reddish uranium pyrochlore (hatchettolite) in the blue sapphire.

Malawi

Corundum, including both ruby and sapphire, some of gem quality, was reported in 1958 from Chimwadzulu Hill in the south of Malawi, Central Africa (Rutland, 1969). In 1967–68 an organization was set up to mine and market the stones but there has never been any great quantity on the market and supply seems to dry up before re-starting again. At the time of writing a few quite good quality rubies of a fine bright red have appeared at gem shows in England. The corundum is found *in situ* in an epidotized amphibolite and also in gravel on the summit plateau of Chimwadzulu Hill which is about 50 miles south of Lake Malawi. It is believed that the amphibolite originated from a peridotite intruded into geosynclinal sediments. Corundum crystals are embedded in a coarse aggregate of rounded hornblende crystals, enclosed in a fine-grained granular grey matrix of epidote and plagioclase. The south-western sides of the hill contain a lot of corundum, probably from eluviation. Yellow, green and blue crystals have been found and some parti-coloured specimens are known. All the crystals show well-developed basal pinacoids and some have triangular growth markings on these faces; the rhombohedral faces are less prominent; the prism faces are rough while the others are smooth. All crystals examined show a prominent parting parallel to the basal plane. RI for a blue stone is in the range 1.770–1.760 and for a red one 1.771–1.762. Luminescence is weak even in red stones, probably due to an appreciable iron content; the colour of the rubies is close, at its best, to good quality Thai rubies which are also rich in iron. Dichroism is strong, being pale brownish-red for the extraordinary ray and deep magenta for the ordinary ray. In the blue stones the colours are pale yellowish-green and purplish-blue. Most stones have fine channels inside.

Colombia, Brazil, Nigeria, China and Nepal

Gem quality blue sapphire has recently been recovered from stream beds and terrace gravels near the Colombian village of Mercaderes, about 143 km south-west of Popayán (Keller *et al.*, 1985). Apatite, rutile and boehmite were found as inclusions and the stones had RI 1.762–1.770, DR 0.008 and SG 3.99–4.02. Crystals were simple hexagonal prisms in the main with pyramidal forms rare. The blue colour resembles that of the Umba Valley material from East Africa. Stones are colour-zoned with blues and greens being the most common; yellows, pinks and some reds have also been found. Many have a brownish-yellow core. Some crystals with a bluish to purplish-red to orange colour show a colour change and chromium is present as seen with the spectroscope. Most sapphires are very iron-rich and titanium-poor. Some lamellar twinning has been observed.

Blue sapphire has recently been reported from Santa Terezinha de Goias, Brazil, a location also celebrated for emerald. The stones have pronounced colour zoning and RI of 1.758–1.760, 1.765–1.767 with a DR of 0.007. The SG is 3.91–4.35, the high figure probably due to included material. Muscovite and margarite have been detected with the help of the microprobe.

Sapphire from Nigeria has RI 1.760–1.764 and 1.770–1.772 with a DR of 0.008–0.010. SG is between 3.96 and 4.00. Two-phase inclusions and negative crystals are reported. Blue sapphires have been reported from Mingxi, Fujian Province, China. The blue is very dark, reminiscent of the darker Australian stones. Most are found in streams but some are recovered from alkali basalt flows with mantle xenoliths. Ruby from Nepal showed tapering curved crystalline inclusions, straight bands of blue sapphire and irregularly curved feathers. SG is 3.99 and RI 1.760–1.768.

Scotland

Blue sapphire has been found as megacrysts in a xenolithic dike at Loch Roag, Isle of Lewis, Scotland (Jackson, 1984). The crystals show striated and truncated hexagonal bipyramidal form and sometimes occur as tabular prisms. They are a fine blue colour and all so far examined have shown fractures into which thin films of secondary iron staining have percolated. Colour zoning has been noted.

Tanzania and Kenya

Ruby crystals of good colour, opaque to translucent, have been found in the Matabuto Mountains of northern Tanzania. The short prismatic crystals occur in a bright green zoisite with black amphibole (anyolite); some water-worn pieces were facetable. Ruby has also been found in the Tsavo National Park area of Kenya; the mining has been sporadic but the finest stones are of very bright rich colour; those of lower quality show marked colour banding. Ruby from Longido, Tanzania, can be a bright orange-red at best.

Corundum from the Umba valley in Tanzania shows considerable colour variation which is attributed to varying proportions of the transition elements iron, titanium, chromium and vanadium. Energy-dispersive X-ray fluorescence analysis could not detect any nickel or cobalt. RI fall in the range 1.761–1.769, 1.768–1.778. SG is in the range 3.98–4.01. Minerals included are zircon, apatite, hematite, mica, monazite, calcite, pyrrhotite, graphite, Healed fractures, negative crystals, colour zoning, intersecting twin lamellae with lath-like boehmite particles and rutile are probably also present.

CROCOITE

$PbCrO_4$. Monoclinic, forming prismatic crystals. H 2.5–3. SG 5.9–6.1. RI (α) 2.29–2.31 (β) 2.36 (γ) 2.66. Biaxial positive. DR 0.270. Orange to blood-red pleochroism. Weak reddish to dark brown fluorescent effects in UV. Distinct absorption band at 555 nm.

Fine transparent orange stones have been cut from crystals found at Dundas, Tasmania, Australia where crocoite occurs as a secondary mineral in the oxidized zones of lead deposits. Named from the Greek word for saffron.

CUPRITE

Cu_2O. Cubic, forming cubes, octahedra and combinations of the two forms. H 3.5–4. SG 6.14 for Namibian gem quality material. RI 2.848.

Cuprite occurs as a secondary mineral in copper deposits. Superb dark red faceted gemstones have been cut from crystals found at the Onganja mine, Namibia; the surface may suffer alteration so that the stone needs repolishing. Named for its composition.

DANBURITE

$CaB_2Si_2O_8$. Orthorhombic, crystals slender with rhombic cross-section, terminated with pyramid and dome forms giving a wedge shape.

Interfacial angles close to those of topaz but H 7. SG 2.97–3.03. No distinct cleavage. RI (α) 1.630 (β) 1.633 (γ) 1.636 with DR 0.006. Biaxial negative for red to green light, positive for blue to violet. In common with other calcium-bearing gemstones, danburite may show a faint rare-earth (didymium) absorption spectrum. Dispersion 0.016. Sky-blue fluorescence under long-wave UV; reddish thermoluminescence and phosphorescence.

Most danburite is colourless but from time to time a deposit in the Malagasy Republic produces stones which compare with the deep reddish-brown of fine quality topaz. Unlike topaz there is no distinct cleavage and though both species are biaxial, the axial plane in danburite is parallel to the basal face rather to a face in the prism zone.

Danburite is named for Danbury, Connecticut, USA but this area has not produced gem quality material which comes rather from Mexico (the Charcas lead mines, San Luis Potosi), the Malagasy Republic, Burma (both colourless and yellow crystals from Mogok), Bungo Kyushu, Japan.

Danburite occurs with feldspar in a dolomite at Danbury and with quartz and calcite at Charcas, where the best crystals are found. The Malagasy danburite provides crystals from which stones up to 50 ct can be cut and is at Maharita on Mt Bity. It has RI of (α) 1.630 (β) 1.633 (γ) 1.636 with a DR of 0.006 and SG of 3.00–3.01. Axinite occurs with the danburite found at the Toroku mine, Miyazake Prefecture, Kyushu, Japan. Burmese crystals show etching.

Danburite is not normally used in jewellery; the colourless stones are quite plentiful, the yellow to orange–brown ones rare, coming on to the market only when collections are sold or old deposits temporarily re-worked. A golden danburite with RI of 1.630, 1.638 and SG of 3.00 is reported from Sri Lanka. Traces of a rare earth spectrum were noted.

DATOLITE

$CaBSiO_4(OH)$. Monoclinic, forming prismatic crystals. H 5–5.5. SG 2.8–3.0. RI (α) 1.622–1.626 (β) 1.649–1.658 (γ) 1.666–1.670. Biaxial negative. DR 0.044–0.047. Some specimens may fluoresce blue in short-wave UV (perhaps from europium impurities).

Datolite, a colourless, pale yellow or pale green mineral, is found as a secondary mineral in basic igneous rocks. Gem stones have been cut from one or two locations in Massachusetts, USA, and from the Habachtal, Austria. Named from a Greek word meaning to divide, in reference to the massive variety's granular texture.

DIAMOND

Diamond is at least 99.5% carbon, the remainder being nitrogen with some hydrogen and oxygen. In a perfect diamond all the carbon atoms are bonded in exactly the same way, each joined to four neighbours, thus giving an octahedral rather than tetrahedral natural growth shape. The crystal system is cubic and diamond crystals most commonly occur as octahedra which are often distorted and which may show rounded edges. Many are flattened through the under-development of some faces so that the crystal is almost tabular. Cubic crystals are commonest in industrial quality stones. The rhombic dodecahedron arises from development of the edges of the octahedron or from development of the edges of the cube; some crystals incline towards the spherical. While octahedra of regular shape and with straight edges are known in the trade as 'glassies', very thin crystals arising from the poor development of parallel faces of an octahedron with strong development of other faces are known as 'flats' when they are of cuttable thickness. Some have been used to cover miniatures and are known for that reason as portrait stones. Twinning is common in diamond; interpenetrant twinning can occur in a variety of forms, one a fairly rare star form where dodecahedral crystals interpenetrate. Contact twins where one individual has been rotated about 180° around a twinning axis occur frequently. Where one crystal rotates through 180° on an octahedral plane, a twinned octahedron is formed, known in the trade as a macle; its plane of rotation is a hexagon and the resulting shape is triangular; these stones often show re-entrant angles which is a sure sign of twinning. Many macles have rounded faces. Some diamond crystals grow together as aggregates, from only three or four up to hundreds of individual crystals. The shape is nearly always irregular.

The carbon atoms in diamond are arranged so that each one is at the centre of a tetrahedron of four other atoms, a discovery made by the Braggs in 1913. Growth takes place in layers, the slowest growing faces persisting in the completed crystal and faster growing ones running out more quickly to points and edges. Octahedral and cube faces are the slowest growing and are therefore seen most frequently. Crystal faces show steps or layers close to and parallel to the edges of octahedron faces; the steps or terraces may occur in lamellae on an octahedron face. On a cube face steps are square and the striations on dodecahedral faces are parallel to the long axis. These lines and similar ones on other faces are known to the diamond cutter as grain. Trigons and squares on octahedral and cube faces, respectively, are due to the growth process: elevated projections are produced by a faster rate and depressions oriented in an opposite sense to the face by slower growth rates. Trigons are equilateral triangles with straight edges and points toward the edges of the face; there may be millions of trigons on a single face, all aligned with each other. Chemical etching produces triangular markings on octahedron faces (aligned with the edges of the face – the reverse effect to that shown by natural trigons), square markings on cube faces and boat-shaped markings on dodecahedral faces. As etching proceeds the edges of the markings (which are pits in the surface) become curved, this showing that growth has been preceded by a layer formation with the layers parallel to the octahedron faces. The layered growth of diamond gives rise to a perfect octahedral cleavage; although other cleavage planes are known the octahedral one dominates and is usefully exploited by the diamond cutter.

Diamond is the hardest of all known substances (10 on Mohs' scale). Using an indenter an average indentation hardness for diamond was given as 10 000 compared with 2200 for corundum, the next hardest gem mineral. Wilks and Wilks discovered that the hardness varies with direction, the hardest being the cube face at 45° to the axis; next hardest is the dodecahedron face at right angles to the axis followed by the octahedron towards the cube, the octahedron towards the dodecahedron and the cube parallel to the crystal axis; the dodecahedron parallel to the crystal axis is the softest direction. The difference in hardness between the hardest and softest directions may be as much as a factor of 100. Only the two softest directions are of real use to the diamond polisher – the others are too hard.

The SG of diamond is constant, ranging from 3.514 to 3.518. Hardness 10. The RI is also constant at 2.4175 in sodium light; the dispersion is 0.044. Although diamond is theoretically isotropic, an apparent birefringence can be seen in many stones as anomalous double refraction between crossed polars. This effect can also be seen in crystals, probably as a result of strain, and can be useful to the cutter who wants to ensure that strain does not cause his stone to break up during the cutting process. For many years it has been recognized that there are different types of diamond and the most recent research places them in four main categories, Ia, Ib, IIa and IIb, with some intermediate types (Davies, 1984). Nitrogen is an impurity in Type I stones, usually at the level of 0.1%. More than 98% of clear, large natural diamonds are Type Ia in which the nitrogen is present as pairs of atoms (IaA) or as larger clusters containing an even number of nitrogen atoms (IaB). Type Ia diamonds also contain platelets, about 100 nm across which do not, as was once thought, contain a significant amount of nitrogen. Type Ib stones are very rare (less than 0.1% of known stones) and contain isolated nitrogen atoms.

All Type I stones absorb in the infrared from 6 to 13 μm and in the UV beyond 300 nm; they usually show a blue fluorescence. They are often yellowish (the colour known as 'Cape', not a true canary yellow) and show a series of absorption lines, the most prominent of which is at 415.5 nm. This line has been designated N3 and probably results from three nitrogen atoms surrounding one carbon atom in a flat configuration. A weaker line at 478 nm and other weaker lines are also seen. Diamonds coloured brownish rather than yellowish have a weak line at 504 nm with other weaker lines, some of the stones having a greenish fluorescence. Since nitrogen has one electron more than carbon it acts as an electron donor; it does not confer electrical conductivity: the donor energy is too large giving rise to strong insulating properties.

There is no nitrogen in Type II diamonds; they do not show characteristic absorptions in the 6 to 13 μm region and they transmit in the UV below about 300 nm. They conduct heat very well. Type IIa stones are colourless, contain virtually no impurities and transmit in the UV to about 225 nm. Type IIb diamonds are blue from their boron content and are rare (about 1% of all diamonds). They transmit in the UV to about 250 nm and will usually phosphoresce bluish under short-wave UV; some grey and brown specimens have been reported. As boron has one electron less than carbon it acts as an electron acceptor; the acceptor energy is very low, allowing electrical conductivity.

The absorption spectrum of diamond is particularly important when the stone is coloured and when there is a possibility that the colour has been artificially induced or altered. Allowing for different authorities quoting slightly different figures for some of the absorption lines, the yellowish Cape stones show absorption lines at 415.5 (much the strongest, although all lines are quite hard to see), 478, 465, 452, 435 and 423 nm. The lines are best seen when the stone is illuminated by light filtered through a flask of copper sulphate in the direction parallel to the girdle plane. Brown diamonds (the brown series) show lines at 504 nm (strong) and weaker lines at 532 and 498 nm. These stones have a green fluorescence. Yellow stones (the true yellows rather than the Cape stones) show no discrete spectrum but sometimes display a weak line at 415.5 nm. No absorption spectrum is shown by the Type IIb blue stones.

Many diamonds fluoresce blue to violet with the colour sometimes appearing in zones, sometimes due to twinning. Yellow stones may fluoresce yellow–green and some Indian pink stones fluoresce orange with a similar phosphorescence. The deep blue Hope diamond shows a red fluorescence. Generally diamond fluoresces more strongly in long-wave UV; short-wave UV effects are similar in colour but weaker. Many stones show bluish-white in short-wave UV and may phosphoresce yellow. Any stone fluorescing sky-blue and phosphorescing yellow must be diamond.

As diamond is formed deep in the earth at high temperatures and pressures there are no liquid inclusions. Diamonds are thought to have been formed in the Upper Mantle among ultramafic rocks which are of igneous origin (although some may have undergone recrystallization); the world-wide distribution of diamond suggests that its first host was a widely spread magma of fairly uniform composition. Diamond contains a large range of mineral inclusions suggesting that complex events preceded its formation, possibly before the formation and emplacement of the serpentinized olivine (kimberlite) in which the diamonds are found. Ultramafic xenoliths in kimberlite include lherzolite with or without some garnet, websterite with or without garnet and dunites. When ultramafic rocks are introduced into the Upper Mantle, only carbon needs to be present for diamond to form so that diamond bearing ultramafic xenoliths may be of many different compositions. They can contain olivine, pyrope garnet, enstatite, diopside (chrome diopside) with some chalcopyrite, graphite and pentlandite, many of which can be seen as inclusions in the diamond crystals.

The other main suite of xenoliths in which diamond is found are the eclogites, igneous rocks consisting of an assemblage of clinopyroxenes and garnet. The principal minerals in eclogites are pyrope–almandine garnet and clinopyroxene, but many others occur as subordinates: corundum (as ruby in some cases), ilmenite, quartz, spinel, graphite, high-sanidine, pyrrhotite and jadeite. It appears that the eclogite is the solidified low-melting fraction of fertile ultramafic rocks (fertile xenoliths may produce basaltic material with a residuum on partial melting) which is precipitated after partial melting whereas the ultramafic xenoliths are probably the result of an early high cataclasis and/or recrystallization. It is generally believed that diamond crystallized by slow precipitation from a molten or partially molten peridotite but the source of the carbon necessary for diamond to form is still uncertain (it may have been concentrated above its saturation level in the magma). Calcite, the most important carbonate mineral is found in some xenoliths of kimberlite and in kimberlite itself but it is uncommon in eclogite and ultramafic xenoliths even though some of these contain graphite and diamond. It is possible that carbonate minerals could not easily form in the diamond-bearing xenoliths so that the available carbon was forced to crystallize as diamond, or that a carbon or diamond phase came about from the reduction of carbon dioxide, perhaps centred around a specific reducing agent such as pyrrhotite which is a common primary inclusion in diamond from several sources.

The facts about diamond formation are still insufficient, the above being just one set of theories which seem to have some substance. It is generally agreed, however, that the diamonds reach the surface or at least higher levels of the crust by sporadic eruptions which gave a breccial type of rock, kimberlite. During these events diamonds became embedded in the kimberlites and in channels below them. The ejection must have taken place at a high speed since there is no sedimentation of the high-density peridotite and eclogite xenoliths and since slow rates would have allowed the resorption or inversion of diamond into graphite. We still do not know why so many kimberlite pipes (the means by which the diamonds are brought up to the surface) were emplaced in the Late Cretaceous era (about 70 million years ago) to which period many are ascribed. Up to 90% of pipes are ascribed to this period.

Returning to diamond's mineral inclusions, the most frequent is olivine. It is seen as pale green crystals which may contain divalent chromium and

Plate 10 The varieties of quartz.

Plate 11 Tourmaline crystal with quartz.

Plate 12 Peridot.

Plate 13 Tanzanite.

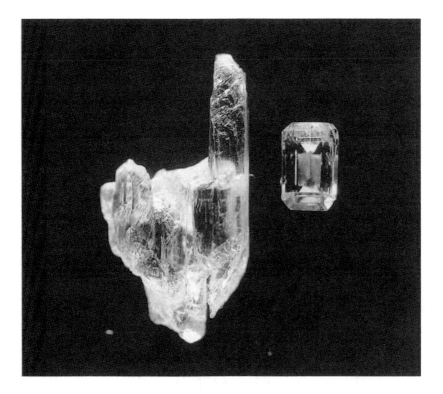

Plate 14 Hiddenite.

Plate 15 Star ruby and sapphire.

Plate 16 Irradiated blue topaz.

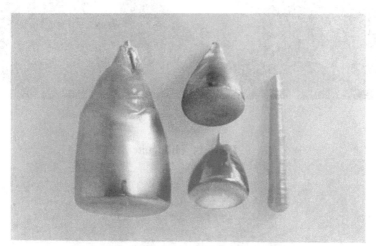

Plate 17 Verneuil-grown ruby and sapphire.

Plate 18 Knischka ruby crystals with varying habits.

sometimes a dark brown grain of chromite. Ultramafic diamond often contains pink to dark-red chromium–rich pyrope garnet, the proportion of trivalent chromium varying and causing differences in shades of red. Where the garnet crystals are reddish-brown to brown an eclogitic origin is indicated and the garnets contain more iron than chromium. Enstatite is found typically in diamond originating from ultramafic mother-rocks; it is usually colourless or pale brown and is about 92% enstatite rather than hypersthene or bronzite. Chrome diopside, an emerald-green colour, is characteristic of diamonds of ultramafic origin. The chromium content may reach 15% by weight. Very pale greyish-green iron-bearing diopside marks eclogitic stones and is especially characteristic of diamonds from the Premier mine in South Africa. Their colour may be masked by the high refractive index of the diamond.

Chromite is found in ultramafic diamonds, forming dark brown to black grains or well-shaped octahedra with a strong metallic lustre. Ruby was only identified as a guest mineral in diamond in the late 1970s (Gübelin, 1982). Reddish inclusions with a weak reddish fluorescence and a fine emission line near 693 nm indicated ruby, a conclusion confirmed by quantitative analysis using the electron microprobe. This indicates an eclogitic origin for the diamond as does rutile which occurs as rare reddish-brown crystals of prismatic habit. Iron-rich biotite is a rare inclusion; sulphide minerals such as pyrrhotite, FeS, and pentlandite, NiS, are more common. Their morphology usually reflects that of the host.

Diamond crystals are very commonly included in diamond and serve to bear out the theory that diamond crystallizes periodically rather than in one continuous process. In many diamonds the diamond crystal inclusions are not in congruent orientation with the host, suggesting that they were overgrown randomly. Sometimes the included diamond is darker in colour, often yellow to brown, than the host and sometimes two or three generations may be found in a younger diamond.

Graphite in diamond is usually the result of subsequent alteration of the diamond form of carbon into the graphite form; this process probably begins preferentially at stress points such as tension cracks and on the contact planes between the guest graphite and the diamond. It is possible that graphite may sometimes be found in diamond as a result of the host having grown over another diamond which was already covered with minute graphite flakes.

India

Diamonds were first found alluvially and even up to recent times these deposits were known as wet diggings. Alluvial diamonds were known in India at least as early as 800 BC; the stones were found in compact sandstones and conglomerates before and after weathering as well as in river beds and terraces. Much of our knowledge of Indian mines comes from the writings of Tavernier who visited the areas between 1630 and 1668. The best-known workings were at Kollur in the gorge of the Krishna river in the south of the kingdom of Golconda (now Hyderabad). A number of celebrated stones were found in this area; it is interesting to note that the geology of the Indian deposits is quite different from that of the diamond-bearing areas of South Africa and Siberia.

Although the area contained a number of diamond deposits the major source of diamonds in India today is the Panna area of the state of Madhya Pradesh where the Jumna river and some tributaries of the Ganges are found. The Panna area contains three types of deposit; primary (pipe) deposits, secondary deposits where the diamonds are found in conglomerates now covered by later deposition and deposits in the beds of streams and rivers. One pipe at least (there are three recorded) was being worked in 1976, as was the Ramkheria alluvial deposit. All this area is part of the Deccan Plateau in which kimberlites occur in the eastern part; the valleys of the rivers in the eastern part of the Deccan Plateau being the earliest places in which diamonds were found. In fact the diamond area of southern India occupies an area about 1000 miles from north to south and about 650 miles from east to west, though the area does narrow to some 250 miles towards the south. It is not precisely known how many kimberlite occurrences there were in the Indian diamond belt since over the centuries erosion has removed many traces of early workings; there may have been only a few pipes which supplied the eastern drainage system.

South Africa

As with Indian diamonds the first discoveries in South Africa were alluvial. The first stones were discovered in that part of the interior of the country occupied by the Griquas who had occupied the land where the Vaal River and the Orange River met. Boers settled in this area after the Great Trek from the Cape in 1836. There is a record of a diamond being found in 1859 on the river bank of the Vaal near Platberg; the stone weighed 5 ct and was sold for £5. The first authenticated account of the finding of a diamond was in December 1866 or February 1867. It was made on a farm just south of the Orange River in Cape Province and the crystal was recognized for what it was after passing through a number of hands (events echoed later on many occasions). This stone was later cut into an oval brilliant of 10.73 ct, named the Eureka and presented to the people of South Africa. The first diamond rush began when a crystal of 83.5 ct was picked up on a farm on the Orange River to the west of the farm on which the Eureka had been found. This stone later became the Star of South Africa after cutting as a 47.96 ct pear-shaped stone.

These and some others were isolated finds; the first actual diggings were at Klipdrift (present-day Canteen Kopje) on the banks of the Vaal River near Barkly West. Digging was carried out haphazardly, some prospectors being lucky, others finding nothing. The nature of the deposits was not understood though as many as 10 000 claims were worked by 1870. The first pipe mines were found in the Orange Free State in that year on the farm named Koffiefontein on the Riet River north-west of Fauresmith. As usual a diamond was picked up on the surface but this time it was still at its place of origin instead of having been transported a great distance from its parent pipe. The diamond had originated in kimberlite, now known to be serpentinized olivine, but appearing then as a yellowish material where it reached the surface and bluish below (later these two types of kimberlite became known as yellow and blue ground, the blue weathering to yellow after exposure). Deep digging eventually took place at the mine. In 1870 a 50 ct diamond crystal was found at the Jagersfontein farm in the gravel of a stream which ran through the farm. This, like the Koffiefontein mine, was also the mouth of a pipe; it was easier to work as the pipe was surrounded by hard basalt which did not, like the kimberlite, continually break up and fall into the workings.

The Jagersfontein mine produced a 972 ct crystal which was called the Excelsior diamond; 21 stones were cut from it many years later. More pipe mines were discovered; the Dutoitspan mine was established on what was originally a farm; the farmer's name was Du Toit and 'pan' signified a shallow, usually dried-up basin used as a reservoir for rainfall draining from the uplands. What were later to become known as the De Beers and Kimberley pipes were found on the same farm, Vooruitzicht, 3 miles from Dutoitspan. The De Beers brothers who owned the farm sold it to claimants; it later became part of De Beers Consolidated Mines and has produced at least £1000 million worth of diamonds. The De Beers brothers took no part in diamond mining, preferring to farm elsewhere. At a depression not far east of the De Beers farm prospectors found diamonds at a place later to become the New Rush mine, subsequently the Kimberley mine; it is now the celebrated Big Hole at Kimberley, now worked out and water-filled.

It should be mentioned that although both Jagersfontein and Koffiefontein turned out to be pipe mines, this was not realized until very much later, the first crystals being picked up from the ground. From 1871 several thousand kimberlite pipes were found in the central plateau of southern Africa and even more kimberlite dikes were found; they are still being found. Few would repay exploitation as the costs are so great and, at the time of writing, sales of diamond are slow. The South African diamond belt extends at least 270 miles from east to west and 350 miles from north to south. Some dike mines go down to 2000 feet and some pipe mines to 2500 and more. It is known that the pipes in the Kimberley area erupted to a surface about 4600 feet above present-day Kimberley before erosion removed that depth of rock. It has been estimated that about 3 billion diamonds have been removed from the pipes by erosion and the amounts found do not equal that figure so that there must be many more to find, probably widely scattered.

Rich deposits of alluvial diamonds have been found along the coastline of South Africa for about 300 miles southwards from the Orange River estuary. These deposits are in Namaqualand and are being exploited on the beaches, in sand and gravels in marine terraces and in river sediments inland.

The central plateau of South Africa rests on basement rocks of granite–greenstone which are some of the oldest rocks exposed on the earth's surface. The oldest exposed stratum is about 3.4 billion years old. The plateau forms part of the larger Southern African Shield which also underlies the eastern part of Botswana, north-eastern Transvaal, and Zimbabwe as well as the Lesotho and Swaziland regions. This is an area of great stability and this stability may be significant when considering the wide spread of diamonds in kimberlites within the plateau. It is thought that there was a diamondiferous intrusion over an extensive area of the plateau about 1.9 billion years ago; the next events are uncertain although diamonds survived the glaciation which took place when the central shield region of what is now South Africa (then part of the Gondwanaland complex) drifted to southern polar latitudes. The glaciers and their associated rubble covered most of the central plateau. The fine-grained clay-like material from the glaciers which also carried pebbles and boulders from the underlying basement rocks consolidated to form the rocks known as the Dwyka Tillite Series. This formation is associated with diamonds in a number of places and it is possible that the Dwyka Tillites moved diamonds about in great quantities south-eastwards across South Africa. The Premier mine, 25 miles east of Pretoria, was the biggest of a cluster of diamondiferous pipes and dikes that erupted more than 1.25 billion years ago. This is the oldest known kimberlite pipe with diamonds *in situ* and produced the largest gem diamond crystal ever found (the 3106 ct Cullinan diamond). Some other large stones are also recorded and many stones are particularly clear and pure. There are 15 varieties of kimberlite at the Premier mine so that there are many varieties of diamonds. Type IIa stones are almost unknown from any other source.

Namibia

In 1863 a Cape Town company obtained concessions for prospecting the coast of South-West Africa (now Namibia) north of the Orange River; no diamonds were found. Diamonds were found, however, in 1908 after the country had become a German protectorate and the German government declared the coastline from the Orange River to latitude 26°S forbidden territory or Sperrgebiet. In 1919 Ernest Oppenheimer negotiated a merger of six German companies into the Consolidated Diamond Mines of South West Africa (CDM) and scientific prospecting and mining were undertaken from that date. The quality of the stones found in this area is very high, nearly all the diamonds being gem quality. As the stones have been transported for hundreds of miles down the Orange River only the best crystals survive. Work on the beaches is carried out by holding back the surf while exploration goes forward. Marine dredging takes place out at sea.

Other African countries

The discovery in 1943 of diamonds concentrated into two areas of southern Ghana led geologists to suspect the existence of a kimberlite body in the vicinity but no pipe was found. It is thought that very old kimberlite bodies did exist and that the smallness of the diamonds indicate that they have been transported over long distances. Ancient kimberlitic activity has been tentatively located at Akwatia, now a mining settlement about 100 miles inland from Accra.

The Ivory Coast lies to the west of Ghana and for many years small diamonds were found in the alluvial deposits of the Bou River. Suspected kimberlites were discovered only in 1943 and diamondiferous dikes were located twenty years later. Diamonds from a cluster of dikes have yielded better quality stones than those from the alluvial deposits for which no kimberlite source has been found. The dikes are about 60 miles south-west of the alluvial deposits (which are sometimes known as Tortiya). A kimberlitic episode in the Ivory Coast is believed to have taken place about 1.4 billion years ago but the diamond deposits have a complicated history which is not yet completely elucidated. It is interesting to note that diamonds from the dikes (the place name is Seguela) are largely dodecahedra compared with the prevailing octahedra of diamond crystals from kimberlites from the Cretaceous (137 million years ago).

There are two diamond sources in Liberia, in only one of which has kimberlitic activity been established. Diamonds are found at Sanniquelli; stones from this deposit may have come from a kimberlite in the Nimba Mountain area which is mostly in Guinea. Kimberlites and dikes are found in the Moro River basin; from here a network of rivers spread the stones over the coastal districts of the country. On the whole these stones are of industrial quality. Diamonds from Sierra Leone, on the other hand, are frequently of high gem quality. The diamond-bearing area is in the south-east part of the country and near the town of Sefadu which gives its name to a cluster of kimberlite dikes. Diamonds are found in rivers, particularly the Moro, Moa and Sewa. Uncontrolled digging has not helped the development of a stable diamond industry in Sierra Leone, from which the

968.9 ct crystal later named the Star of Sierra Leone was recovered. This was found at Yengema, part of a cluster of kimberlites, in 1972. No one is certain that the source of the Sierra Leone diamonds is a large pipe though it is fairly clear that some major source in addition to the various clusters must have existed at one time.

The Aredor mining project in Guinea is situated near the Sierra Leone border and in 1983 its Kerouane plant was expected to produce at least 90% gem quality material in sizes up to 0.53 ct.

Diamonds have been found in the kimberlites of the Drakensburg Mountains, particularly in the Maluti Range. This lies in Lesotho, a mountainous enclave within South Africa. There may be up to 400 kimberlite occurrences but no serious prospecting and mining was carried out until the 1950s. One of the kimberlites, Letseng-la-Terai, about 40 acres in extent and over 10 000 ft above sea level, was found to be diamondiferous and a mine was established there in 1977. The age of the kimberlite at Letseng-la-Terai has been measured radiometrically at 93 million years. Fine stones have come from it, including a crystal of 601 ct. Other pipes in Lesotho are diamondiferous though at present they are not worth exploiting.

Although it had long been rumoured that Botswana had great mineral wealth the greater part of the country is covered by the Kalahari Desert. De Beers began prospecting in 1955 but no diamondiferous kimberlite was found until 1967 when the Orapa pipe was located. The mine was established in 1971. It is the second largest kimberlite pipe to be found, the largest being the Mwadui pipe in Tanzania. It is one of a cluster of 28 pipes and is rich in diamonds, about 15% being of gem quality. The age of the pipe is estimated at 93.1 million years. Orapa lies about 130 miles west of Francistown and is part of a district named Letlhakane, a name also given to the mine which exploits another of the pipes. It is thought that only the original cone of the pipe has been eroded compared to an estimated 4000 ft of erosion at Kimberley. If this is the case the Orapa pipe may contain great depths of kimberlite.

Botswana forms part of the 'Southern Shield' and in the diamond belt other kimberlites have been found, the most important of which so far is near Jwaneng, about 225 miles south of Orapa and about 70 miles west of the capital Gaborone. Several kimberlites occur in the Jwaneng cluster, at least one containing diamonds. Smaller occurrences of diamond have been found in a kimberlite at Dokolwayo, Swaziland and there have been reports of diamond in Zimbabwe and Mozambique.

After the deposits in South Africa, those of the various countries of West Africa are of highest importance; Ghana has been the second largest of the world's diamond producers (although stones are small in general) and Sierra Leone produces gem material of very high quality. The whole area is beset by smuggling and it is not always possible to say where a particular stone or consignment comes from. Diamond found in Guinea occurs north-east of the Sefadu kimberlite cluster of Sierra Leone, in the Kissidougou–Beyla–Macenta region. The deposits are thought to date from the Cretaceous period, like those of Sierra Leone. Diamondiferous kimberlites are also found in Mali though they have not been seriously exploited. A similar story is true of Upper Volta.

Diamond deposits in Angola were exploited by the Belgians from 1907 when the mining company Forminiere was set up. Until the late 1950s mining was entirely an alluvial operation and by 1970 Angola was the seventh largest diamond producer, now under the auspices of the company DIAMANG. There are hundreds of kimberlite bodies in Angola and the gemstone proportion of the recovered diamonds is very high.

Although diamonds have been reported from Congo (Brazzaville) and from Gabon, the stones probably originated elsewhere. In Zaire alluvial diamonds were first discovered in the Tshikapa district of West Kasai in 1907. Before this date kimberlites had been found in the Belgian Congo close to the border with what is now Zambia. Some of the pipes were diamondiferous but the first occurrences were not considered worth mining at that time. Very large deposits of diamonds were found in the Lulua River basin in the Bakwanga area of West Kasai and along the banks of the Bushimaie River. Most stones were of industrial quality, only about 5% being usable as gemstones. The origin of these deposits was found in the 1950s to be major kimberlite pipes in an area about 5 miles square – a very large concentration.

Two widely separated diamond areas are found in the Central African Republic; both are in the Oubangui River but are about 450 miles apart. No trace of a kimberlite source has been found – both deposits give a high proportion of gem to industrial material. There may be as many as 400 kimberlite pipes in Tanzania but only one, the Mwadui pipe, has produced any great quantity of diamonds and this is now run-down. It is the largest diamondiferous kimberlite known and it is associated with the name of John T. Williamson who mined it from 1940 until his death. Probably the best-known stone to come from Mwadui is the Williamson pink diamond in the personal jewellery collection of Her Majesty the Queen. The Mwadui pipe and alluvial deposits close by in the Alamasi valley have provided virtually all the diamonds known from Tanzania.

North America

North America is poor in diamonds, although crystals have turned up in many places (Sinkankas, 1959, 1976), up to 75 000 over the continent as a whole. Diamondiferous pipes have been located in Arkansas and stones have been found along the eastern slopes of the Appalachians in Alabama, North and South Carolina, Georgia and Virginia; they are also known from Tennessee and from glacial tillites in Wisconsin, Michigan, Indiana and Ohio. They have occurred in association with kimberlite diatremes in Colorado and Wyoming and with the gold washings in northern California.

North America was at one time attached to the land mass that is now North-west Africa, on the super continent of Pangaea. The two were split by rifts; a similar process separated the once-joined continents of South America and South Africa (the old land mass of Gondwanaland). It is not surprising that the same geological formations are found in all these places, although events after their separation have been different. Some areas are open to the public who may search for diamonds on payment of a fee; a number of stones have been found in this way.

The most important *in situ* diamond deposit is at Murfreesboro, Arkansas. Here peridotite was recognized as far back as 1842 but prospecting only began after a farmer had found two crystals on his property in 1906. At Murfreesboro the kimberlite is not in the form of pipes but occurs as rocks which are, however, recognizable because they are identical to those in South Africa. Mining is in shallow surface workings only.

Most other diamond occurrences are in streams and reports are sporadic. Several quite large stones have been found in the Great Lakes region, especially from Wisconsin and Indiana; some were in gold-bearing gravels in the former state. In California crystals are uniformly small and most have a yellowish tinge, although they are comparatively unflawed (Sinkankas, 1959, 1976).

USSR

One of the most important kimberlite and diamond occurrences is in the Siberian platform of the USSR. Many hundreds of kimberlites have been located, but conditions militate against the possibility of ever exploiting them all. There are similarities between the central Siberian shield and the diamondiferous regions of southern Africa; V. S. Sobolev was one of the first to suggest that diamonds might occur in Siberia. First attempts at prospecting in the basin of the Vilyui River in the heart of the Siberian shield were hampered by the extreme conditions and the teams therefore extended their area of operations to include the mouths of the central Siberian rivers along the Arctic coastline, the rivers themselves and then their tributaries. The first diamond was found in the basin of the Vilyui River in 1953. The first kimberlite (in the Yakutia region) was located by Larisa Popugaieva in 1954. This was the Zarnitsa pipe which did contain diamonds as it happened although exploitation was not feasible.

The famous Mir pipe was discovered in 1955 during an extensive exploration of the Vilyui River basin; this was rich in diamonds. Very soon other pipes were found, including the Udachnaya pipe 250 miles to the north. By 1956 40 pipes had been found and since that time hundreds more. It is thought that at least three pipes ejected their diamonds into the Vilyui River basin, thus accounting for the extraordinary wealth of this area. The gemstones are small but clear.

Australia

The most important diamond discovery of recent years lies in the north-west of Australia. Here there are rocks of Precambrian age overlain by Cambrian age rocks in which lamproites (similar to kimberlites but originating from shallower depths in the mantle) have been recognized for some years. However, true kimberlite was not discovered until the 1970s in an area near Lake Argyle south of Kununurra. The lamproites were discovered in an area known as the Ashton venture and it was teams associated with this enterprise that eventually located the true pipe. From the pipe which has an area of about 45 hectares, alluvial deposits extend for at least 19 miles to Lake Argyle. The kimberlite penetrates the Cambrian rock which is the top covering of the Australian shield but is thought to be relatively young, perhaps 150 million years.

During 1985 a larger than usual number of pink diamonds came into the GIA gem laboratory in New York and it is thought that the stones came from Australia. Pink grain lines, linear patterns of bright interference colours coinciding in strength with the grain lines and a fairly intense colour are thought to be indicative of Australian origin.

Indonesia

Diamonds have been found in what is now Indonesia in that part of Borneo called Kalimantan. A translation of this might be 'river of diamonds' which suggests that the stones are found alluvially. It is thought that the kimberlites from which the stones originated yielded their diamonds to sandstones and conglomerates which have, like the kimberlites themselves, been eroded so that the diamonds are found in gravels. However, the supply seems to have dwindled during this century so that few now reach the markets. The kimberlites were probably located in the mountains in the north of Borneo.

South America

Brazil, Venezuela and Guyana are the main diamond-producing countries in South America. Although Brazil once formed part of Gondwanaland, like southern Africa, no really large kimberlite pipes have yet been found. The rocks of Brazil are very old, however, so that sophisticated prospecting methods, perhaps with the assistance of remote sensing, might yet reveal important deposits. Diamonds were first discovered alluvially in 1725 by Portuguese gold prospectors in the bed of the Rio dos Marinhos, a tributary of the Rio Pinheiro, in the state of Minas Gerais. Attempts to locate kimberlites subsequently have turned up quite a large number but so far none of them have been found to be diamondiferous. Minas Gerais, a town on the Jequitinhonha River, has for long been a centre of diamond working and it took the name of Diamantina on this account. The extreme west of Minas Gerais, the Estrela do Sul region, has produced the largest stones found in Brazil. Mining is proceeding in a number of places from the Rio Blanco plateau on the Venezuelan border to the north of Minas Gerais and Matto Grosso do Sul.

North Brazil, Venezuela and Guyana share a source of diamonds known as the Roraima deposit which has been eroded by many rivers flowing in different directions. It is probable that the mountain watershed once contained a large number of kimberlites; production on the Venezuelan field was similar to that in Brazil; stones were picked up from many scattered alluvial deposits. However in the 1970s a diamond region in Bolivar province 300 miles south-west of Ciudad Bolivar began to produce stones; the area is named Guanaimo. It was a jungle area, quite unexplored with respect to mineral resources and a diamond rush developed. Mining is still reported to be uncontrolled and many stones are smuggled into Colombia and Brazil; production is also said to have declined. The rocks of the Roraima area are very old and there is evidence of a good deal of igneous activity.

Diamonds began to appear on the London market from Guyana (then British Guiana) in 1900; by 1919 regular shipments were established. It is thought that the stones came from the same ancient and lost kimberlites that supplied the stones for the Venezuelan deposits and for the Roraima deposits of Brazil. Stones are found in an area known as the Pool below the Pacaraima Mountains; they are picked up from gravels and are often of high quality. The average weight is half a carat and the miners have traditionally been known as pork-knockers.

DIASPORE

$AlO(OH)$ with some manganese. Orthorhombic, forming elongated platy crystals and some thicker ones from which faceted transparent stones have been cut. There is one direction of perfect cleavage. H 6.5–7. SG 3.3–3.5. RI (α) 1.702 (β) 1.722 (γ) 1.750. Biaxial positive. DR 0.048. Gem quality material from Turkey shows three absorption bands in the blue (in similar positions to those in green and blue sapphire).

Until the report of pale yellow gem quality material from Turkey in

1977 diaspore was unknown as a gemstone. It is found in metamorphosed limestones, in some schists and altered igneous rocks as well as in bauxite deposits. Diaspore of a light greenish-brown and with a colour change from this colour in daylight to a pinkish-brown in incandescent light is reported. Absorption bands are at 471, 463 and 454 nm with a fine sharp line in the red near 701 nm. RI (α) 1.702 (β) 1.722 (γ) 1.750 with DR 0.047–0.049 and SG 3.394. What may be at the time of writing the world's largest diaspore is from the locality of Mamaris, Yatagan in the province of Mugla, Turkey. The deposit is said to occur in a 200–300 m vein in aluminium-rich rocks. The stone which weighs 157.66 ct changes from green in daylight to pinkish-red in incandescent light. Named from a Greek word meaning to scatter, referring to diaspore's habit of disintegrating in the blowpipe flame.

DIOPSIDE

$CaMgSi_2O_6$ with a complete series to hedenbergite, $CaFeSi_2O_6$. Monoclinic, forming stubby prismatic crystals often well developed with a perfect cleavage in one direction. H 5.5–6.5. SG 3.22–3.38 with higher values on increase of iron. RI (α) 1.664–1.695 (β) 1.672–1.701 (γ) 1.695–1.721. Biaxial positive. DR 0.024–0.031. Absorption bands at 505, 493 and 446 nm for pale green stones; chrome-rich (emerald-green) stones have bands at 508, 505, 490 with a doublet at 690 and woolly bands at 670, 655 and 635 nm. Some stones give a blue or creamy-white fluorescence in short-wave UV and also an orange-yellow; some specimens fluoresce mauve in long-wave UV and others may phosphoresce in a peach colour. Four-rayed stars are quite common and chatoyant stones are also known. A chrome diopside stated to be from the USSR had RI of between 1.670 and 1.700 with a DR of 0.030. This accords with the literature. Inclusions were very like those seen in stones grown by the flux-melt method but electron microprobe analysis showed no elements that could prove a man-made origin; rumours of synthetic chrome diopside have been in the trade for many years but so far nothing has been proved.

Diopside occurs in calcium-rich metamorphic rocks and in kimberlites with diamond (notably the chromium-rich green diopside). Most diopside used as gemstones is dark to light green or emerald green for the chromium-rich stones. Yellow (Burmese) and violet (violane; found at St Marcel, Piedmont, Italy) are less common. Fine greens (chromium) come from the USSR and from Outukumpu, Finland and are found with diamond at Kimberley, South Africa; star stones and cat's-eyes come from India; Burma provides chatoyant material with a very sharp eye. Other ornamental diopsides are found in the Malagasy Republic, in the Zillerthal, Austria (green transparent stones), De Kalb, New York, USA (fine large transparent green crystals). Named from Greek words meaning double appearance.

DIOPTASE

$CuSiO_2(OH)_2$. Trigonal, forming crystals with a trigonal axis and centre of symmetry only, the habit being stubby. There is a perfect cleavage in one direction. H 5. SG 3.28–3.35. RI (ω) 1.644–1.658 (ϵ) 1.697–1.709. Uniaxial positive. DR 0.053. Dispersion 0.036. Absorption band in the yellow and green with its centre at 550 nm; the blue and violet are strongly absorbed.

The rich emerald-green colour of dioptase would make it a valuable gemstone were it not for the small size of most crystals and the preference of most people for the spectacular crystal groups. Dioptase occurs in the oxidized zone of copper deposits and the finest material comes from the Altyn-Tübe area of the Khirgiz steppes, USSR and from Zaire. The Tsumeb area of Namibia also produces some fine crystals. The name refers to the ease of observing the cleavage directions through the transparent crystal and is taken from a Greek word.

DOLOMITE

$CaMg(CO_3)_2$. Trigonal, forming rhombohedral crystals often with curved faces and with one direction of perfect cleavage. H 3.5–4 varying with direction. SG 2.85, rising to 2.93 with a high impurity content. RI (ω) 1.679–1.703 (ϵ) 1.500–1.520. Uniaxial negative. DR 0.179–0.185. Various luminescent effects reported.

Clear crystals of dolomite have been faceted and are marked by their high birefringence. Dolomite occurs in sedimentary rocks or in altered magnesium-rich igneous rocks. Faceted crystals are usually from the deposit at Eugua, Navarra, Spain. Named from D. Dolomieu, a French mineralogist.

DUMORTIERITE

$Al_7O_3(BO_3) (SiO_4)_3$. Orthorhombic, crystals rare; occurs more commonly as masses with a cleavage in one direction. H for massive varieties near 7. SG 3.41. RI (α) 1.686 (β) 1.722 (γ) 1.723. Biaxial negative. DR not usually observed in gem material. May show blue fluorescence under short-wave UV (French material); Californian variety may show blue–white to purple under short-wave UV. Transparent blue, blue–green and green dumortierite is reported from Brazil with SG 3.31, H 6–7, RI (α) 1.670 (β) 1.672 (γ) 1.669 and DR 0.015–0.016.

The massive blue and violet dumortierite is hard and well suited to ornamental use. It is found in aluminous metamorphic rocks or in pegmatites, most gem material occurring in France. A few reddish-brown transparent stones have been cut from Sri Lanka material. Dumortierite impregnating quartz, found in Arizona, USA, has SG 2.8–2.9 and RI 1.54–1.55. Much of this material is polished. Named for Eugene Dumortier, a paleontologist.

DURANGITE

$NaAl(AsO_4)F$. Monoclinic, forming orange–red crystals. H 5. SG 3.87–4.07. RI (α) 1.662 (β) 1.695 (γ) 1.712, biaxial negative. Strongly pleochroic with two rays colourless, the third orange–yellow.

Durangite, named from its location of Durango, Mexico is found associated with cassiterite and hematite at the Barranea tin mine.

EKANITE

(Th, U) (Ca, Fe, Pb)$_2$Si$_8$O$_{20}$. Tetragonal, forming elongated crystals; also found as pebbles. H 5–6.5. SG 3.28–3.32. RI (ω) 1.573 (ϵ) 1.572. Uniaxial negative. DR 0.001. An emerald-green ekanite has been reported to have an RI of 1.590–1.597; a light brown stone with similar index is also known.

Ekanite, which is metamict and radioactive, is green, dark brown to black and is found in the gem gravels of Sri Lanka, particularly at Eheliyagoda. It was discovered in 1953 by F. L. D. Ekanayake for whom it is named.

EMERALD see BERYL

ENSTATITE

$MgSiO_3$ forming series with iron silicates, bronzite, hypersthene and orthoferrosilite, a member of the pyroxene group. Orthorhombic, forming prismatic crystals, commonly twinned in lamellae. There is a good cleavage in one direction. H 5–6. SG (for enstatite) 3.20–3.30. RI (α) 1.650–1.665 (β) 1.653–1.671 (γ) 1.658–1.680. Biaxial positive. DR 0.007–0.009. Strong absorption line at 506 and another at 547.5 nm. Most stones in the overall series show a pink to green pleochroism.

Green and brown stones from India and Brazil are usually in the bronzite range with higher constants than the enstatites which are also brown and green, occurring in Tanzania and Sri Lanka. A chrome-green stone is found with diamond at Kimberley, South Africa; both four- and six-rayed star stones are known, many from India. A yellowish-green enstatite was found in an alluvial deposit at Marimba Hill, Kenya. The SG was 3.23 and the RI 1.652 and 1.662 with a DR of 0.010. The colour was attributed to 1.45% FeO and 0.22% Cr_2O_3. Associated with the enstatite was an emerald-green actinolite whose fragments were too small to cut. A colourless enstatite from Sri Lanka had SG 3.25 and RI (α) 1.658 (β) 1.661 (γ) 1.668 with a strong absorption band at 506 nm with other fainter bands. The material was not absolutely without colour since a faint brownish pink could be seen along some directions. The stone was found to be rich in magnesium. Orthoferrosilite has little ornamental use and most stones in the series tend towards dark colours (the star stones appear virtually black). The emerald green chromium-rich stones are very attractive and rare. The group occurs in basic and ultrabasic rocks and in a number of other environments. Enstatite altered to a near-serpentine composition and leek-green in colour has been called bastite. It has been cut into cabochons; occurrence in Burma and Baste, Harz Mountains, Germany. SG is 2.6. The name enstatite is taken from a Greek word meaning opponent, with reference to its high melting point; bronzite refers to colour and lustre; hypersthene (a certain amount of gem hypersthene is found in Baja California, Mexico) is from Greek words with the sense of tough; orthoferrosilite refers to crystal structure and chemical composition.

EOSPHORITE

(Mn, Fe)$AlPO_4(OH)_2 \cdot H_2O$ series to childrenite. Monoclinic, forming pseudoorthorhombic crystals, prismatic, frequently twinned. H 5. SG 3.05 for pure manganese end-member. RI (α) 1.638–1.639 (β) 1.660–1.664 (γ) 1.667–1.671. Biaxial negative. DR 0.029–0.035 (less than that of childrenite). Pleochroism yellow, pale pink, pink. Strong absorption line at 410 nm with a weaker one at 490 nm; best seen in brownish-pink stones.

Eosphorite occurs in granite pegmatites and the pink stones are quite attractive. Facetable material is found at Itinga, Minas Gerais, Brazil. Named from a Greek word meaning dawn-bearing, alluding to the colour.

EPIDOTE

$Ca_2(Al, Fe^{3+})_3(SiO_4)_3(OH)$ forming a series with clinozoisite; epidote and clinozoisite are monoclinic, zoisite is orthorhombic and dimorphous with clinozoisite. Crystals prismatic or tabular, sometimes massive, twinned or striated. There is a perfect cleavage in one direction. H 6–7. SG clinozoisite 3.21–3.38, epidote 3.38–3.49, zoisite 3.15–3.38. RI clinozoisite (α) 1.670–1.715 (β) 1.675–1.725 (γ) 1.690–1.734. Biaxial positive. DR 0.005–0.015. RI Epidote (α) 1.715–1.751 (β) 1.725–1.784 (γ) 1.734–1.797. Biaxial positive. DR 0.015–0.049. Zoisite (α) 1.685–1.705 (β) 1.688–1.710 (γ) 1.697–1725. Biaxial positive. DR 0.004–0.008.

Epidote shows pleochroism: colourless, pale yellow, yellow-green/greenish yellow/yellow-green. Zoisite also shows pleochroism: deep blue/purple/green for the tanzanite variety. Epidote shows a strong absorption line at 455 nm but on the whole the group is not identifiable by the absorption spectra alone.

Minerals of the epidote group are typically found in low- to medium-grade metamorphic rocks; clinozoisite and epidote are also found in igneous rocks, zoisite in calcareous rocks, including metamorphosed dolomites and regionally metamorphosed calcareous shales. Epidote is usually dark green (pistachio green); clinozoisite can be an attractive yellowish-brown. The blue dichroic variety of zoisite, trade named tanzanite, occurs in a number of colours in the rough, which is then cut and heated to develop a fine sapphire blue. Many epidote group minerals show chatoyancy from fibrous inclusions. Tawmawite is an emerald-green chromium-bearing rock, found in Burma; unakite is a rock composed of pink feldspar and green epidote. From Tanzania comes a deep green (chromium-bearing) rock with ruby crystals embedded in it. Most facetable epidote comes from the Untersulzbachthal, Austria though there are other locations, including Minas Gerais, Brazil, which provides some yellowish-green facetable crystals; unakite comes from Zimbabwe and some of the southern states of the USA. The best crystals of clinozoisite come from Gavilanes, Baja California, Mexico and the best zoisite from Lalatema, Tanzania. A red transparent facetable clinozoisite is found at Arendale, Norway. It has SG 3.32, RI 1.715 and 1.731 with a DR of 0.016. The colour is from manganese. The massive pink zoisite, thulite, is named from old name for Norway where it occurs. The hardness is 6 and the SG 3.10. A vague shadow edge at 1.70 can be seen on the refractometer. The colour is from manganese. Epidote is named from a Greek word meaning to increase; referring to the base of the prism which has one side longer than the other. Zoisite is named for Baron von Zois, who possessed some early specimens: clinozoisite refers to the monoclinic form of the crystals; unakite is named from the Unaka range in the United States.

EUCLASE

Be(Al, OH)SiO$_4$. Monoclinic; crystals prismatic, often showing numerous smooth faces. Perfect cleavage parallel to the plane of symmetry. H 7.5. SG 3.10. RI (α) 1.650–1.651 (β) 1.655–1.658 (γ) 1.671–1.676. DR 0.020.

Some greenish-blue stones may show a chromium absorption spectrum with two vague bands in the blue and a doublet at 705 nm. Fine sapphire-blue euclase from the Miami district of Zimbabwe showed a marked transparency region between 500–400 nm (though no absorption bands could be seen). This observation was made with a spectrometer. The stones had 0.12% Fe_2O_3. This euclase from the Miami area of Zimbabwe comes from six claims in pegmatite; the colour is an intense dark indigo or cobalt-blue in the prism zone. It occurs with beryl which it may replace, forming large imperfect crystals with well-developed faces. Some doubly terminated crystals have been reported.

Named for its perfect cleavage, euclase presents considerable difficulty to the lapidary. Most stones are a pale blue to blue–green and the mineral is to be found in granite pegmatites. The spectacular deep blue Zimbabwe material comes from old rocks; specimens with a high birefringence are reported from Santana de Encoberto, Minas Gerais, Brazil and the Ouro Preto region of this State also produces euclase. Smaller deposits are found in the southern Urals and in the Morogoro area of Tanzania.

EUDIALYTE

$Na_4(Ca, Ce)_2(Fe^{2+}, Mn)ZrSi_8O_{22}(OH, Cl)_2$, crystallizing in the trigonal system forms transparent rose-red crystals at Kippaw, Quebec, Canada. The RI is (ϵ) 1.600–1.601 (ω) 1.596–1.597. Uniaxial positive. DR 0.004. SG 2.8–3.00. H 5–5.5. The colour is thought to be due to manganese, since absorption bands at 870, 800 and 745 nm have been observed. Named from the Greek with reference to its easy solubility in acids.

FELDSPARS

The feldspars are divided into two main groups, the potassium and the plagioclase feldspars. The potassium feldspars all have the composition $KAlSi_3O_8$ but have different structures; orthoclase is monoclinic as are sanidine and anorthoclase but the two last differ in structure between themselves and from orthoclase. Microcline is triclinic.

The plagioclase feldspars form a solid solution series from albite, $NaAlSi_3O_8$ to anorthite, $CaAl_2Si_2O_8$. The series has long been divided into six species; albite, oligoclase, andesine, labradorite, bytownite and anorthite. Divisions are according to the relative percentages of albite against anorthite (Fig. 8.1).

Orthoclase monoclinic, occurring in some places as fine large crystals with

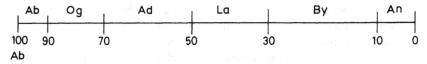

Figure 8.1 After Arem, 1977.

rounded faces and two directions of cleavage intersecting at 90°. H 6–6.5. SG 2.55–2.63. RI (α) 1.518–1.529 (β) 1.522–1.533 (γ) 1.522–1.539 with DR 0.008, biaxial negative. Most orthoclase is yellow and shows an iron spectrum with a strong line at 420 nm, weaker ones at 445 and 420 nm. It may fluoresce weak blue in long-wave UV or orange in short-wave UV. Under X-rays it glows white to violet.

Orthoclase is found in alkalic and plutonic acid rocks, granites and pegmatites. Fine transparent yellow stones are found at Itrongahy, Malagasy Republic and some sodium-bearing specimens, known as adularia, are found in Switzerland. An orange orthoclase sunstone is found at Tvedestrand, Norway. Both yellow and colourless chatoyant stones are reported from Burma and Sri Lanka.

Sanidine/Anorthoclase is monoclinic, usually occurring in glassy lumps. H 6. SG 2.52–2.62. RI (α) 1.518–1.527 (β) 1.522–1.532 (γ) 1.522–1.534. DR 0.008. Biaxial negative. Strongly resembles smoky quartz when faceted. Occurs in acid igneous rocks, most faceted stones having come from the Eifel region of West Germany.

Albite forms triclinic twinned platy crystals in white, yellow, pink, green, grey, reddish and colourless varieties. It forms at low temperatures and is common in pegmatites and granites. It also occurs in marbles. Fine colourless material occurs at the Rutherford mine, Amelia, Virginia, USA (the crystals are a platy variety known as cleavelandite). Colourless crystals, some with a blue or yellow tinge come from Kenya and have H 6–6.5. SG 2.63, RI 1.535, 1.539, 1.544. Facetable albite is found in the Malagasy Republic and has SG 2.62, RI 1.530, 1.532, 1.540. Albite moonstones are common. A low-temperature albite from Kioo Hill, Kenya, has the peristerite structure and is a milky blue to straw-yellow. It occurs in the central parts of larger common feldspar crystals in a kyanite-bearing pegmatite. SG 2.623, RI (α) 1.531 (β) 1.535 (γ) 1.539, DR 0.004.

Moonstone is the name given to feldspars with wide variations in composition. For a feldspar to merit the name it must contain finely dispersed plates of one type within another as a result of unmixing on cooling. Orthoclase moonstone contains albite within orthoclase; where the albite crystals are very fine a blue sheen results; where the albite plates are thick the sheen is white. Body colour may be white, yellowish, brown, reddish-brown or greenish. Goethite inclusions are responsible for the red. The sheen of moonstone is sometimes called adularescence and may in some stones be concentrated into a near-chatoyant band. SG is usually 2.56–2.59 with Sri Lanka stones being at the low end and Indian ones at the high end. RI is 1.520–1.525 with DR 0.005.

Burma and Sri Lanka moonstone is adularia and with a white to blue sheen. A fine sanidine moonstone with a blue sheen comes from Grant County, New Mexico, USA. An orthoclase moonstone from Virginia has RI 1.518–1.524 and is of good quality. Some moonstone is found in Tanzania. Moonstone shows characteristic surface cracks; these are fissure systems along incipient cleavages caused by pressures from exsolution. They show as short parallel cracks with shorter cracks arising perpendicularly from the length of the parallel fissures. Rectangular dark areas due to stress cracking and negative crystals may be seen. Burmese stones show oriented needles.

Oligoclase may be found as colourless to pale green transparent crystals as at the Hawk mine, Bakersville, North Carolina, USA (RI 1.537–1.547, SG 2.651).

Andesine is found in a number of places but not usually in gem form.

Bytownite is found in basic plutonic rocks and in some metamorphic rock. It is sometimes reddish (Arizona and New Mexico locations, USA).

Anorthite has the highest calcium content of the plagioclases; small faceted gems are sometimes obtainable.

Labradorite is colourless to yellow but many other colours arise from inclusions such as hematite, which produces a sheen of golden to red spangles known as aventurescence (sunstone). Zircon and magnetite are also known as inclusions. The best-known labradorite is translucent to opaque but shows blue, green and golden schiller.

A labradorite moonstone is reported from the Malagasy Republic (SG 2.70, RI 1.550–1.561). The material displays a blue sheen. Pale yellow transparent stones from Australia have RI 1.556, 1.564, SG 2.695. Oregon labradorite (albite:anorthite 32:68) has SG 2.71–2.73, RI 1.559–1.563, 1.569–1.573 with DR 0.008.

Sunstone is the name used to describe material with hematite or goethite inclusions which reflect light in parallel orientation giving a golden sheen. It may be oligoclase or labradorite; the best material comes from quartz veins at Tvedestrand and Hitterö, Norway. Attractive sunstone is found in the Harts Range, Northern Territory, Australia. SG is 2.57 and RI 1.52–1.527. The colour ranges from a pale pink to deep reddish-brown with spangles. Crystals show no striations on the major cleavage face (there are two cleavages at approximately right angles to each other) and have a perthitic or microperthitic intergrowth structure.

Microcline is triclinic and occurs as well-formed crystals, often twinned (see below) and with the two directions of cleavage shared by all the potassium-feldspars. H 6–6.5. SG 2.54–2.63. RI (α) 1.514–1.529 (β) 1.518–1.533 (γ) 1.521–1.539. Biaxial negative; usually opaque. May show yellow–green fluorescence in long-wave UV and green in X-rays; inert in short-wave UV.

Often known as amazon-stone from the Amazon River basin in South America, microcline occurs in acidic plutonic rocks, also in pegmatites and granites, syenites and schists. Fine greenish-blue crystals with whitish streaks are found at Amelia, Virginia and Pike's Peak, Colorado, USA; fine specimens occur in Brazil and in the USSR. Although the blue–green stones are known most widely microcline does occur in other colours such as colourless, white, pink, yellow, red or grey.

Plagioclase All have hardness of 6–6.5.

SG	RI
albite 2.57–2.59	1.527, 1.531, 1.538
oligoclase 2.62–2.67	1.542, 1.546, 1.549
andesine 2.65–2.69	1.543, 1.548, 1.551
labradorite 2.69–2.72	1.560, 1.563, 1.572
bytownite 2.72–2.75	1.561, 1.565, 1.570
anorthite 2.75–2.77	1.577, 1.585, 1.590

DR varies from 0.007 to 0.013. Albite, labradorite and some andesine are biaxial positive, the others are negative.

Perthite is an intergrowth of albite and oligoclase with some orthoclase or microcline. It usually forms brown and white lamellae and the white feldspar shows a golden or white schiller. Good quality material is found at Dungannon Township, Ontario, Canada.

Peristerite shows a whitish or bluish schiller. Ornamental material is found in Ontario, Canada and in Kenya.

Feldspar crystals show a number of habits shared by the monoclinic and triclinic systems, the most interesting of which are the Carlsbad, Manebach and Baveno twins. Carlsbad twins are the most common in orthoclase, the other types being rare. In sanidine this is also the case though some specimens of the other types are known. Microcline shows a combination of albite and pericline twinning and albite twins on either of the Carlsbad, Baveno or Manebach laws but more often sodium feldspars twin on the albite, pericline or other systems. A combination of pericline and albite twinning also occurs in anorthoclase.

The compositions of feldspar minerals are complicated by the possibility of potassium entering the plagioclase structure or sodium entering the orthoclase structure. The resulting compositions are known as ternary feldspars. It is also possible for unmixing to occur at low temperatures in, for example, the plagioclases where high temperature mixed compositions are stable. The unmixing may result in a segregation of the potassium and sodium molecules into separate feldspar phases, one distributed within the other. A mixture of albite with oligoclase or orthoclase is called perthite and the well-known moonstone is an albite–oligoclase mixture, as are sunstone and peristerite. Lamellae of feldspar in another feldspar give a play of colour known as schiller, best seen in labradorite where a number of different colours may be observed.

Lamellar twinning is not seen in the potassium feldspars though their crystals are commonly twinned in other ways. Lamellar twinning is common in the plagioclase group which are all triclinic and have two directions of good cleavage like the orthoclases.

Some feldspar names not yet explained are: microcline, referring to the cleavage being close to but not quite 90°: sanidine, named from a Greek word meaning board with reference to the flattish crystals: anorthoclase, also from the Greek, refers to the cleavage not being at 90°. Feldspar itself is taken from the German word for rock, fels. Plagioclase refers to the oblique cleavage of the mineral: albite from the Latin word for white: perthite is named for Perth, Ontario, Canada and labradorite from Labrador in the same country. Adularia comes from the Adular-Bergstock, Switzerland: peristerite is from the Greek word for pigeon from the resemblance of the mineral to the neck feathers of the bird: oligoclase is from a Greek word meaning little break, alluding to a supposedly less perfect cleavage than that shown by albite. Andesine is named for the Andes mountains of South America, bytownite for Bytown, Canada. Anorthite is from two Greek words meaning 'not straight', alluding to the oblique angle at which the crystal faces meet.

FLUORITE

CaF_2. Cubic, occurring as cubic crystals, often interpenetrating, or as masses. Octahedra usually occur as the result of cleavage applied after recovery. Perfect octahedral cleavage. H 4. SG 3.18. RI 1.432–1.434. Green specimens may show absorption bands at 634, 610, 582 and 445 nm and large pieces may show a stronger band at 427 nm. Strong sky-blue to violet–blue fluorescence under long-wave UV (this may be due to traces of divalent europium) and some stones show a green fluorescence (perhaps from divalent ytterbium) though the Blue John variety is inert. Under X-rays fluorescence effects are similar and may be excited even in Blue John; under short-wave UV all effects are weak; X-ray bombardment may give rise to a persistent phosphorescence.

Fluorite occurs in hydrothermal deposits, in sedimentary deposits or around hot springs. It is usually associated with sulphide ore emplacement. Gem varieties are pink, blue, green, purple, yellow, brown, colourless, some stones showing colour-zoning. Pink stones, usually small, are found in Switzerland and Zimbabwe; fine greens are found at Westmoreland, New Hampshire, USA (these crystals can be large); emerald-green fluorite is found in Namibia. The mine is thought to be near Otjiwarongo. Fine groups are found in Weardale, Durham, Cleator Moor and the Alston area of Cumbria. The banded mauve and white form, with colourless and yellow markings, has traditionally been found in the Treak Cliff outlier of the Mam Tor range of hills in Derbyshire, England. This has for long been known as Blue John and many elaborate carvings have been made from it, despite the cleavage. The area in which Blue John is found was originally mined for lead. Fine multi-coloured crystals from Illinois, USA show prominent vicinal faces on the cube faces giving a striated effect. Named from a Latin word meaning to flow, alluding to its use as a flux in iron smelting. Fluorspar is an admissible alternative name.

FRIEDELITE

$Mn_8Si_6O_{15}(OH, Cl)_{10}$. Trigonal, usually occurring in fibrous masses though tabular crystals are known. Perfect cleavage in one direction. H 4–5. SG 3.04–3.07. RI (ω) 1.654–1.664 (ϵ) 1.625–1.629. Uniaxial negative. DR 0.030. May show reddish fluorescence under both types of UV; green and yellow also reported.

Faceted stones and cabochons are orange to brownish-red. The mineral is found in manganese deposits and most gem material comes from Franklin, New Jersey, USA. Gem-quality friedelite has been found in the Kalahari manganese field near Kuruman, South Africa. It is said to occur in three discrete, near-horizontal veins within low-grade manganiferous ironstone. One form of friedelite is evenly coloured and translucent, rose-red to carmine-red; the other variety is of a similar colour but is streaked with pink and white veinlets of rhodochrosite and other carbonates. All material so far found has been cryptocrystalline. RI is (ω) 1.657 (ϵ) 1.630 but a mean figure (all that can be seen on the refractometer) is 1.64. An absorption band extends from 595 to 535 nm and from 480 to 400 nm.

Hardness is 4.5–5 and SG 3.06–3.11 (the latter figure for matrix samples).
Named after Charles Friedel, a French mineralogist.

GARNET

The garnet group of minerals has the common formula $A_3B_2Si_3O_{12}$ where
'*A*' can be iron, calcium, manganese or magnesium; '*B*' may be aluminium,
iron, titanium or chromium. All have very similar structures and there is
complete solid solution between some species but not between others. All
are members of the cubic system in which they commonly form rhombic
dodecahedra, icositetrahedra or combinations of the two forms; some
garnets occur as flattened slices or lumps with no crystal faces visible. There
is no cleavage but stones may chip easily with a marked conchoidal
fracture. Pyrope, almandine and spessartine have a hardness of 7 to 7.5;
grossular and uvarovite are 6.5–7.5 and 6.5 to 7 in andradite.

Pyrope has an SG 3.65–3.87 and RI of 1.730–1.766. In the absorption
spectrum stones from Kimberley, South Africa, Bohemia (Czechoslovakia) and Arizona show a broad band in the yellow–green centred near
575 nm which covers two of the three main almandine bands which would
otherwise be visible in that position, due to the invariable presence of some
of the almandine molecule in pyrope. The third band at 505 nm can be seen
faintly. Narrow chromium lines are rarely seen in the red. The composition
of pyrope is $Mg_3Al_2(SiO_4)_3$ but pure pyrope is unknown in nature,
almandine and spessartine components being invariably present. Some
pyrope found in East Africa is pyrope–spessartine with SG 3.816 and
RI 1.757; they may show a colour change from greenish–blue in daylight
to magenta in tungsten light. Some Norwegian stones with RI 1.747 and
SG 3.715 change from violet in daylight to wine red in tungsten light but
stones are very small. Inclusions in pyrope are not of great importance in
identification; circular aggregates of quartz crystals can be found in some
stones and Arizona material showing octahedra and small needle–like
crystals. The so–called Bohemian garnets contain chromium and occurred
in volcanic breccias (rock structures composed of an accumulation of
angular fragments) and tuffs (deposits of volcanic ash and dust compacted
into rocks) and in conglomerates near Trebnitz, Czechoslovakia. Most
pyrope is found in peridotites, kimberlites and serpentines, also in eclogites
and basic igneous rocks. Most stones classed as pyrope are almandine with
some of the pyrope component.

Main localities for pyrope are Czechoslovakia; Kimberley, South Africa
where the stone is associated with diamond and occurs as an inclusion
within diamond; Otteroy, Norway (the colour-change material); Arizona, USA; the Umba Valley, East Africa (colour-change stones). Garnets

from the Umba valley of Tanzania contain a variety of inclusions; yellow apatite, quartz, zircon, pyrite, rutile are all reported. The garnets consist of solid solution series of the end members pyrope, almandine and spessartine with low grossular components. SG and RI of the pyrope-almandine garnets vary between 3.71 and 4.04, 1.738 and 1.787. Those of the solid solution pyrope-spessartine vary in SG from 3.82–4.05 and in RI from 1.751–1.788.

Almandine $Fe_3Al_2(SiO_4)_3$ has an SG 3.95–4.3 and RI of 1.75–1.83; most stones are above 1.78. The absorption spectrum shows three broad strong bands in the yellow at 576 nm, the green at 527 nm and the blue–green at 505 nm with other weaker bands. The 505 nm band is persistent and can be seen in almost all pyrope as well as in many spessartines, particularly in those from Sri Lanka. As inclusions almandine contains zircon crystals with haloes, rutile needles, usually short and crossing at 110° and 70°. Sri Lanka stones contain densely packed hornblende rods; fibrous asbestiform needles of augite or hornblende run parallel to the dodecahedral edges and may give a four-rayed star in cabochons; apatite, quartz, ilmenite, spinel and biotite inclusions are also known.

Almandine is a constituent of metamorphic rocks; it may also be found in igneous rocks, contact metamorphic zones and alluvially. Most gem almandine on the market comes from the mica schists of the Jaipur area of India; Trincomalee in Sri Lanka yields fine large stones; star garnets are found along Emerald Creek, Benewah County, Idaho, USA: here they occur in a greenish-grey chloritic schist and a muscovite mica schist. Fine deep red crystals are found in a grey micaceous schist on the Strikine River estuary near Wrangell, south-eastern Alaska.

Rhodolite is a name given to a garnet intermediate in composition between almandine and pyrope with a magnesium to iron ratio of 2:1. The colour is a distinctive purplish-red but many stones sold as rhodolite are really pyrope–almandine; there is no agreement as to what constitutes a true rhodolite and for this reason alone the use of the name should be discouraged; it can also be confused with rhodonite. Stones from the gravels of Cowee Creek, North Carolina, USA are the true rose-red rhodolites and have an SG of 3.837–3.838 with RI of 1.760–1.761. Apatite crystals form common inclusions and almandine bands are always present in the absorption spectrum. Rhodolite is also found in the North Pare Mountains of Tanzania; this material may show a colour change from blue in daylight to a purplish-red in tungsten light; it has an SG of 3.88 and an RI of 1.765. Stones from this locality reach quite large sizes; those from Cowee Creek are usually only 1–2 ct.

Spessartine $Mn_3Al_2(SiO_4)_3$ has an SG of 4.12–4.20 and RI of 1.79–1.81;

stones from Amelia, Virginia have 1.795 and Brazilian ones 1.803–1.805. The almandine band at 505 nm is often visible but the bands caused by Mn are at 495 and 485 nm, a stronger band at 462 nm and a very strong one at 432 nm. There are also narrow bands at 424 and 412 nm, the latter very intense. Spessartine contains wavy feather-like structures with a shredded appearance, formed of liquid droplets.

Spessartine is found in granite pegmatites, also in gneisses, quartzites and rhyolites. The colour ranges from a yellowish-orange through a fine orange to dark red, according to how much of the almandine component is included. Magnificent orange spessartines are found in a pegmatite at the Rutherford mines, Amelia, Virginia, USA and stones of a very fine colour are also found at several localities in California, particularly in San Diego County where fine orange stones come from the pegmatite at the Little Three mine near Ramona. A relationship between colour and refractive index for the Rutherford no. 2 mine spessartine has been calculated; dark brownish-red iron rich stones have RI 1.809 and and pale orange (manganese-rich) stones have 1.795. Attractive almandine–spessartine orange–red garnet from Clearwater county, Idaho, USA, contains manganese (shown by absorption band at 432 nm).

Fine spessartines come from Brazil, particularly from the pegmatite of Poço dos Cavalos, Ceará and from several deposits in Minas Gerais; here the finest spessartine is found at the Ribeirão Lava-Pés pegmatite near Poaiá but important deposits exist at Barbacena, Entre-Rios, Queluz and Morro da Mina near Lafeite. Dark red spessartine is found in Norway at a number of places, including Fosbaek, Iveland, where it is found in muscovite. It is named from the original locality of Spessart in north-west Bavaria, West Germany but this locality does not produce gem-quality material. Spessartine is found in the gem gravels of Burma and Sri Lanka. Very large orange crystals of spessartine are found in the Shingus-Dusso pegmatites of Gilgit, north Pakistan. (Almandine is found with them.)

Uvarovite $Ca_3Cr_2(SiO_4)_3$ usually has an SG near 3.71–3.77 and RI of 1.74–1.87. Although it forms exceptionally fine emerald–green crystals they contain few areas sufficiently transparent to facet; nevertheless, some cabochons have been cut. It occurs in chromites and serpentines. Fine specimens are found at Outukumpu in Finland. Crystals of fine colour are also found in the Urals at Mount Saranovskaya, Kyshtym and various places in Miass. Named for Count S. S. Uvarov, a Russian statesman.

Grossular $Ca_3Al_2(SiO_4)_3$ usually has an SG near 3.65, although the range is 3.4–3.71. RI is usually near 1.73–1.74. Pale stones show no absorption spectrum; green stones with a trace of chromium may show a faint chromium spectrum. Massive material may show a weak line at 461 nm or a band at 630 nm. Green massive stones from Pakistan show a line at

697 nm with weak lines in the orange; there is also a strong band at 630 nm and diffuse lines at 605 and 505 nm. All massive grossular gives an orange under X-rays and so do some faceted stones.

Grossular occurs in metamorphosed impure calcareous rocks, especially in contact zones; also in schists and serpentines. There are many varieties of ornamental importance. Hessonite is an orange-brown colour, often known as cinnamon-stone. It has a characteristic granular appearance from included apatite and zircon and also a treacly appearance. Most hessonite comes from Sri Lanka and has SG 3.65 and RI 1.743; the spessartine molecule is sometimes found in hessonite and has the effect of raising the constants and giving manganese bands in the blue and violet. Fine specimens are found with green diopside at Eden Mills, Vermont, USA; at many places in California; fine crystals are found at the Jeffrey mine, Asbestos, Quebec, Canada. The grossular occurs in irregular dikes which occur in peridotite and dunite; the commonest dike rocks are diorites characterized by saussuritized plagioclase and biotite. Many of the intrusions of granite and diorite are associated with calcium-rich and aluminium-rich mineral assemblages (including grossular, idocrase and diopside) of hydrothermal origin. The colour of the grossular varies from colourless through shades of white, pink, orange and green. Very fine orange hessonite crystals show striated icositetrahedron faces and these combine with the rhombic dodecahedron form. A green grossular containing chromite cores comes from the same locality.

A yellow form of transparent grossular from Tanzania gives an orange fluorescence under X-rays and has SG 3.604, RI 1.734. A fine transparent green garnet found in graphite schists consisting mainly of quartz, potassium feldspar and graphite occurs at Lualenyi, Kenya. This material, named tsavolite after the Tsavo National Park, has a mean SG of 3.61 and RI 1.743. It contains 0.68 to 3.3% V_2O_3 and a smaller amount of chromium; vanadium is held responsible for the colour. Rod-like light green crystals of actinolite are characteristic inclusions, thus distinguishing tsavolite from demantoid which contains byssolite fibres. A lime green transparent grossular from Tanzania contains apatite inclusions. A yellowish-brown grossular from the Umba Valley in Tanzania had an SG of 3.68; it was a member of the pyrope–spessartine solid solution group with a high proportion of the grossular component. The composition was 28.5% pyrope, 24.0% grossular, 43.6% spessartine, 3.3% almandine and 0.6% goldmanite (the calcium–vanadium–aluminium–iron garnet). The vanadium content was only 0.2% so that the V^{3+} absorption pattern was too weak to give an alexandrite effect.

The variety *hydrogrossular* occurs in a fine green massive form from the Transvaal, South Africa and is also found in Pakistan and Tanzania. Some material may be coloured by chromium; other varieties of hydrogrossular, containing manganese, are pink (SG 3.27, RI 1.675–1.705). The jadelike green stones have an SG of 3.48 and an RI of 1.728. Some hydrogrossular is grey and contains zoisite. Fine green material shows a chromium absorption spectrum. A pink grossular found in a marble at Lake Jaco and Morelos, Mexico, is variously known as landerite, xalostocite or rosolite, names which would probably be better dropped. Hydrogrossular from New Zealand is known as rodingite (SG 3.35, RI 1.702). Pakistan green stones have SG 3.63, RI 1.738–1.742. Californite is a mixture of idocrase and grossular and is usually pale to medium green. It is found in California, Pakistan and South Africa. A colourless grossular is found in several places, including Burma and California at Georgetown; SG 3.50, RI 1.737.

Andradite $Ca_3Fe_2(SiO_4)_3$ has an SG of 3.7–4.1 and RI of 1.88–1.94. The most important variety, demantoid, usually has an SG of 3.82–3.85 and RI of 1.88–1.89. A black variety, melanite, contains from 1–5% titanium oxide; the variety topazolite (undesirable name) is yellowish-green with RI 1.887. The RI of melanite is near 1.89. The dispersion is 0.057 which is even larger than that of diamond but the effect is masked by the body colour of the stone; nonetheless, the dispersion in demantoid gives it a life which is not shown by tsavolite whose dispersion is only 0.027. In the absorption spectrum a strong band acting as a cut-off at the violet end of the spectrum at 443 nm is characteristic for demantoid; this band is due to ferric iron. Where chromium is present a doublet is seen very deep in the red at 701 nm with weaker diffuse lines at 640 and 621 nm. Demantoid shows red through the Chelsea colour filter. Demantoid contains characteristic inclusions of byssolite (a fibrous amphibole); these resemble horse tails and are usually referred to by that name; their presence is diagnostic for demantoid.

Andradite is found in schists and serpentines and in alkali-rich igneous rocks (melanite). Demantoid and the yellow–green andradite are characteristic minerals of the former group although some brown and green stones occur in metamorphosed limestones and contact zones. The finest demantoid comes from the Ural Mountains of the USSR where it was first found in the platinum washings at Nijni Tagil as greenish to white pebbles. Later it was found in the Bobrovka stream in the Sysert district near the village of Poldnevaya, both as loose grains in gold washings and as rounded nodules *in situ*, the mother rock being a serpentine of greenish-grey colour which is penetrated by veins of olivine. Minerals occurring with demantoid include dolomite and magnetite. Nodules with a typically greasy appearance may range up to 2 inches across; they consist of a large number of irregular veins cemented together by serpentine which is in a thin film. Crystals of recognizable habit are very rare.

Some fine chrome-green andradite has been reported from Korea and the yellow–green stones occur in the Ala valley, Piedmont, Italy. Some demantoid as well as green and brown andradite is found in Zaire. Melanite is found at a number of places in Italy, including Vesuvius. Yellow, green and brown andradites are found in serpentine outcrops and alluvially in San Benito County, California, USA. Some chatoyant material, the only known garnet of this kind, has also been reported, the colour golden.

GRANDIDIERITE

$(Mg, Fe^{2+})Al_3(BO_4)(SiO_4)O$. Orthorhombic, forming elongated crystals with two directions of cleavage, one perfect. H 7.5. SG 2.97–3.00. RI (α) 1.590–1.602 (β) 1.618–1.636 (γ) 1.623–1.639. Biaxial negative. DR 0.037. Strong pleochroism, dark blue–green, colourless and dark green.

Blue green cabochons of grandidierite are cut from material found in the pegmatites of Andrahomana, Malagasy Republic. Named for the French explorer Grandidier.

GYPSUM

$CaSO_4 \cdot 2H_2O$. Monoclinic, forming large and often perfect crystals, also rosettes; twinning common. There are three directions of cleavage, one of them perfect. H 2. SG 2.32. RI (α) 1.520 (β) 1.523 (γ) 1.530. Biaxial positive. DR 0.010.

Gypsum (often referred to as alabaster) is found in sedimentary rocks, saline lakes, the oxidized portions of ore deposits, and in volcanic deposits. The massive granular alabaster has been fashioned into cabochons. Selenite is often used to describe large colourless transparent twinned crystals with swallow tails. The name refers to their moonstone-like surface glow. Much ornamental alabaster is quarried in Tuscany, particularly at Castellina. Some ornamental alabaster is quarried in England, particularly at Chellaston, Derbyshire.

Gypsum comes from the Greek word for what is now known as plaster; alabaster from a Greek word referring to a stone from which ointment containers were made.

HAMBERGITE

$Be_2BO_3(OH, F)$. Orthorhombic, prismatic habit with easy cleavage parallel to a face in the prism zone. H 7.5. SG 2.35. RI (α) 1.55 (β) 1.59 (γ) 1.63. Biaxial positive. DR 0.072.

Hambergite is colourless and is remarkable only for its comparative rarity and its high birefringence. Those crystals which have been found suitable for faceting come from Anjanabanoana in the Malagasy Republic. Named for Axel Hamberg, the Swedish mineralogist who first drew attention to the mineral, it is found in granite or alkali pegmatites.

HAUYNE

$(Na, Ca)_{4-8}Al_6Si_6(O, S)_{24}(SO_4Cl)_{1-2}$. Cubic, usually found as grains but crystals may be dodecahedral or octahedral. One direction of distinct cleavage. H 5.5–6. SG 2.44–2.50. RI 1.496–1.505. Some material from the Laacher See, West Germany, gives an orange fluorescence under long-wave UV.

A member of the sodalite group of minerals, hauyne is one of the constituents of lapis lazuli. Some specimens however, though very small, are transparent enough to be faceted and the magnificent bright blue colour makes them desirable gemstones. Found in alkaline igneous rocks. Named for Abbé Haüy, a celebrated French mineralogist.

HEMATITE

Fe_2O_3. Trigonal, forming characteristic reniform masses and occasionally crystal groups with a bright surface which is prone to tarnish. H 5–6.5. SG 4.95–5.26. RI (ω) 3.22 (ϵ) 2.94. Uniaxial negative. DR 0.280. Red streak.

Hematite is a major ore of iron and takes its name from a Greek word for blood, referring to its colour when powdered. Bright crystals are found on the Isle of Elba; rosette-like masses are found in the Ticino and Grisons cantons of Switzerland and have been given the name iron roses. Much hematite is cut and polished for setting in jewellery where the bright lustre may suggest a black diamond. Material from Cumbria, England, has been fashioned in Idar-Oberstein, West Germany. Hematite occurs chiefly as sedimentary deposits. Hematine or hemetine is an artificial imitation of hematite and is a mixture of stainless steel with chromium and nickel sulphides; it is magnetic whereas hematite is not.

A new type of hematite seen in the gem cutting industry consists of granular hematite as a main constituent with magnetite, martite and gangue minerals as minor constituents. Iron content is 69% and the material comes from Brazil. The colour is bluish-grey with a dark brown streak and there is some magnetism, in contrast to other varieties of hematite. (SG of 2.33 compared to 5.2–5.3 for the natural material is shown by an imitation of unknown composition tested in Germany during 1984.)

HEMIMORPHITE

$Zn_4Si_2O_7(OH)_2 \cdot H_2O$. Orthorhombic, forming tabular crystals with a perfect cleavage in one direction. H 4.5–5. SG 3.4–3.5. RI (α) 1.614 (β) 1.617 (γ) 1.636. Biaxial positive. DR 0.022.

Blue crystals and masses have been faceted and cut into cabochons; the blue is a delicate sky-colour. Hemimorphite occurs as a secondary mineral in the oxidized zones of ore deposits; principal localities for the facetable material are Mapimi, Durango and Santa Eulalia, Chihuahua, Mexico. Named for its crystal form.

HERDERITE

$CaBePO_4(F, OH)$. Monoclinic, forming stout or prismatic crystals. H 5–5.5. SG 2.95–3.02. RI (α) 1.594 (β) 1.613 (γ) 1.624. Biaxial negative. DR 0.030. Some stones fluoresce pale green under both long-wave and short-wave UV and give an orange glow with strong phosphorescence under X-rays. A green cut stone in the US National Museum has much lower refractive index and DR and this may be due to its fluorine content. Violet and colourless specimens described from Maine, USA, have RI of 1.587–1.621. A brown herderite from Brazil belongs to the hydroxyl-herderite group and has very high RI of 1.610, 1.630 and 1.642. The fluorine content is 0.20%. A colourless to yellow hydroxyl-herderite from Brazil has RI (α) 1.610 (β) 1.631 (γ) 1.645 with DR 0.035 and SG 3.08.

Herderite occurs as a late stage hydrothermal mineral in granite pegmatites. Most gem quality material comes from Minas Gerais, Brazil; specimens from Maine are paler. At its best herderite is pink or green with some violet specimens recorded. Named for S. A. W. von Herder, a mining official from Freiberg, Germany.

HODGKINSONITE

$MnZn_2SiO_5 \cdot H_2O$. Monoclinic, forming pyramidal or prismatic crystals. One direction of perfect cleavage. H 4.5–5. SG 3.91–3.99. RI (α) 1.720 (β) 1.741 (γ) 1.746. Biaxial negative. DR 0.026. Dull red fluorescence under long-wave UV. Pleochroism lavender and colourless.

One of the minerals known only from the metamorphosed limestone at Franklin, New Jersey, USA, hodgkinsonite, a pink to lavendar mineral, is named for H. H. Hodgkinson, its discoverer. Faceted stones are bright and attractive, though very rare since no material is recovered now.

HORNBLENDE

The common hornblende is a silicate of calcium, potassium and sodium with magnesium, iron and aluminium and is a member of the monoclinic system. It is a common constituent of igneous and metamorphic rocks, named from German words meaning to deceive. Used to denote dark minerals found with ores but not itself containing them. Most hornblende is far too dark to make ornamental specimens.

Greenish-brown gem quality hornblende from Baffin Island, NWT, Canada, is found in a dolomitic marble and has SG 3.09, H 5.5–6, RI (α) 1.625–1.627, (β) 1.632–1.633, (γ) 1.644–1.647 with DR 0.020. Veils of healing feathers and unidentified colourless crystals are included.

HOWLITE

$Ca_2B_5SiO_9(OH)_5$. Monoclinic, commonly forming nodular masses. H about 3.5. SG 2.45–2.58. RI (α) 1.583–1.586 (β) 1.596–1.598 (γ) 1.605. Biaxial negative. DR 0.022. Howlite may fluoresce a brownish-yellow in short-wave UV and some Californian material gives a deep orange in long-wave UV.

Howlite, which is white, is frequently dyed blue to imitate turquoise. However, spheres and cabochons are occasionally cut. Howlite is found in arid regions and borate deposits, reported from Lang, Los Angeles County, California, USA and from the Mohave Desert in the same state. Named after H. How who described a mineral with similar composition.

IDOCRASE see VESUVIANITE

IOLITE see CORDIERITE

JADEITE

$Na(Al, Fe^{3+})Si_2O_6$ a member of the pyroxene group of minerals. Monoclinic, occurring as masses of interlocking crystals and found as pebbles or boulders. H 6.5–7. SG usually near 3.33. Shows a marked splintery fracture but is very tough. RI: shadow edge near 1.66; crystals have (α) 1.640 (β) 1.645 (γ) 1.652–1.657. Biaxial positive. DR 0.012–0.020. Rich green material coloured by chromium shows a strong line in the absorption spectrum at 691.5 nm with others at 655 and 630 nm; there is a diagnostic line at 437 nm and weaker ones at 450 and 433 nm. Pale coloured jadeite may fluoresce a dim white in long-wave UV; X-rays may excite an intense blue–violet glow in pale yellow and mauve material.

The brighter-coloured of the two minerals which may be classed as jade (nephrite is the other), jadeite occurs in serpentine derived from olivine rocks but is found more often in alluvial conditions as boulders and pebbles. The finest material ('Imperial jade') is a deep emerald green, translucent and with no hint of white streakiness. This variety is found in Upper Burma at Tawmaw and nearby places where it occurs with albite forming veins in a country rock of serpentine. Fine sectioned boulders, the larger weighing 80 kg, are in the Geological Museum, London. So far there appears to be no deposit of jadeite with ornamental potential in China. An attractive mixture of bright green jadeite with the sodium–chromium silicate, ureyite, is known as maw-sit-sit. Some Burmese jade not of Imperial quality is none the less translucent and green; the name Yunnan jade is often given to it. A material known as chloromelanite is opaque dark green to black.

A study by W. Campbell Smith of many prehistoric jade axe heads found in Britain showed that they were all jadeite although no source for the material exists in Britain and there is no source of any significance in Europe. The Pre-Columbian civilizations of Central America used jadeite for a variety of implements and jadeite is found in California in San Benito County in its parent rock of serpentine. The area is often referred to as Clear Creek from a small river running through it and boulders of jadeite are found in the river. On the whole the colour of this jadeite is poor but it is an important find for investigating its origins: it is rare for jadeite to be found *in situ*.

Jadeite is also found in Mexico as finished objects, not (so far) *in situ*. The colour is green though not quite so intense a green as that shown by Burmese jadeite. A pearl-grey colour is more common, ranging through dark green to nearly black. Deposits of jadeite occur in Guatemala in the valley of the Rio Grande Motagua in the south-central part of the country. The mineral is also found *in situ* at a site near Manzanal. Guatemalan jadeite is characterized by its extreme toughness coupled with a typical coarse granular structure. No chromium is present but albite very often is and serves as a point of identification. Some apple-green jadeite is reported from the USSR.

Jadeite is named from the Spanish piedra de ijada (stone of the loins) referring to its presumed healing power over ailments of the kidneys. A material of a dark green colour with lighter coloured veins found in New Guinea is a chrome-rich jadeite intergrown with picotite (chrome spinel), quartz, opal and limonite. The SG is 3.35. It is thought to be derived from an olivine rock and has been called astridite, after a Belgian queen.

The material saussurite which may be mistaken for one of the jade minerals is the product of the decomposition of plagioclase feldspar. It has a tough compact structure and a splintery fracture. The colour is greenish-grey to white. SG 3.0–3.4, H 6.5–7. In its altered form the mineral zoisite is predominant in saussurite, giving an RI of 1.70. It was originally found near Lake Geneva and named for H. B. de Saussure; it also occurs in Cornwall and in Italy.

JEREMEJEVITE

$Al_6B_5O_{15}(OH)_3$. Hexagonal, elongated and tapering crystals. H 6.5. SG 3.28–3.31. RI (ϵ) 1.640 (ω) 1.653. Uniaxial negative. DR 0.007–0.013. Namibian material shows pleochroism of cornflower blue to colourless or straw.

Jeremejevite whose name is taken from that of Pavel V. Jeremejev, a Soviet mineralogist, was first discovered as tiny grains in a granitic debris in Eastern Siberia; later, gem quality crystals were found at Cape Cross near Swakopmund, Namibia. They are blue–green and allow for cut gemstones of up to 5 ct to be fashioned. The extent of the deposit is thought to be limited.

KAKORTOKITE

Kakortokite is an ornamental rock consisting of white nepheline syenite with red crystals of eudialyte and black arfvedsonite. SG 2.7–2.8. Naujaite is a whitish nepheline syenite containing microcline nepheline with small amounts of albite, analcime, acmite and sodium amphiboles. Both materials are found at Julianehaab, Greenland.

KAMMERERITE

This mineral is a chromian variety of clinochlore, itself a member of the chlorite group. Crystals of a fine dark red or reddish-purple, large and transparent enough to facet, are sometimes found in a chromium deposit at Kop Krom, Kop Dǎglari, Turkey. The SG is 2.70 and the RI 1.588 and 1.594 with a DR of 0.006; lower figures are sometimes found. The hardness is 2–2.5. Clinochlore has the composition $(Mg, Fe^{2+})_5Al(Si_3Al)O_{10}(OH)_8$ and is monoclinic.

KORNERUPINE

$Mg_3Al_6(Si, Al, B)_5O_{21}(OH)$. Orthorhombic, forming columnar crystals with two directions of perfect cleavage. H 6.5. SG 3.28–3.34. RI (α) 1.665 (β) 1.678 (γ) 1.682. Biaxial negative. DR 0.013. Pronounced pleochroism, green, yellow and reddish-brown. Most kornerupines show an absorption band at 503 nm.

Most kornerupine is a dark brownish-green to greenish-yellow; some vanadium-bearing material from the Kwale district of Kenya is a beautiful bright apple-green and some blue specimens have been found at Itrongahy, Malagasy Republic; the colour is close to that of pale aquamarine. Grey to brown prismatic crystals are found at Betroka and Inanakafy in the same country. Brown chatoyant stones are fairly often seen on the market. In both Burma and Sri Lanka rolled pebbles of a green to brown colour can be found: some very rare asteriated stones have come from the Mogok area. Some East African kornerupine shows a bluish-green to green colour, unlike that found in the Kwale district of Kenya. The blue–green stones show an intense emerald-green in polarized light parallel to the c-axis: a weak blue is seen at right angles to this direction. RI (α) 1.662–1.663 (β) 1.673–1.674 (γ) 1.674–1.675 with DR 0.012–0.013. Green kornerupine from Burma shows a yellow fluorescence under both long-wave and short-wave UV; this glow is stronger in some of the East African kornerupines. Sri Lanka material is inert.

Kornerupine from the Elahera district of Sri Lanka has been found to have RI 1.668 and 1.680 with a DR of 0.012. The SG is between 3.34 and 3.36 and the colour is brownish-green. Apatite crystals and zircon crystals with haloes are reported as inclusions. Named for the Danish geologist A. N. Kornerup.

KUNZITE see SPODUMENE

KYANITE

Al_2SiO_5, trimorphous with sillimanite and andalusite. Triclinic, forming bladed crystals, generally flattened and elongated. Perfect cleavage in one direction. H 4–7.5, varying with direction; 4 along length, 7.5 across. SG for gem material near 3.68. RI (α) 1.712–1.718 (β) 1.721–1.723 (γ) 1.727–1.734. Biaxial negative. DR 0.017. Strong pleochroism, violet, colourless and cobalt blue; shades of yellow and green in some stones. May show chromium absorption spectrum with line in deep red at 710 nm, two others in deep blue with a dark edge near 600 nm. May fluoresce dim red in long-wave UV.

Kyanite occurs in schists, gneisses and granite pegmatites. Transparent crystals in blue and green with stripes of blue are found in Brazil and blue stones with green banding in the Machakos district of Kenya; this area also produces colourless crystals and elsewhere in Kenya fine blue transparent crystals are found. A blue kyanite with a change to reddish in incandescent light from a blue–green in daylight had SG 3.676 and RI (α) 1.714 (β) 1.724 (γ) 1.731. The colour was found to be attributable to the chromophore pairs (chromium plus iron) or to chromium alone. The location was East Africa and the stone contained long growth tubes. An emerald-green kyanite from Tanzania has SG 3.68, RI (α) 1.714 (β) 1.725 (γ) 1.732. The colour is caused by chromium; iron, vanadium and titanium are also present. Named from a Greek word meaning blue.

LAPIS LAZULI see LAZURITE

LAWSONITE

$CaAl_2Si_2O_7(OH)_2 \cdot H_2O$, dimorphous with partheite. Orthorhombic, forming colourless to pale-blue crystals. H 7–8. SG 3.08–3.09. RI (α) 1.665 (β) 1.669 (γ) 1.684. Biaxial positive. DR 0.019.

Lawsonite is a mineral of low temperature metamorphic rocks, especially of schists. Found in California, it is named for A. C. Lawson of the University of that state.

LAZULITE

$MgAl_2(PO_4)_2(OH)_2$, forming a series to scorzalite $(Fe, Mg)Al_2(PO_4)_2$ $(OH)_2$. Monoclinic, with acute pyramidal crystals. H 5.5. SG near 3.08 at the lazulite (Mg) end of the series. RI (α) 1.604–1.635 (β) 1.633–1.653 (γ) 1.642–1.662 (figures for lazulite). Biaxial negative. DR 0.031–0.038. Strong pleochroism, colourless to blue to dark blue. Transparent light and dark blue specimens of lazulite–scorzalite had RI (α) 1.626 (β) 1.654 (γ) 1.663 with DR 0.037 and SG 3.22. The composition was $Mg_{0.5}Fe_{0.5}Al_2$.

Lazulite is found in granite pegmatites and metamorphic rocks. Fine blue transparent crystals have been found in the state of Minas Gerais, Brazil and others have come from the Bhandara district of India and from the Yukon area of Alaska. Named from a German word meaning blue stone.

LAZURITE

$(Na, Ca)_{7-8}(Al, Si)_{12}(O, S)_{24}[(SO_4), Cl_2, (OH)_2(OH)_2]$ cubic, forming very rare dodecahedral crystals, more commonly found massive. The ornamental material lapis-lazuli is a rock composed of lazurite, hauyne, sodalite and nosean, which form members of the sodalite group. Lazurite is chemically a variety of hauyne, rich in sulphur. Lapis-lazuli contains calcite and pyrite, the best material having virtually no calcite and only a small amount of disseminated pyrite. H 5–6. SG 2.7–2.9, going higher if much pyrite is included. RI near 1.50. May show orange streaks in long-wave UV, especially material from Chile; pinkish in short-wave UV. H_2S gas with characteristic smell of bad eggs results if a drop of HCl is placed on a specimen.

Lazurite and lapis-lazuli occur as contact metamorphic materials in limestones and granites. The finest material comes from Badakshan, Afghanistan; the Chilean Andes furnish less fine lapis and it is also found in Mongolia. Lapis-lazuli from Sar-e-Sang, Badakshan, Afghanistan, is unquestionably the finest known. The complex consists of strongly metamorphosed rocks with gneisses, cipolins, skarns, crystalline schists, amphibolite, veins of leucocratic granites with pyroxenite and hornblendite dikes which are found as large boulders in the bed of streams. The lapis-bearing rock is cipolins overlying gneiss and the lapis occurs in skarns which form bands about 40 m wide. The lazurite is nearly always found with pyrite and calcite, dolomite, forsterite, diopside and scapolite are also in association. Since the Soviet invasion of Afghanistan it is believed that no mining is taking place (Wyart, 1981). Named from a Persian word meaning blue.

LEGRANDITE

$Zn_2(OH)AsO_4 \cdot H_2O$. Monoclinic, prismatic crystals. H 4.5. SG 3.98–4.04. RI (α) 1.675–1.702 (β) 1.690–1.709 (γ) 1.735–1.740. Biaxial positive. DR 0.060. Pleochroism colourless and yellow.

Legrandite, which occurs in vugs in limonite, has a bright yellow colour and some crystals yield areas transparent enough to facet. The chief producer is the Ojuela Mine, Mapimi, Mexico. Named for a M. Legrand, a Belgian mine manager who collected the first specimens.

LEPIDOLITE

$K(Li, Al)_3(Si, Al)_4O_{10}(F, OH)_2$. Monoclinic, a member of the mica group, forming tabular crystals or platy masses with a perfect cleavage. H 2.5–4. SG 2.8–3.3. RI (α) 1.525–1.548 (β) 1.551–1.585 (γ) 1.554–1.587. Biaxial negative. DR 0.018–0.038.

Occurring in granite pegmatites, the lilac to purple massive lithium mica, lepidolite has been used ornamentally for such objects as paperweights. Named from a Greek word referring to the scaliness of the massive material.

LEUCITE

$KAlSi_2O_6$. Tetragonal, pseudocubic, forming icositetrahedral crystals. H 5.5–6. SG 2.47–2.50. RI about 1.50 (if uniaxial the material is optically positive). Faint anomalous birefringence may be seen. Fluoresces medium-bright orange in long-wave UV (Italian crystals) and bluish in X-rays.

Leucite occurs in potassium-rich basic lavas and virtually all facetable crystals are found in the Alban Hills near Rome, Italy. Leucite is colourless though it often displays interference colours. Named from a Greek word meaning white.

LINARITE

$PbCu(SO_4)(OH)_2$. Monoclinic, forming prismatic crystals and crusts, the crystals with a perfect cleavage in one direction. H 2.5. SG 5.30. RI (α) 1.809 (β) 1.838 (γ) 1.859. Biaxial negative. DR 0.050. Pleochroism various shades of blue.

Linarite occurs in the oxidized zones of lead-copper deposits and is a very fine dark blue. There is insufficient transparent material to allow faceted stones to be cut. Fine large crystals have been found at the Mammoth mine, Tiger, Arizona, USA. Named from Linares, Spain.

LUDLAMITE

$(Fe^{2+}, Mg, Mn)_3(PO_4)_2 \cdot 4H_2O$. Monoclinic, forming tabular, wedge-shaped crystals with a perfect cleavage in one direction. H 3.5. SG 3.19. RI (α) 1.650–1.653 (β) 1.667–1.675 (γ) 1.688–1.697. Biaxial positive. DR 0.038–0.044.

Apple-green ludlamite may be cut into cabochons but the mineral is rare; fine crystals are found at the Blackbird mine, Lemhi County, Idaho, USA and at Hagendorf, Germany. It is found as a secondary mineral in the oxidized zone of ore deposits. Named for Henry Ludlam, English mineralogist and collector.

MAGNESITE

$MgCO_3$. Trigonal, forming rhombs with a perfect rhombohedral cleavage. H 3.5–4.5. SG 3.0–3.1. RI (ω) 1.700–1.717 (ϵ) 1.509–1.515. Uniaxial negative. DR 0.022. In short-wave UV shows blue, green or whitish glow sometimes with greenish phosphorescence. Effervesces in warm acids.

Fine clear colourless rhomb-shaped crystals are found at Brumado, Bahia, Brazil which is the only source of facetable material. Magnesite occurs as an alteration product of magnesium-rich rocks and is also found in sediments and as a gangue mineral (i.e. that part of an ore not economically desirable) in hydrothermal ore deposits. Named from its composition.

MALACHITE

$CuCO_3(OH)_2$. Monoclinic, crystals rare and small. More commonly found as masses, frequently banded; sometimes stalactitic. Perfect cleavage in one direction. H 3.5–4.5. SG 3.6–4.05. RI (mean) 1.85. Effervesces in warm acids and colours Bunsen burner flame green.

Fine green banded malachite is found with azurite and cuprite in the oxidized zones of copper ore deposits; superb quality masses are found in the USSR at Mednorudyansk and Nizhne-Tagilsk and large crystals are found at Tsumeb, Namibia. Large quantities are also found in Zaire. Care needs to be taken when fashioning malachite since the masses consist of closely-packed radial fibres. Named from a Greek word alluding to its colour.

MANGANOTANTALITE

(Mn, Fe) $(Ta, Nb)_2O_6$ with a manganese to iron ratio of 3:1. It forms two series, with manganocolumbite and with ferrotantalite. Orthorhombic, forming short prismatic crystals with a distinct cleavage in one direction. H 6–6.5. SG 8.00. RI (α) 2.19 (β) 2.25 (γ) 2.34. Biaxial positive. DR 0.150.

Manganotantalite is found in granite pegmatites, crystals being brownish to red. Facetable crystals are found at Alto Ligonha, Mozambique and in Minas Gerais, Brazil. A stone from the Morrua area of Mozambique had H 5.5–6, SG 7.73–7.97 and RI (by real and apparent depth) 2.16. Pleochroic colours were deep red and pale pink. Negative crystals with two-phase fillings were seen to be parallel to the c-axis. Named for its composition.

MARBLE

Marble is a crystalline limestone in the form of massive calcite. Many varieties are used as ornamental building stones since impurities give a large and attractive range of colours. A variety consisting mainly of fossil shells is known as lumachella, or fire marble, referring to the play of colour which can be seen when the material is viewed at certain angles. It can be distinguished from opal matrix by its effervescence with acids. The best material comes from Bleiberg, Carinthia, Austria and from Astrakhan in the USSR.

MARCASITE

FeS_2, dimorphous with pyrite. Orthorhombic, forming abundant tabular or pyramidal crystals, often with curved faces with a distinct cleavage in one direction. A spear-shaped habit is common as is the 'cockscomb pyrite', an aggregate of flattened twinned crystals. The mineral marcasite is not the marcasite of the jeweller, which is pyrite. Marcasite is the less stable form of FeS_2 and has a hardness of 6–6.5. SG 4.85–4.92.

Marcasite forms at low temperatures, often in sedimentary environments and in low temperature veins. It occurs world wide and has been imitated in jewellery by steel which is magnetic (marcasite is not). The colour is a pale brassy yellow and surfaces may show iridescence. Named from an Arabic (?Moorish) word; the name was used for crystallized pyrite up to about 1800.

MELIPHANITE

$(Ca, Na)_2Be(Si, Al)_2(O, OH, F)_7$. Tetragonal, forming thin and tabular crystals with a perfect cleavage in one direction. H 5–5.5. SG 3.0–3.03. RI (ω) 1.612 (ϵ) 1.593. Uniaxial negative. DR 0.019. Distinct pleochroism, yellow and red.

Meliphanite (which has been known as melinophane) is sometimes cut into small yellowish to reddish cabochons. It occurs in nepheline syenites and skarns, most ornamental material coming from the Julienehaab area of Greenland and the Langesundsfjord area of Norway. Named from Greek words meaning having the appearance of honey (from the colour).

MELLITE

$Al_2C_6(COO)_6 \cdot 18H_2O$. Tetragonal, forming prismatic crystals showing pyroelectricity. H 2–2.5. SG 1.64. RI (ω) 1.539 (ϵ) 1.511. Uniaxial negative. DR 0.028. May show dull white fluorescence in short-wave UV, sometimes a light blue. A cut golden mellite had SG 1.58, RI 1.509, 1.541 with a DR of 0.032. It came from the brown coal deposits of Thuringia.

Occurring in brown coals and lignites as a secondary mineral, mellite, a honey-yellow aluminium mellitate is an organic mineral formed by inorganic processes. It is found in the Bitterfeld region of Germany and in the Paris Basin, France. Named from a Latin word for honey.

MESOLITE

$Na_2Ca_2Al_6Si_9O_{30} \cdot 8H_2O$. Monoclinic, fibrous crystals which when packed together give a silky effect. H 5. SG 2.29. RI (mean) 1.50.

A member of the zeolite group of minerals, mesolite produces white cabochons with a silky lustre. The mineral is found in amygdaloidal basalt and occurs at a number of locations. Named from a Greek word alluding to its position mid-way between natrolite and scolecite.

MICROLITE

$(Na, Ca)_2Ta_2O_6(O, OH, F)$. Cubic, forming octahedral crystals with an indistinct octahedral cleavage. H5–5.5. SG about 5.5. RI 1.93–1.94. Some anomalous birefringence may be caused by metamictization.

Microlite, occurring as a primary mineral in granite pegmatites, may be cut into cabochons of a pale yellow to brown colour. Very occasionally a Brazilian green crystal may be faceted. Fine crystals also occur at the Rutherford mine, Amelia, Virginia, USA. A garnet-red cut stone is recorded. Named from a Greek word for small, in allusion to the small crystals.

MILLERITE

NiS. Hexagonal, forming masses which may be fashioned. Crystals acicular, too thin for ornamental use. There are two directions of perfect cleavage. H 3–3.5. SG 5.3–5.6. Brassy yellow faces subject to tarnish to a greenish-grey.

Millerite, named for mineralogist W. H. Miller of Miller indices fame, occurs as a low-temperature mineral in limestones and dolomites, in serpentines and ore deposits in carbonate rocks. Good material, a fine yellow, comes from Timagami, Ontario, Canada where it occurs in large cleavable masses. A cloudy yellowish-green stone has been recorded in cut form from Rössing, Namibia.

MIMETITE

$Pb_5(AsO_4)_3Cl$. Monoclinic, commonly forming acicular or globular crystals. H 3.5–4. SG 7.24 (may be lower when calcium replaces lead). RI

(ω) 2.147 (ϵ) 2.128. Biaxial negative. DR 0.019. Tsumeb stones fluoresce orange–red in long-wave UV.

Cabochons of a bright orange or yellow have been cut from globular masses of mimetite from Chihuahua, Mexico. Transparent crystals found at Tsumeb, Namibia but occurrence even there is sparse. Mimetite occurs as a secondary mineral in the oxidized zone of lead deposits. Named from a Greek word, meaning imitator, since the crystals resemble those of pyromorphite.

MONTEBRASITE see AMBLYGONITE

MOONSTONE see FELDSPAR

NAMBULITE

$NaLiMn_8Si_{10}O_{28}(OH)_2$. Triclinic, forming wedge-shaped, flattened prismatic crystals, frequently twinned. Two directions of distinct cleavage. H 6.5. SG 3.51. RI (α) 1.707 (β) 1.710 (γ) 1.730. Biaxial positive. DR 0.023.

Although nambulite is a rare mineral the few facetable crystals that have been found (notably at the Kombat Mine, Tsumeb, Namibia) would make superb gemstones since the orange–red colour is outstanding. Nambulite occurs in veins cutting manganese oxide ores; in addition to the Tsumeb location it is also found at the Tunakozawa Mine, Japan. Named for Professor Matsuo Nambu of Tohoku University, Japan (a notable student of manganese minerals).

NATROLITE

$Na_2Ca_2Si_3O_{10} \cdot 2H_2O$. Orthorhombic, forming massive fibrous crystals with a perfect cleavage in two directions. H 5.5. SG 2.20–2.26. RI (α) 1.473–1.483 (β) 1.476–1.486 (γ) 1.485–1.496. Biaxial positive. DR 0.012. Fibrous varieties have silky lustre. Some specimens may give orange–yellow fluorescence under long-wave UV.

Natrolite is one of the zeolite family of minerals and is found in basaltic cavities and in other dark igneous rocks. The mineral is white but some crystals from Bound Brook, New Jersey, USA are colourless and can be faceted. Named from the Latin natron, referring to the presence of sodium.

NATURAL GLASS

Obsidian

Obsidian is formed from the cooling of volcanic lava. The cooling takes place too quickly for the lava to form into a rock; had a rock formed it would have consisted mainly of quartz, feldspar and mica. Obsidian is a mixture with a variable chemical composition, although up to 77% can be silica, with 10–18% alumina. Obsidian is generally opaque and black, but yellow, red, greenish-brown and grey colours are known. The texture shows signs of incipient crystallization, usually with small vesicles; these are variably-shaped cavities formed by the entrapment of gas bubbles during solidification. Slight double refraction due to strain is often seen. The hardness is usually about 5 and the SG 2.33–2.47. The RI is usually between 1.48 and 1.51. Basalt glass has higher constants at 1.58–1.65 and 2.70–3.00. A texture known as perlite and consisting of a network of curving cracks is characteristic of obsidian and disintegration along the curves gives bead- or ball-like pieces which have been given the name marekanite from a type locality in Siberia. A spherulitic textured obsidian with radiating feldspar fibres is called peanut obsidian and comes from Sonora, Mexico. Obsidian is often simulated by bottle glass with a higher SG and RI; it is also more transparent than obsidian. Obsidian was used in the ancient Mexican civilizations for its ability to be fashioned into a sharp cutting edge. With white flecks it is known as snowflake obsidian and is found in the USA.

Tektite

Small pieces of glass were found near the Moldau River in Bohemia as far back as the eighteenth century and named moldavite after the river. The true origin of the glass was uncertain and it is only relatively recently that a meteoritic origin has been postulated. Button-shaped glass pieces, known as australite, have been found in south Australia and Tasmania. Billitonite was found on Billiton Island between Sumatra and Borneo. Not all reported occurrences are necessarily true tektite; some may be obsidian. Tektites are composed of silica (70–77%) with alumina (15–10%). RI is 1.48–1.52 and the SG is 2.30–2.50 with a hardness of 5 (figures for transparent bottle-green moldavite). Some Borneo tektites are dark brown and those from Colombia are sometimes colourless. All have surface markings and are button-shaped. A tektite much richer in silica (86–90%) is reported from mining areas in western Tasmania. This is poorer in alumina (8–6%). The colour ranges from colourless to black through olive-green. The SG varies from 1.85 to 2.30 and the name darwinite or queenstownite is taken from the Jukes-Darwin mining field near Queenstown, Tasmania. Moldavite in Czechoslovakia can be divided into three groups. Those from the Radomilice area of southern Bohemia are spherical or oval, not very flattened; in these the bubbles are infrequent and the colour pale or bottle green. Of all the Czech moldavites these are richest in silica but have a relatively low aluminium and iron content. There is hardly any lechatelierite (closely approaching silica glass in composition) present. From other areas of southern Bohemia the colour of the moldavite is mostly bottle green; the stones are more flattened and contain profuse bubbles. They have a high lechatelierite content. The silica, aluminium and iron content is intermediate between the first and third types. The other group comes from Moravia and brown colours dominate; olive green is also found. Bubbles are infrequent and specimens are spherical, richer in aluminium and iron than the other types. Calcium and magnesium oxides are low. Four occurrences are recorded: from Upper Miocene sediments; from Pliocene and Pleistocene sediments; from slope lavas probably from the Pleistocene; and from Holocene or Pleistocene alluvia. A recent report (1985) shows that moldavite from Chlum nad Maisi in southern Bohemia, Czechoslovakia, is worth using in jewellery and a fashion has sprung up in that country.

Silica-glass

This glass may contain up to 98% silica. It was reported from the Libyan Desert in 1932 in lumps up to 5 kg, pale yellowish-green in colour. The RI is 1.462 and the SG 2.21. The hardness is 6.

NEPHELINE

$(Na, K)AlSiO_4$. Hexagonal, forming stumpy crystals and masses. H 5.5–6. SG 2.55–2.66. RI (ω) 1.529–1.546 (ϵ) 1.526–1.542. Uniaxial negative. DR 0.004. Some German material shows medium light-blue fluorescence and some Ontario material dull orange in long-wave UV.

Nepheline is found in plutonic and volcanic rocks and in pegmatites associated with nepheline syenites. Most material comes from Canada. A variety called elaeolite is red, green, brown or grey and may show some kind of chatoyancy from inclusions. In general nepheline ranges from colourless through grey, yellow and green to a brickish-red. Named from a Latin word for cloud, referring to the mineral turning cloudy when immersed in acid. Elaeolite is from a Latin word meaning oil and refers to its greasy lustre.

NEPHRITE

$Ca_2(Mg, Fe)_5(Si_4O_{11})_2(OH)_2$. Monoclinic, the fibrous variety of actinolite (q.v.). Nephrite is extremely tough, being made up of masses of fibrous crystals densely matted together. Colours are quiet but varied. H 6–6.5. SG usually near 2.95. RI close to 1.62 but for the individual crystals it is (α) 1.600–1.627 (γ) 1.614–1.641 with DR 0.027. Biaxial negative. The strong pleochroism is not obvious due to the fibrous nature of nephrite. There is no luminescence but there are absorption bands at 498 and 640 nm with a doublet at 689 and a sharp line at 509 nm.

The colours of nephrite which, with jadeite, is one of the two minerals which may be called jade, are quieter than those of jadeite. They range from a creamy-beige, traditionally known as mutton-fat jade, through green (from iron) to brown (from oxidized iron). There may be a dark brown skin or rind. Yellowish-green and black are also known. Nephrite occurs as boulders with a rounded outline. Most Chinese jade artefacts before about the eighteenth century are nephrite and there are many examples of nephrite with a brown staining attributable to burial. No deposit of nephrite has yet been reported from China itself, most nephrite used there coming from the south and west of Sinkiang in East Turkestan where the areas of Kashgar, Yarkand and Khotan produced alluvial boulders. In the nineteenth century nephrite was discovered in Siberia near to the south-west end of Lake Baikal. Large boulders were discovered by the explorer Alibert in rivers south of Irkutsk (large ones are in a number of museums, including the British Museum (Natural History) which weighs 524.5 kg). Most of this jade is a fine dark green with black graphitic spots.

There are several nephrite locations in North America; fine green boulders and pebbles were found in Wyoming in 1936. Nephrite is a product of metamorphism and occurs in schists, gneisses, serpentines and metamorphosed limestones but is so much more durable than the enclosing rocks that it is rare to find it *in situ*. The point at which pebbles and boulders are found may be long distances away from its place of formation. Wyoming material has undergone much surface alteration and it is only possible to discover the true value of the nephrite when the boulder has been sliced. The main occurrences are in south-central Wyoming where pebbles and boulders are retrieved from desert areas; there are few in place exposures. In some of the exposures that have been located, however, an association with quartz has been noted, some of the quartz occurring as inclusions in the nephrite and some being replaced by nephrite. Fine green nephrite, some appearing chatoyant, is found in Alaska and in California the Cape San Martin region on the Pacific coast in Monterey County produces pebbles on the beach and *in situ* material in grey schist outcrops where there are intruded peridotite rocks altered to serpentine. The nephrite occurs as lenses and nodules next to the serpentine. The colour of the nephrite is a greyish-green and in specimens from some other localities in the vicinity a silvery reflection from minute inclusions makes identification easy. Alaskan nephrite locations cluster round the Kobuk

River watershed; the chatoyant material has been found to be composed of very thin elongated crystals of tremolite–actinolite in a parallel position thus giving the eye effect. Some of this material does not show the matting (felting) characteristic of most nephrite and it has been suggested that cat's-eye tremolite would be a more appropriate name, tremolite forming a series with actinolite and having the composition $Ca_2(Mg, Fe^{2+})_5Si_8O_{22}(OH)_2$. Fine nephrite is found in British Columbia, Canada, being found along the gravel bars of the drainage region of the Fraser River in the south of the province. A large industry is based on these finds. Quite a lot of the nephrite is found *in situ* with serpentines. Some creamy-white nephrite is found in Poland, some of it showing patches of green; location is in the Jordansmuhl area. Dark green nephrite is also found in Taiwan, New South Wales, Australia and in Zimbabwe. A fine green nephrite is found in New Zealand and has long been used by the Maoris for ornamental and technical purposes. It is found *in situ* in parts of the Southern Alps but more commonly is found as alluvial boulders and pebbles in river valleys of north Westland, South Island.

Nephrite is named from the Greek word for kidney, referring to the rounded almost kidney shape of the boulders.

NICCOLITE

NiAs. Hexagonal, usually massive, crystals rare. H 5–5.5. SG 7.78.

Peach-red cabochons of niccolite have been cut from masses which are found as vein deposits in basic igneous rocks. Large masses are found in the Sudbury area of Ontario, Canada. The metallic lustre is distinctive. Named from the Latin word for nickel.

ODONTOLITE

Odontolite can be a fine blue, close to that of good-quality turquoise which it may simulate. It consists of the fossilized bones or teeth of extinct animals but the organic material is replaced by minerals. Constants vary but may show an SG characteristic of apatite which is the major replacing mineral in many specimens. The minerals calcite and vivianite (ferrous phosphate) are also found, the latter giving at least some of the colour. Material of a good blue will have an SG near 3.00 and RI 1.57–1.63. The hardness is close to 5. The best odontolite is found at Simmore, Departement Gers in southern France.

OLIVINE

Mg_2SiO_4—Fe_2SiO_4, a complete solid solution with end members forsterite and fayalite. The gem variety, peridot, has an ideal iron content of about 12–15%. Peridot with a six-rayed star, formed by needles and biotite platelets has been reported. More than 15% iron gives a rather dark and muddy colour. Some very bright green stones with a trace of chromium are known and some stones contain manganese. Orthorhombic, forming striated prisms, rolled pebbles or nodules in volcanic areas. H 6.5 (fayalite) to 7 (forsterite); distinct cleavage parallel to the vertical axis of the crystal. SG of forsterites from a number of locations range from 3.22 to 3.45 (green Burmese material and yellow material from Kenya respectively; the Kenya material is $Fo_{90}Fa_{10}$). Most green peridot is in the range 3.3–3.4; brownish peridots are close to 3.5. RI (Burma) (α) 1.654 (β) 1.671 (γ) 1.689 with DR 0.036. (Kenya yellowish) (α) 1.650 (γ) 1.686 with DR 0.036. Pleochroism is generally weak and the iron content inhibits luminescence. Absorption spectrum shows three main bands due to iron at 493, 473 and 453 nm with weaker bands at 653 and 529 nm. Characteristic inclusions are glass blebs in Hawaiian stones, chromium-spinel crystals in some stones from the south-western United States and petal-like liquid inclusions surrounding chromium-spinel crystals; this last type of inclusion is traditionally known as a 'lotus leaf'. Biotite grains are also reported. A virtually colourless olivine from Sri Lanka had the very low RI of (α) 1.640 (γ) 1.675 with an SG of 3.20. FeO content was low at 3.6% by weight. A yellowish stone from the Eifel region of West Germany is reported to have an RI of (α) 1.655 (β) 1.683 (γ) 1.690 with a DR of 0.035 and an SG of 3.34.

Forsterite is named for the mineralogist J. Forster, fayalite for the Fayal Islands in the Azores. Olivine is named from the Latin for olive and peridot from a French word of uncertain origin.

Olivine occurs in a variety of environments. Forsterite is found in magnesian limestones altered by igneous intrusion; fayalite is quite rare, intermediate compositions are found as a main constituent of basic igneous rocks; gem mining takes place in basic and ultrabasic rocks. In Burma fine peridot is found north of Mogok on the northern slope of Kyaukpon. It occurs as bright green euhedral crystals loose in a weathered serpentine. Burmese stones have a 'sleepy' appearance due to a profusion of crystallites. Probably the world's finest peridot comes from the island of Zabargad (various Romanizations are possible) sometimes known as the Island of St John situated about 60 miles south-east of the Râs Banâs peninsula in the Red Sea. The island is small, only 3.2 km long and 2.4 km wide. It and its peridot occurrences are described by Gübelin (1981). The rocks are derived from magmatic activity with associated metamorphism of existing sediments, all being later exposed through tectonic uplift and erosion. Mafic igneous rocks of deep-seated origin form the bulk of the island. These have low silica content and are mostly lherzolites with abundant olivine, pyroxene and amphibole. The metamorphic rocks consist of serpentines, granulites, schists and slates. Of more recent origin are alluvial sediments and a large deposit of gypsum. Peridot has been found wherever the peridotites outcrop; the largest crystals were found on the eastern slopes of a feature known as Peridot Hill and occurred in vein-like areas of the serpentinized peridotite. It is presumed that the peridot crystals originally lined open cracks and fissures and were broken from their sites by tectonic movements; Gübelin feels that this could be true only if the transformation of the peridotite into serpentinite occurred metasomatically and before or during the formation of the peridot crystals, otherwise they would have been serpentinized too. Many of the crystals show no signs of this; they are notably fresh or only slightly etched so that they could easily have developed at the same time or later.

Mining was originally manual excavation of individual veins, but methods were improved in the years before the first world war; in a four-year period more than US$2 million worth (at today's values) was recovered. The mines have been nationalized since 1958 and it appears that little or no mining is currently taking place, perhaps for military reasons.

With the cessation of peridot from Red Sea deposits, those at San Carlos in Arizona, USA have gained in importance. The San Carlos Apache Reservation is in Gila County, Arizona, and the main deposit of peridot is known as Peridot Mesa. This is capped with basalt which is fine-grained and vesicular, perhaps extruded as the result of a single geologic event. Peridot occurs as spherical, ovoid, semi-angular masses in the vesicular basalt, the masses ranging from 1 cm or less to 30 cm or more in the longest dimension. These masses are composed of granular olivine with most of the peridot crystals no larger than sand grains so that faceted stones from this area are small; colour ranges from a brownish-green to a bright lime green with less iron. Chromite and chrome spinel inclusions are found together with negative crystals, 'lotus leaves' and glass blebs. Crystals of chrome diopside and biotite are also reported. A smoke-like veiling, perhaps due to some kind of solid solution unmixing occurring as the peridot comes to the surface, is also noted in a major paper by Koivula (1981). Exclusive rights to the mining are held by various Apache families and it is not always easy to assign a particular stone to a particular deposit in an area in which there are many neighbouring ones. It is interesting to note that the San Carlos peridot, like some others, contains a trace of nickel which may play a part in its colouration.

Peridot of gem quality comes from Ameklovdalen, Sondmore, Norway; stones are a bright yellowish-green and are low in iron. Green and brown grains from which stones can be cut are found in the state of Chihuahua in northern Mexico.

Peridot is found in China in the Zhangjikou-Xuanhua area of Hebei Province. There is no directional hardness like that shown by some Arizona material; SG is 3.36, RI (α) 1.653 (β) 1.670 (γ) 1.689 with DR 0.036. Hardness is near to 7. Chromite crystals and cleavage discs like lotus petals are included.

OPAL

$SiO_2 + nH_2O$; water content usually 6–10% in precious opal. Amorphous and formed of minute spherical particles, constituting a gel; opal takes a variety of habits, including stalactitic, reniform, botryoidal; lustre tends to waxy. Common opal (i.e. without a play of colour) may be white, yellow and various pale to dark colours; a nickel-bearing variety is an apple-green. H 5.5–6.5. SG (black and white opal) 2.10; orange–red varieties 2.00. RI 1.44–1.47; Mexican opal usually 1.42–1.43 though 1.37 is recorded. Some opal gives a green fluorescence from included uranium minerals. An Australian report classifies opal into three types:

• Near-amorphous (most precious opal and its associated potch).
• Opal with a structure based on a disordered interlaying of cristobalite and tridymite. (Opal associated with volcanic environments and almost all common opal.)
• Opal with a structure of low-cristobalite (rare and usually associated with volcanic environments).

A cat's-eye opal with an orange–brown body colour showed the presence of cristobalite by X-ray diffraction; the pattern for cristobalite was superimposed upon an amorphous background. A slightly radioactive, strongly fluorescent yellow-green cut opal, reportedly from Mexico, gave an absorption spectrum like that recorded for uranium glass. Bands were observed at 495, 460 and 430 nm (the latter a cut-off). SG and RI were 2.18 and 1.455, respectively. A pink opal shown to the Santa Monica laboratory of the Gemological Institute of America was found to consist of palygorskite (a clay mineral often found in this type of opal) and cristobalite with the opal. Testing was by X-ray diffraction.

The play of spectral colour in opal is a diffraction effect produced from a three-dimensional array of silica spheres and voids which have to be regularly stacked for diffraction to take place; irregular stacking is found in common opal. The size of the spheres ranges from 150 to 400 nm and it is upon the size that the particular colour seen depends. Opal may be white or black, according to whether the background to the play of colour is dark or light. Fire opal is orange–red and may or may not show a play of colour. Water opal has a play of colour in a limpid colourless host. The varieties named above are the most important ones in opal, the multiplicity of names referring in the end to one or another of these types. A few related substances can be mentioned; siliceous sinter is a massive glassy opal found near hot springs; pseudomorphs after organic materials such as wood, shell and fossils are often shown by opal and can be very beautiful. Hyalite is transparent colourless or grey opal which is sometimes faceted. Hydrophane is opaque and light in colour but when immersed in water becomes transparent with a play of colour (it is vital to ensure that any water in which opal is immersed is absolutely clean since opal is porous). An interesting variety which is so porous that it sticks to the tongue is the bluish-white cacholong. Tabasheer is found in the joints of bamboo. Girasol is almost transparent and shows a moonstone-like effect. Contra luz opal is a beautiful variety of opal from Mexico; it resembles water opal in its play of colour set in a water-like host but in this case the colours can also be seen by transmitted light as well as by the more customary reflected light. This variety and fire opal are often faceted but the white and black stones are almost invariably cut as cabochons.

Opal occurs in sedimentary rocks such as sandstones or in other locations where low-temperature silica-bearing solutions can percolate through rocks.

The only opal known in classicial times was the so-called 'Hungarian' opal. This was found as seams in greyish-brown andesite near the village of Czerwenitza (now in Czechoslovakia). The play of colour is strong against a milky-white background.

The finest opal comes from various sites in Australia, where opal was discovered in about 1872 in Queensland. Boulder opal is hardened sandy clay with opal layers sandwiched within the clay layers; Yowah nuts are walnut-sized concretions found in layers with the kernel being the opal (it does not show on the outside so all specimens have to be cut open in some way). Seam opal describes opal of differing thicknesses found as seams in a sandstone matrix (usually smaller stones).

The finest black opal is found at Lightning Ridge in New South Wales; this occurrence was first exploited in about 1905. At first the opal, which occurs in nodules, was disregarded by European dealers but soon became greatly sought after, as it still is. Fine white opal is found at Coober Pedy in South Australia in a sandstone or clay matrix; this field, still in production, was discovered in about 1915. Also in South Australia is the Andamooka field where the opal is also white but may be brownish in colour. Some of this opal is artificially darkened to give the appearance of black opal.

Mexican opal is found in siliceous volcanic lavas, in cavities and in a number of places. Yellow and red fire opal is found at Zimapan in the state of Hidalgo; the rock is a trachyte porphyry. Water opal is found at San Luis Potosi in the state of Chihuahua and opal also occurs in the state of Queretaro in a number of places. The finest Mexican opal comes from the Queretaro area 200 km north-west of Mexico City; although the opal has been known since the eighteenth century, deposits were not developed until the end of the nineteenth century. Mining is in open-pit quarries, the opal occurring in a series of thinly bedded rhyolite lava flows which contain many gas cavities. The opal is found as a secondary filling in the cavities as well as filling any other free space in the lava. It has lower constants than most opal from Australia with an RI in the range 1.42–1.43 and SG about 2.00. It is particularly transparent and shows a good deal of red and green.

Some bright orange opal without a play of colour is found. A variety of mineral inclusions are found in the opal.

Opal is found at a number of places in Honduras where it occurs as veins in a dark reddish to black trachyte. Opal is also found in Indonesia in thin seams in a dark rock. Water opal is a common variety and an interesting variety shows red patches of colour against a coffee-brown background which is quite translucent. Opal was found in the state of Piauí, Brazil in a sandstone; the opal shows bright but very small colour patches; it is notably insensitive to heat and will stand up to hard wear. The water content is low. The opal in Piauí state is only of economic value from one mine area, the Pedro II where there are a number of separate workings. The opal at the Boi Morto mine occurs in films and veins in a quartzitic sandstone and at other places in the area it is found eluvially and alluvially; alternatively it can be recovered from solid sandstone. It is thought that quartz–dolerite intrudes a country rock such as a sandstone; the intruded rocks form a sill which then decomposes to provide dissolved silica at the right conditions for spheres to form. Surface weathering then promotes decomposition, eliminating magnesium and most iron and calcium from the intrusive rock and leaving clays. Alternating wet and dry seasons may accelerate decomposition and vary the water-table and tectonic stability to allow the spheres to settle in a reservoir extending over a wide area (Jobbins, 1980). Much Brazilian opal has a whitish background with the play of colour appearing as small patches. This is quite attractive; some orange to orange–brown opal is reported but the colour is not as deep as the Mexican fire opal.

Fine opal pseudomorphs, in which the initial shape of wood is taken by a silica replacement, are found at Virgin Valley, Humboldt County, Nevada, USA. Discovered about 1900, the opal is of superb quality but is prone to cracking due to loss of some of the high water content. Idaho also produces similar material and a yellow transparent variety which is often fashioned. Some Idaho opal shows a fine play of colour and in certain cases the colour patches give an almost asteriated effect. A rose-coloured opal is reported from Idaho, USA with an SG of 1.95 and RI 1.44–1.45.

Opal with a play of colour has been reported from the Coromandel Valley in the North Island of New Zealand. The occurrence is in a rhyolite. The opals are structurally similar to those from other volcanic areas such as Indonesia; some are flaky. Three types are found: a jelly or 'crystal' opal (almost transparent) with a fairly good play of colour, a jelly opal with violet and a green flash, and a milky opal with a rolling flash in some directions only. The second type appears to have come from a quartzitic conglomerate and the third from a rhyolitic tuff. No rock was found when the first sample was examined.

A green opal from Tanzania examined by the GIA had SG 2.125, RI 1.452. Under long-wave UV the opal gave a bluish-white glow where there were whitish areas on the stone, the green part being inert. A medium reaction of the same kind took place under short-wave UV. The absorption spectrum showed general absorption in the red and a cut-off in the blue. The colouring element is nickel. Opal is found at several places in Indonesia and is thought to be of volcanic origin. Some stones show a play of colour, rich in red, against a coffee-brown background. Inclusions in Mexican opal are flow structure, goethite needles, acicular unnamed crystals, chalcedony and hornblende. Some Tanzanian yellow opal contains dendrites of manganese oxides.

Because much opal occurs as very thin seams, it is not always possible to fashion a solid stone from material which is otherwise of fine quality. It is possible, however, to cement the seams on to a backing such as obsidian or black onyx (the backing also enhances the colours). This is called an opal doublet. When the doublet is capped (for protection and to enhance the colours by making the cap serve as a lens) it is called a triplet. The best triplets have a rock crystal cap but glass and plastic are commonly used. Floating opal is a name given to a container (usually plastic) in which opal fragments are suspended in water or glycerine.

PAINITE

$CaZrBAl_9O_{18}$. Hexagonal, prismatic. Hardness just over 8. SG 4.01. RI (ω) 1.81 (ϵ) 1.78 with DR 0.028. Uniaxial negative. Dichroism pale brownish-orange and deep ruby red. Chromium absorption spectrum.

Only three specimens of painite are known at the time of writing in late 1984. The original crystal was sent by Mr A. C. D. Pain (for whom the stone is named) from Burma to the London Chamber of Commerce laboratory in 1952. Further tests at the British Museum (Natural History) established that the species was new to science. Mr Pain, who had long worked in Upper Burma, sent a number of fine stones to London over a long period and his collection, known by his name, is now displayed in the British Museum. The miner who found the first crystal soon found another and this too went via Mr Pain to the Museum. In 1979 the Gemological Institute of America reported the discovery of a third crystal from a box of supposed red spinels from Mogok. The weight of the first crystal was (before testing) 1.6897 g; that of the third 0.34 g.

PECTOLITE

$NaCa_2Si_3O_8(OH)$ with some manganese, forming a series through manganpectolite to serandite. Triclinic, forming acicular crystals or radial, globular masses; perfect cleavage in one direction. H 4.5–5. SG 2.74–2.88.

RI (α) 1.595–1.610 (β) 1.605–1.615 (γ) 1.632–1.645. Biaxial positive. DR 0.036. Orange–pink in long-wave UV (material from Bergen Hill, New Jersey, USA); specimens from other areas may show cream or green as well as orange in short-wave UV. Faint yellow phosphorescence shown by specimens from Paterson, New Jersey.

Pectolite occurs in cavities in basaltic rocks in association with zeolites; a blue–green variety from Alaska resembles jade and fine prismatic crystals, often twinned and some of them facetable, occur at the Thetford mines, Quebec and at Asbestos in the same province of Canada. The finest material, blue mixed with white, at its best dark and translucent, comes from the Dominican Republic where it is locally called larimar. Named from a Greek word meaning congealed, referring to the mineral's translucency.

PERICLASE

MgO. Cubic, forming octahedral crystals with a perfect cleavage in one direction. H 5.5. SG 3.56. RI 1.736. May show pale yellow fluorescence in long-wave UV.

Found in marbles from high-temperature contact metamorphism. Most faceted stones are synthetic and colourless; rare natural material from Vesuvius, Italy; Texas and New Mexico, USA. Named from its cleavage.

PERIDOT see OLIVINE

PETALITE

$LiAlSi_4O_{10}$. Monoclinic, usually found as masses. Hardness just over 6. Perfect cleavage in one direction. SG 2.3–2.5. RI (α) 1.503–1.510 (β) 1.510–1.521 (γ) 1.516–1.523, with a DR of 0.012–0.014. Biaxial positive. Absorption band at 454 nm recorded for one stone which also showed weak orange fluorescence; the same stone showed bright orange with persistent phosphorescence under X-rays.

Petalite is a characteristic glassy colourless gemstone with only a few specimens reaching a pale yellow. It occurs in granite pegmatites, most transparent specimens coming from Brazil. Colourless and light yellow petalite from Brazil may show slight chatoyancy but a mixture of petalite and analcime, pinkish-brown in colour, shows a much stronger eye. The material comes from Zimbabwe and has SG 2.39 and RI (α) 1.505 (γ) 1.517 with a DR of 0.012. GIA reported in 1986 that a chatoyant petalite of a pink colour had been found in South Africa. SG is approximately 2.34 and RI approximately 1.51. The hardness is about 6.5. The mineral takes its name from the Greek word for leaf, referring to its cleavage.

PHENAKITE

Be_2SiO_4. Trigonal, giving prismatic, acicular or rhombohedral crystals. H 7.5–8. SG 2.93–2.97. RI (ω) 1.654 (ϵ) 1.670 with a DR of 0.016. Uniaxial positive. Dispersion 0.005. May show pale greenish fluorescence under UV but under X-rays a blue colour is reported.

Phenakite takes its name from a Greek word meaning 'to cheat' since it was often mistaken for quartz in early times. Almost all specimens are colourless and somewhat glassy, although wine-yellow crystals have been reported from the USSR and a cut stone coloured greenish-blue with SG of 3.00 and RI 1.654–1.670 was reported in 1963. This stone gave a light blue fluorescence under UV and had pronounced dichroism with peacock-blue and violet–red colours. A vanadium-doped stone in the author's collection also appears a fine peacock-blue and was grown at Bell Laboratories, New Jersey. Phenakite occurs most frequently in pegmatites and mica schists; fine crystals are found with emerald in the area of the Takovaya River north-east of Sverdlovsk, USSR and others (some reported as pale red) from São Miguel do Piracicaba, Minas Gerais, Brazil. Other clear crystals have been found at the Kiswaiti Mountains, Tanzania. Phenakite of gem quality occurs in the Habachtal, Austria; it varies from colourless to yellowish or orange–brown and fluoresces a dark red to light green. SG is 2.95–2.98, RI (ω) 1.654 and (ϵ) 1.670 with a DR of 0.016.

PHOSGENITE

$Pb_2CO_3Cl_2$. Tetragonal, forming thick prismatic crystals with a distinct cleavage in one direction. H 2–3. SG 6.13. RI (ω) 2.114–2.118 (ϵ) 2.140–2.145. Uniaxial positive. DR 0.028. Strong yellowish fluorescence in UV and X-rays. Very weak red–green pleochroism may be seen in some thick crystals.

Phosgenite may be a fine yellow–brown or pale shades of green, pink, yellow or white. Some colourless material may be faceted but most phosgenite is cut as cabochons. The finest yellow–brown stones come from Sardinia where they occur in the lead ore deposits at Monte Poni. The name refers to phosgene (carbonyl chloride); phosgenite contains the same elements.

PHOSPHOPHYLLITE

$Zn_2(Fe^{2+}, Mn)(PO_4)_2 \cdot 4H_2O$. Monoclinic, forming well developed prismatic or tabular crystals. Perfect cleavage in one direction. H 3–3.5. SG 3.08–3.13. RI (α) 1.595–1.599 (β) 1.614–1.616 (γ) 1.616–1.621. Biaxial negative. DR 0.021–0.033. Violet fluorescence in short-wave UV.

Crystals from Potosí, Bolivia are a very fine blue–green and occur in massive sulphide deposits. It is also found in granite pegmatites at Hagendorf, Germany. The name is a compound of two Greek words meaning phosphoric and possessing cleavage. Gemstones as such are rare but the occasional cabochon may be encountered in large collections.

POLLUCITE

$(Cs, Na)_2Al_2Si_4O_{12} \cdot H_2O$, forming a series with analcime. Cubic, forming rare cubic crystals. H 6.5–7.5. SG 2.85–2.94. RI 1.518–1.525. Orange to pink fluorescence in UV and X-rays.

Pollucite occurs in granite pegmatites, the colourless crystals occasionally providing areas large enough for faceting. Most of these crystals come from Newry, Maine, USA. Named from the Greek Pollux, brother of Castor and of Helen of Troy. Pollucite, first discovered in Italy in 1846, was originally called pollux and was associated with another mineral then called castor; this latter mineral was later renamed petalite.

PREHNITE

$Ca_2Al_2Si_3O_{10}(OH)_2$. Orthorhombic, occurring most commonly in masses; sometimes stalactitic. There is a distinct cleavage in one direction. H 6–6.5. SG 2.88–2.94. RI (α) 1.611–1.632 (β) 1.615–1.642 (γ) 1.632–1.665. Biaxial positive. DR 0.021–0.033. Some specimens may show faint brownish-yellow fluorescence under UV and X-rays.

The pale green, blue–green and yellow prehnites make fine cabochons and some of the yellowish Australian material has been faceted. A chatoyant prehnite is reported from the Prospect quarries near Sydney, New South Wales, Australia. The mineral occurs as botryoidal masses and cavity fillings in a small complex analcite dolerite sill intruded between shales and sandstones. The mineral is found in basaltic rocks with zeolite minerals, having been deposited from ground-water at low temperatures; it is also found in serpentine rocks and acid igneous rocks. Named for Colonel Prehn who found the mineral on the Cape of Good Hope. Fine green material comes from the Fairfax Quarry, Centreville, Virginia, USA and there is an occurrence at Barrhead, Renfrewshire and other places in Scotland.

PROSOPITE

$CaAl_2(F, OH)_8$. Monoclinic, forming crystals and granular masses. H 4.5. SG 2.88. RI (α) 1.501 (β) 1.503 (γ) 1.510; massive material usually gives a reading near 1.50.

Prosopite may be confused with turquoise (though this has an RI of 1.62). Found at Santa Rosa, Zacatecas, Mexico in association with azurite. Named from its deceptive appearance (with reference to its propensity to form pseudomorphs).

PROUSTITE

Ag_3AsS_3. Trigonal, forming rhombohedral or prismatic crystals with a distinct cleavage in one direction. H 2–2.5. SG 5.57–5.64. RI (ω) 3.088 (ϵ) 2.792. Uniaxial negative. DR 0.296. Strong pleochroism in shades of red.

Superb dark red colour darkens on exposure to light (but surface alteration can be removed). The finest crystals come from the Dolores Mine, Chanarcillo, Chile; facetable material occurs here and near Freiberg in Saxony, East Germany. Proustite, which is named for the French chemist J. L. Proust, is a primary silver mineral found in low-temperature ore deposits or in the upper portions of vein deposits.

PUMPELLYITE

$Ca_2MgAl_2(SiO_4)(Si_2O_7)(OH)_2 \cdot H_2O$. Forming series with ferropumpellyite and with julgoldite. Monoclinic, often in fibrous masses. Distinct cleavage in two directions. H 5–6. SG 3.1–3.5 (chlorastrolite). RI (α) 1.674–1.702 (β) 1.675–1.715 (γ) 1.688–1.722. Mean RI 1.70. Biaxial positive and negative. DR moderate. Distinct pleochroism colourless through pale green to brownish-yellow.

The variety of pumpellyite called chlorastrolite has been cut into cabochons which are sometimes a fine deep green. Others have a green and white patterning resulting from mixtures with other minerals. The pattern resembling a turtle shell is most popular and the best chlorastrolite which can be a fine emerald or even Imperial jade green is rare. The mineral occurs in a number of different environments, the best gem material being found in the Lake Superior area of the USA. Named for the Michigan geologist Raphael Pumpelly.

PURPURITE

$(Mn, Fe)PO_4$, forming series to heterosite. Orthorhombic, usually as masses with a good cleavage in one direction. H 4–4.5. SG 3.69. RI (α) 1.85 (β) 1.86 (γ) 1.92. Biaxial positive. DR 0.007. Strong pleochroism, grey through rose to purple.

Cabochons of purpurite can be a very fine deep purple; material occurs from the oxidation of phosphates in granite pegmatites, the best quality coming from Usakos, Namibia. Named from a Latin word for purple.

PYRITE

FeS_2 dimorphous with marcasite. Cubic, occurring commonly as cubes, although a form known as a pyritohedron is especially common; in this form there are four triad axes but no diad axes. The principal axes at right angles to the cube faces are no longer tetrad but diad. The striations on adjacent faces are thus at right angles to each other. Metallic lustre. H 6–6.5. SG 5.0–5.3. Greenish-black streak.

Pyrite (often called pyrites or iron pyrites) is found in nearly all rock types and is the commonest of all the sulphides. It is cut in the rose style with a flat back; cutting can be difficult as pyrite is brittle and heat-sensitive. The marcasite of the jeweller is usually pyrite since this is more stable than true marcasite. Found world-wide. Named from the Greek word for fire, alluding to pyrite's property of emitting sparks when struck.

PYROPHYLLITE

$Al_2Si_4O_{10}(OH)_2$. Monoclinic, forming tabular crystals or foliated masses with a perfect cleavage in one direction and sectile. H 1–2. SG 2.65–2.90. RI (α) 1.534–1.556 (β) 1.586–1.589 (γ) 1.596–1.601. Biaxial negative. DR 0.050. The variety agalmatolite may fluoresce a weak cream to white in long-wave UV (Chinese material).

Pyrophyllite is found in schists or in hydrothermal veins with quartz. It is very like talc with a soapy feel and colours ranging from brownish-green to white. South African wonderstone (or koranna stone) is dark grey and comes from the Transvaal; its SG is 2.72, RI is near 1.58, about 86% pyrophyllite. Carvings have been made from North Carolina material. Named from Greek words for fire and leaf, referring to the sheet-like appearance and thermal properties of pyrophyllite. Agalmatolite means figure stone and refers to its use in carvings.

PYROXMANGITE

(Mn, Fe)SiO_3. Triclinic, forming rare tabular crystals with a perfect cleavage in two directions. H 5.5–6. SG 3.61–3.80. RI (α) 1.726–1.748 (β)

1.728–1.750 (γ) 1.744–1.764. Biaxial positive. DR 0.016–0.020. Pleochroic in shades of red and pink.

Fine red crystals of pyroxmangite have been faceted, usually from material found in Honshu, Japan. The mineral is found in metamorphic rocks rich in manganese. Fine crystals are also found at Broken Hill, New South Wales, Australia in association with rhodonite. Named for a manganiferous form of pyrites, once thought to be the mineral's composition.

QUARTZ

SiO_2, a member of the trigonal crystal system – prismatic crystals developed from a hexagonal lattice – characterized by the presence of one axis of three-fold symmetry with three axes of two-fold symmetry perpendicular to it and separated from each other by angles of 120°. There is no plane or centre of symmetry. Most crystals show a combination of prism, positive rhombohedron and negative rhombohedron, the positive one usually being better developed. Quartz crystals show enantiomorphism, i.e. there are right- and left-handed forms which are true mirror images and not superimposable. Quartz is twinned about a pair of prism faces, the two individual crystals interpenetrating and giving the appearance of hexagonal symmetry. Twinning about an intermediate pair of prism faces is also common and this kind of twinning gives rise to the alternating amethyst and colourless sectors observed in a crystal section cut at right angles to the trigonal axis.

Important examples of twinning in quartz include Dauphiné twinning in which the individuals are geometrically related by a rotation of 180° about the c-axis and are of the same hand. The crystal axes are parallel but piezoelectricity is reversed in the two parts. Dauphiné twins are almost always interpenetrant; twinning can be revealed by etching the crystal with acid. In Brazil twinning the parts are related by reflection over the form {1120} and are of opposite hand. Some amethyst has been found to be composed of regular polysynthetic Brazil twins as thin lamellae parallel to the terminal faces. In Japan law twinning, the best-known type of twinning in quartz, the two individuals have c-axes inclined at 84° 33', the composition plane being {1122}. Some examples have one individual amethyst and the other citrine or rock crystal.

The spiral or helical structure revealed by X-ray studies, allows quartz to rotate the plane of polarization of a beam of light traversing a section cut at right angles to the prism edge. The angular direction varies from right to left with the orientation of the path of the beam in accordance with the right- and left-handed nature of the crystal. The amount of rotation depends upon the thickness of the section and the wavelength. The lower wavelengths such as those in the UV region are rotated far more than those at the red end of the spectrum.

The rotational property of quartz gives a unique interference figure shown by a section cut at right angles to the prism edge. When viewed in convergent light between crossed polars a black cross with coloured rings at the centre is seen, quite like the normal uniaxial figure. With quartz, however, the arms of the cross do not pass within the first ring and the central area is coloured, the tint depending on the thickness of the section. If the analyser is rotated on its own the tint of the area will change and the order of the change will show the 'handedness' of the section; a right-handed section will show the colour passing from blue through purple to red on a right-hand rotation of the analyser; a left-handed section gives a similar effect on a left-handed rotation. The low symmetry of quartz allows pyroelectric effects and applying a mechanical stress causes the separation of electrical charges on the surface of the crystal. This effect (piezoelectricity) is the reason for the use of quartz crystals in electronics.

There are several clues to the character of quartz crystals: a positive rhombohedron will develop a brighter lustre than a negative rhombohedron; the positive rhombohedron will also show the small, flat three-sided pyramids known as vicinal faces which occur as rectangular pyramids on the prism faces. Growth hillocks are also common and give a cobbled appearance to a face (these are particularly noticed in synthetic quartz). Prism faces are almost always horizontally striated. In right-handed crystals the trigonal pyramid is situated in the upper right corner of the prism face below the positive rhombohedron; left-handed crystals have this face in the left corner of the same face.

Quartz shows marked internal zoning, the zones being arranged concentrically parallel to the external crystal faces. Sectioning a coloured crystal shows zonal colouration and X-ray irradiation turns different zones smoky in rock crystal. In amethyst the colour concentrates in the tips of the rhombohedra. Some crystals show alternate development of amethyst and citrine.

Quartz is the stable modification of crystallized silica at temperatures up to 870°C. At about 573°C the structure of quartz which is stable at ordinary temperatures (low quartz) (α-quartz) changes to high quartz (β-quartz). At 870°C, quartz passes into tridymite with a considerable change of volume and at 1470°C it changes to cristobalite. Quartz has a hardness of 7 and very regular constants, 2.651 for the SG and RI (ω) 1.544 (ϵ) 1.553. DR 0.009. Uniaxial positive. The dispersion (for rock crystal) is low at 0.013. Pleochroism is only strong in rose quartz where it is shades of pink. Quartz shows a range of luminescent colours which on the whole are not useful for identification; X-rays give a faint blue glow in rose quartz.

Rock crystal is colourless and transparent, although it may contain a variety of inclusions; red, golden or yellow rutile crystals (Venus' hair stone); green tourmaline crystals darkening to black; a variety of needle-like crystals with the generic name sagenite; mossy greenish chlorite; yellow to orange fragments and fibres of goethite; blood-red platelets of hematite; blue–green chrysocolla and dumortierite. Some chatoyant material is known though the eye is coarser than that shown by chrysoberyl; colours are yellowish-brown or pale green.

Quartz forms pseudomorphs after other minerals; a particularly attractive example being tiger's eye, the result of quartz replacing the blue asbestos mineral crocidolite after its decomposition with retention of its fibrous structure. The colour is from iron oxides (gold, yellow) – where the original blue colour remains it may be called hawk's-eye. Where blue and gold are found together the name zebra tiger-eye may be used.

Some vein quartz may show milkiness from minute cavities and bubbles filled with carbon dioxide or water. The name milky quartz is sometimes used. Smoky quartz is coloured by the effects of natural radioactivity (possibly from radioactive hydrothermal fluids) and is known as morion or cairngorm from Scottish localities; the smokyness can vary from faint yellowish-grey to a deep brown. The golden-yellow variety of transparent quartz is called citrine from the pale yellow to orange to deep golden orange to brown colours. Some is obtained by the heat-treatment of amethyst. The colour of citrine comes from ferric iron Fe^{3+}; Fe^{2+} gives a green coloured mineral, prasiolite. Amethyst, violet or purple coloured, is probably the most valuable variety of transparent quartz; the colour is produced by irradiation of iron-bearing quartz; iron in both the Fe^{2+} and Fe^{3+} states gives amethyst on irradiation, provided that growth has occurred in certain critical directions such as in the positive or negative rhombohedron. The colour is attributed to the operation of a colour centre. Heating produces a yellow colour if Fe^{3+} was present before irradiation; if Fe^{2+} was present the prasiolite colour develops. Most citrine in the trade is obtained by heating amethyst to about 450°C. Commercial amethyst has been traditionally called Siberian when the stone shows red flashes against a purple background. A pale lilac colour is known as Rose de France. Characteristic inclusions in amethyst are thumb print-like markings which may be due to twinning; the name rippled fracture has been used. Prismatic crystals, negative cavities and marked colour zoning are also seen. The name is taken from a Greek source meaning 'not drunken' as the stone was once attributed with powers to prevent intoxication.

Rose quartz, which may be nearly transparent but is more commonly translucent, probably owes its colour to a charge transfer between the Ti^{4+} state of titanium and the Fe^{2+} of iron, induced by irradiation. Work is still in progress on rose quartz; it is possible that a colour centre involving a phosphate impurity produces another type of rose quartz which is likely to fade. Rose quartz can show a six-rayed star which is best displayed by transmitted light (diasterism). Many occurrences are reported, some of the finest material coming from the Alto Feio in Rio Grande do Norte, Brazil and from places in the Jequitinhonha valley of the same country. Good quality material is also found in the Malagasy Republic and in the USSR. A multi-star quartz cabochon reported in *Gems and gemology* weighed 170 ct and showed several separate stars; the effect is thought to be caused by sillimanite.

Some of the finest amethyst comes from the Mursinska district of the Ural Mountains of the USSR. The crystals are found below the turf in drusy cavities of quartz veins crossing weathered granite. Other fine amethysts come from the state of Rio Grande do Sul, Brazil, the deposit spreading over the border with Uruguay. There are many other important sources. Stones of a dark violet–red are found in the Malagasy Republic at Ambatomanga, Ampangabe and Tongafeno, where they occur in pegmatites. Further good crystals come from Tsunagi, Echigo, Japan and some fine dark stones have come from Western Australia and from geodes in the Deccan trap-rocks of India.

Some of the best rock crystal comes from Herkimer County, New York, USA; stones from this deposit have an exceptional brightness and clarity which has caused them to be called Herkimer diamonds. Most commercial rock crystals comes from Brazil or the Malagasy Republic. Smoky quartz is an important mineral of the central Alpine regions of Switzerland where it occurs in cavities, sometimes as very large crystals. Colourless quartz crystals with a phosphorescent powder inserted into a cavity drilled from the base which was then capped by yellow metal bell caps are reported by GIA. The crystals are apparently sold for some supposed occult significance. Quartzite is the name given to a rock made up of tightly packed quartz grains formed at high temperature and pressure during metamorphism. When this rock contains light-reflecting crystals it is known as aventurine; the most pleasant colour is green aventurine from included crystals of the chrome mica, fuchsite. A faint chromium absorption spectrum may be seen. SG is usually 2.64–2.69. Green aventurine quartz from India has been found to be composed of a complex array of four crystalline materials. Interlocking, irregularly-shaped grains of quartz formed the bulk; fuchsite, rutile and almandine were associated.

Dumortierite quartz is a deep blue to violet and is made up of crystalline quartz coloured by the mineral dumortierite. A light blue star quartz is sometimes encountered, usually from Brazil. A transparent light green quartz was found in 1979 on the Nevada–California border in the United States. Crystals were discovered in pebbly talus below steep basalt-andesite cliffs. It is thought that the quartz was deposited after the cooling and microcrystallization of the basalt–andesite. During solidification of the lava, trapped gases passed upwards producing vesicles: as aqueous solutions passed through the porous rock they became enriched with silica which first crystallized out as chalcedony; later as crystallization became slower, crystalline quartz was formed. Amethyst was probably formed first as a result of low-grade radiation exposure; later metamorphism with a major intrusion of rhyolite may have forced the amethyst colour centre into an unstable state with alteration of colour to green or citrine. The green quartz absorbs at about 442 nm. A variety of quartz in which the amethyst and citrine components occur regularly and appear alternately in faceted stones has been called ametrine. It appears that this effect can be obtained artificially but natural material is reported from Bolivia.

Crypto-crystalline quartz usually takes its colour from impurities such as iron oxides and those of manganese, nickel, titanium or chromium. Some varieties are formed by deposition from percolating ground-waters which deposit silica over long periods, giving rise to banding in varieties such as agate. Other types are formed as gelatinous masses which slowly dehydrate and crystallize. Some agates are deposited within spherical cavities like gas pockets in basalt and this gives banding. The commonest type of crypto-crystalline quartz is made up of minute fibres of silica and is known as chalcedony. Chert stained various colours by oxides is known as jasper. Banded varieties or those with dendritic inclusions are classed as agates. Ordinary chalcedony is often greyish-blue unless dyed and is a compact form of silica. It occurs world-wide. An amethystine chalcedony is reported from central Arizona, USA. It occurs in a weathered sedimentary host rock. The RI is 1.54 and SG 2.61. Dark brownish-red spherules are scattered through it and the colour is described as reddish-purple. Heat treatment will alter some of the material to orange.

Cornelian (the name carnelian is also used) is red–orange or brownish-red and translucent to opaque. It is coloured by iron oxide. Many types of chalcedony can be turned red by heating since they contain disseminated iron compounds which are thus oxidized. Cornelian can be found on beaches in the east of England but most commercial material comes from Brazil. Sard resembles cornelian but is more opaque and tends more to a brownish colour; again Brazil and Uruguay supply commercial needs. Plasma is opaque and a deep green. It contains densely packed actinolite crystals; the colour may be due to chlorite, celadonite, microfibres of amphibole or other green silicates. A green to yellowish-green chalcedony is known as prase; dark green plasma with red (orange) spots from iron oxides is known as bloodstone or heliotrope. India and Brazil are commercial producers. Onyx is banded black and white chalcedony and sardonyx is also banded but with red and white colours. Chrysoprase is a fine bright apple green from nickel. The best material comes from the Marlborough mine in Queensland, Australia but Silesia, Poland also provides good quality specimens. It shows green through the Chelsea colour filter.

Agate displays layers which are concentric or flat bands. It may also contain dendritic (plant-like) inclusions and sometimes resembles landscapes. Moss agate contains mineral inclusions with a moss-like appearance and lace agates show intricate lacy patterns. Fire agate shows an attractive play of colour resulting from platy crystals of iron oxide layered with chalcedony. Turritella agate is made up of shells of the gastropod *Turritella*. All these agate varieties are distributed widely throughout the world, many coming from the United States. Fire and lace agate are found in Mexico, where they are handled commercially.

Jaspers are usually masses of minute silica crystals coloured by various impurities. Shades of brown, yellow, red and green are the commonest. The variety orbicular jasper contains spherules of banded agate in the jasper, and scenic (picture) jasper shows landscape-like patterning. Chrysocolla in quartz is a siliceous material consisting of fine particles of blue–green chrysocolla disseminated in silica. The material is hard and takes a good polish. Coloured varities of agate replacing (pseudomorphous after) wood are known as petrified wood. The structure of the wood can still be seen and colours are bright; there is a lot of this material in the United States. Dinosaur bone is what the name suggests though the bone is replaced by silica; it is an attractive patterned brown and is found in the western United States.

Chalcedony is named after a place, perhaps a seaport in Asia Minor; agate is named from the Greek name for a Sicilian river in which agate was found; onyx comes from a Greek word for nail or claw; sard is from Sardis, an early locality for the stone. Cornelian is taken from the Latin word for flesh with reference to its colour; plasma comes from a Greek word meaning something moulded; this refers to the use of the material in the manufacture of intaglios (engraved designs). Prase comes from a Greek word meaning leek and refers to the colour. Heliotrope is also from the Greek and derives from two words meaning sun and turn; this refers to the red reflection seen when the stone is turned to face the sun while immersed in water.

REALGAR

AsS. Monoclinic, forming striated prismatic crystals with a good cleavage in one direction. H 1.5–2. SG 3.56. RI (α) 2.538 (β) 2.684 (γ) 2.704. Biaxial negative. DR 0.166. Strong pleochroism, colourless, pale yellow, dark red. May be adversely affected by lengthy exposure to light.

Rare cut stones of realgar are a superb red but their softness and fragility militate against ornamental use. The mineral occurs in low-temperature hydrothermal vein deposits, with lead and silver ores. Fine gem quality crystals are found in the state of Washington, USA. Named from an Arabic word meaning powder of the mine.

RHODIZITE

$CsAl_4Be_4B_{11}O_{25}(OH)_4$. Cubic, forming dodecahedral or tetrahedral crystals or masses. H 8.5. SG 3.44. RI 1.694. Some material may glow a weak yellow in short-wave UV and a stronger yellowish-green, with phosphorescence, under X-rays.

Found in pegmatites, colourless to pale yellow crystals of rhodizite can be faceted, stones from Antandrokomby, Malagasy Republic, being the most common. Named from a Greek word meaning rose coloured, referring to the red colour given to the blowpipe flame. Rose-coloured material has been reported from the Sverdlovsk area of the USSR.

RHODOCHROSITE

$MnCO_3$ with some iron and calcium, forming a series to siderite (with iron substitution) and to calcite (with calcium substitution). Trigonal, forming rhombs and elongated crystals; some material occurs as scalenohedra and as banded masses. There is a perfect rhombohedral cleavage. H 3.5–4. SG 3.4–3.6. RI (ω) 1.786–1.840 (ϵ) 1.578–1.695. Uniaxial negative. DR 0.201–0.220. Absorption bands at 551 and 410 nm with weak lines at 565 and 535 nm. Will effervesce in warm acids. Medium pink fluorescence under long-wave UV for Michigan material, dull red to violet in short-wave UV, Argentine and Colorado material.

Rhodochrosite occurs as a gangue mineral in hydrothermal veins and as a secondary mineral in ore deposits. Fine material comes from Alma, Colorado and in recent years superb orange–pink crystals have come from the Hotazel and N'Chwaning mines in the Kalahari area of South Africa. Further fine rhodochrosite is found at San Luis, Catamarca Province, Argentina; some of this is stalactitic with a concentric structure which makes attractive ornaments when sectioned. Most massive rhodochrosite is banded pink and white. Named for the colour.

RHODONITE

$MnSiO_3$ (with some calcium up to a maximum of 20%). Triclinic, forming tabular crystals or masses. Perfect cleavage in one direction. H 5.5–6.5. SG 3.57–3.76, 3.67 in crystals. RI (mean) 1.73. Biaxial positive, altering to negative with the addition of impurities. Broad absorption band at 548 nm, strong narrow line at 503 nm with a more diffuse and weaker band at 455 nm.

Found in manganese ore bodies, rhodonite is difficult to facet on account of its cleavage; facetable material is limited to crystal edges found with galena, especially those crystals found at Broken Hill, New South Wales, Australia, where rhodonite occurs with pyroxmangite. Much massive rhodonite is veined with black manganese oxides; some good quality material of this kind has been found at Daghazeta, Tanzania; older sources include the Sverdlovsk area of the USSR (fine pink- or rose-coloured masses) and Honshu, Japan. Named from its colour.

RUBY see CORUNDUM

RUTILE

TiO_2, polymorphous with anatase and brookite. Tetragonal, forming prismatic vertically striated crystals and frequently groups of contact twins; distinct cleavage in one direction. H 6–6.5. SG 4.2–5.6 over the complete range of compositions which sometimes include niobium, tantalum or iron. RI (ω) 2.62 (ϵ) 2.90. Uniaxial positive. DR 0.287. Dispersion (of interest only in synthetic rutile) 0.280.

Rutile occurs in a variety of environments, including gneisses and schists, pegmatites, crystalline limestones and igneous rocks. Found in quartz as slim golden to brown needles (rutilated quartz); most natural rutile is dark red to dark brown and on this account is not commonly faceted. The synthetic rutile had a long vogue as a diamond simulant but its higher birefringence and softness make its continued use uneconomic compared with later simulants such as cubic zirconia. A dark brown cat's-eye rutile weighing 1.43 ct was examined by the GIA – the stone was natural and thought to be from Sri Lanka. Named for the Latin word for red.

SAPPHIRE see CORUNDUM

SARCOLITE

$(Ca, Na)_4Al_3(Al, Si)_3Si_6O_{24}$. Tetragonal, forming equant crystals. H 6. SG 2.92. RI (ω) 1.604–1.640 (ϵ) 1.615–1.657. Uniaxial positive. DR 0.011–0.017.

The only known locality for this pink to flesh-coloured stone is Monte Somma, Mount Vesuvius, Italy where it occurs in volcanic rock. All known faceted stones are small. Named from Greek words alluding to its colour.

SCAPOLITE

$Na_4Al_3Si_9O_{24}Cl$ (marialite); $Ca_4Al_6Si_6O_{24}(CO_3, SO_4)$ (meionite). Gem scapolite forms part of the isomorphous series between meionite and marialite and belongs to the tetragonal system, in which it forms large prismatic crystals with a distinct cleavage in two directions. H 6. SG 2.50–2.74. RI (ϵ) 1.540–1.555 (ω) 1.546–1.600 over the whole series. Uniaxial negative. Pink and white scapolite may have an SG of 2.63 and RI 1.549 (ω) and 1.540 (ϵ), with a DR of 0.009. One violet stone is reported to have SG 2.634 and RI 1.560 (ω) and 1.544 (ϵ) with DR of 0.016. Clear yellow stones have SG in the region of 2.70 and RI (ω) 1.568 (ϵ) 1.548 with DR of 0.020. Yellow stones from pegmatites in Entre Rios, Mozambique have RI (ω) 1.554–1.556 and (ϵ) 1.540–1.541 with a DR of 0.015. Tanzanian yellow stones have SG 2.671 on average with RI 1.567 and 1.548, DR 0.019. Pleochroism is fairly well marked, with pink stones showing dark to light colours, as do violet stones; yellow stones show colourless to pale yellow. Pink and violet stones show absorption bands at 664 and 652 nm (from chromium), with strong absorption in the yellow. Burmese scapolites fluoresce yellow to orange in long-wave UV, pink in short-wave UV; stones from Tanzania may show strong yellow in long-wave UV. Massive material from Quebec, Canada, fluoresces a strong yellow with phosphorescence (long-wave UV) and some yellow stones have been found to show lilac in short-wave UV and a strong orange in X-rays.

Scapolite occurs in contact zones, altered basic igneous rocks and metamorphosed rocks. Fine yellow, pink and rare purple stones are found in Kenya and Tanzania; superb violet stones, pink and colourless material, with some pink and blue cat's-eyes come from Burma; yellow stones are found in the Malagasy Republic and from the state of Espirito Santo, Brazil. The deep yellow Quebec material contains sulphur and makes fine cabochons. A violet scapolite believed to be from East Africa was found to be 89.2% marialite (sodium) and showed a strong dichroism parallel to the c-axis (violet) to near-colourless at right angles to this direction. SG was 2.65. The material can be distinguished from amethyst by the latter being optically positive. A cat's-eye scapolite with a strong eye had RI of (ω) 1.583 (ϵ) 1.553 with a dark grey body colour. The composition was 69% meionite and the eye was caused by pyrrhotite crystals parallel to the c-axis. A very dark yellowish greenish-brown scapolite cat's-eye had an RI of 1.56 and an SG of approximately 2.70. Under short-wave UV there was a very faint red fluorescence. The stone was found, by X-ray diffraction analysis, to be the variety mizzonite. A reddish-brown cat's-eye scapolite is reported from Kenya; it has an RI of 1.57 and an SG of 2.73. It shows a strong red fluorescence in short-wave UV. It resembles some copper-red chatoyant beryl. Named from a Greek word alluding to the stumpiness of the crystals.

SCHEELITE

$CaWO_4$ with some molybdenum. Tetragonal, forming octahedrally-shaped or tabular crystals; also occurring as masses. Distinct cleavage in one direction. H 4.5–5. SG 5.9–6.1. RI (ω) 1.918–1.920 (ϵ) 1.934–1.937. Uniaxial positive. DR 0.016. Dispersion 0.038. May show faint didymium absorption lines in yellow and green. Bright sky-blue fluorescence in short-wave UV.

Scheelite is found in contact metamorphic deposits, in hydrothermal veins and in pegmatites and placer deposits. Fine facetable crystals have been found at Santa Cruz, Sonora, Mexico and in Arizona and California. Scheelite from the village of Colombage-Ara near Ratnapura in Sri Lanka, pale yellowish-blue or colourless crystals, have RI of 1.920–1.922 and 1.930–1.935, with DR 0.012–0.015. The SG is 5.94–6.30, the hardness is in the usual range for the material. Stones show a clear white or chalky-blue under short-wave UV; Californian stones fluoresce a very strong blue. Stones show a rare earth absorption spectrum. Some bright yellow stones could be taken for yellow diamond, but the birefringence gives them away; the colourless stones with an almost adamantine lustre have also been used as diamond simulants. Scheelite is synthesized by crystal pulling. Named for K. W. Scheele, a Swedish chemist who discovered the existence of tungsten in scheelite.

SCORODITE

$Fe^{3+}AsO_4 \cdot 2H_2O$, forming a series with mansfieldite in the variscite group. Orthorhombic, crystals pyramidal or tabular, prismatic. H 3.5–4. SG 3.28–3.29. RI [Tsumeb material] (α) 1.785 (β) 1.796 (γ) 1.812. Biaxial positive. DR 0.027. Tsumeb stones show intense purple to blue pleochroism, with an absorption line at 450 nm and broad absorption in the green.

Scorodite is most commonly yellowish to reddish-brown or blue–green but some Tsumeb (Namibia) material is a fine intense purple–blue. Scorodite is found as a secondary mineral resulting from the oxidation of arsenious ores. It is named for its garlic-like smell when heated, from the Greek.

SENARMONTITE

Sb_2O_3, isomorphous with valentinite. Cubic, forming octahedral crystals or occurring as masses. The colour is grey to white. The lustre is resinous. H 2–2.5. SG 5.5. RI 2.087. Senarmontite is formed by the alteration of stibnite, Sb_2S_3. It occurs in Inyo County, California and in Canada, France and Germany. Named for Henri de Senarmont, a French mineralogist. A stone reported in 1987 was believed to come from Brazil. It was colourless and had an RI of 2.08.

SEPIOLITE

$Mg_4Si_6O_{15}(OH)_2 \cdot 6H_2O$. Compact white and fibrous, mixed with an amorphous mineral of similar composition. H 2. SG 2. RI 1.53. Highly porous and will float on water.

The ornamental variety of sepiolite is meerschaum, a white material used, among other things, for pipe bowls. It is found near Eskisehir in Turkey and in various other European and North American locations. Named from a German word meaning sea foam in reference to its appearance.

SERANDITE

$Na(Mn, Ca)_2Si_3O_8(OH)$. Triclinic, forming prismatic crystals, often well formed and with a perfect cleavage in one direction. H 4.5–5. SG 3.32. RI (α) 1.660 (β) 1.664 (γ) 1.688. Biaxial positive. DR 0.028.

Rose-pink to salmon coloured crystals have been found in nepheline syenite rocks at Mont Saint Hilaire, Quebec, Canada and some of these crystals have recently been reported as facetable. Named for J. M. Serand a West African collector who was involved in finding serandite on the island of Rouma, Republic of Guinea.

SERPENTINE

A group of minerals with the general formula $A_3Si_2O_5(OH)_4$ where A is magnesium, iron (Fe^{2+}) or nickel. Forms part of the kaolinite-serpentine group. Monoclinic, usually forming masses of fibres (never crystals) with a perfect cleavage in one direction. H 2.5 (bowenite 4–6). SG for ornamental serpentines 2.44–2.62; bowenite is 2.58–2.62.

Serpentine is formed by the alteration of basic and ultrabasic rocks. Bowenite may be yellow, yellow–green, blue–green or dark green and is translucent; fine dark green material may be mistaken for nephrite (one variety from New Zealand has SG 2.67). The ornamental variety of antigorite serpentine is williamsite, which is apple-green and sometimes shows a weak green in long-wave UV. A chrome-green variety is found at Rock Springs, Maryland, USA and has SG 2.6–2.62. RI 1.56. H 4.5.

A mixture of the lizardite variety of serpentine and carbonates with greenish-white banding is the well-known Connemara marble which has an SG of 2.48–2.77 and mean RI 1.56. Various colours of lizardite are found at the Lizard peninsula, Cornwall, England. A green serpentine with calcite veins is known as Verd-antique and is found in Greece, Egypt and Vermont, USA. Pseudophite, sometimes incorrectly known as Styrian jade, is an aluminous serpentine with H 2.5, SG 2.69 and RI 1.57; it is found in Styria, Austria. A faceted antigorite from Pakistan was yellowish-green with RI 1.559–1.561 and DR 0.001–0.002.

Serpentine is named from the serpent-like markings seen in the marble varieties; antigorite from the Antigorio valley, Piedmont, Italy. Williamsite is named for its discoverer L. W. Williams and bowenite for G. T. Bowen, an early student of the material from Rhode Island, USA (although he wrongly identified it as nephrite).

SHATTUCKITE

$Cu_5(SiO_3)_4(OH)_2$. Orthorhombic, forming slim prismatic crystals or masses with a good cleavage in two directions. H about 5. SG 3.8–4.1. RI near 1.75. Found as an oxidation product of secondary copper minerals, shattuckite has been cut into cabochons of a dense blue. Sometimes the mineral is mixed with quartz which alters the properties. Planchéite, often confused with shattuckite, is a distinct species with the composition $Cu_8(Si_4O_{11})_2(OH)_4 \cdot nH_2O$.

Shattuckite is found at the Shattuck mine, Bisbee, Arizona, USA. Pseudomorphs after malachite are sometimes found.

SIDERITE

$FeCO_3$. Trigonal, usually massive but single crystals may be rhomb-shaped. Perfect rhombohedral cleavage. H 3.5–4.5. SG 3.83–3.96. RI (ϵ) 1.633 (ω) 1.873. Uniaxial negative. DR 0.240.

Some of the fine brown crystals from Panasqueira, Portugal, have been faceted and material from Ivigtut, Greenland may also be of ornamental value. Siderite is found widespread in sedimentary deposits and also occurs in hydrothermal ore veins and in pegmatites and basaltic rocks. Named from the Greek word for iron.

SILLIMANITE

Al_2SiO_5, polymorphous with kyanite and andalusite. Orthorhombic; rarely in prismatic crystals, more commonly in fibrous masses. Perfect cleavage in one direction. H 6.5–7.5. SG 3.23–3.27; 3.14–3.18 for compact varieties. RI (α) 1.654–1.661 (β) 1.658–1.662 (γ) 1.673–1.683. Biaxial positive. DR 0.020. Sometimes may show strong pleochroism, pale brown, pale yellow to green; greenish or brownish; dark brown or blue. Some blue Burma stones show weak reddish fluorescence.

Sillimanite occurs in metamorphic rocks, including schists and gneisses, also in granites. Material from Burma, pale blue and transparent, has been cut despite the cleavage and fibrous nature of the material; pale blue to colourless material has been found in Kenya and can be faceted; Sri Lanka produces violet–blue, green and blue material as does Burma. Some chatoyant material is known. Named after Benjamin Silliman, Yale University mineralogist; an alternative name, fibrolite, refers to the nature of the mineral.

SIMPSONITE

$Al_4Ta_3O_{13}(OH)$. Hexagonal, forming tabular or prismatic crystals. H 7–7.5. SG 5.92–6.84. RI (ϵ) 1.976 (ω) 2.034. Uniaxial negative. DR 0.058. Under short-wave UV may show bright blue-white, pale yellow or light blue fluorescence.

A very rare gemstone found in granite pegmatites. Only the bright orange–yellow material from Western Australia is of gem quality though if more specimens were found its hardness and lack of cleavage would make it commercially viable. Named for Dr E. S. Simpson, a government mineralogist of Western Australia.

SINHALITE

$MgAlBO_4$. Orthorhombic, found as pebbles, very few of them showing vague crystal form. H 6.5–7. SG 3.47–3.50. RI (α) 1.665–1.676 (β) 1.697 (γ) 1.705–1.712. Biaxial negative. DR 0.038. Distinct pleochroism, pale brown, greenish brown and dark brown. Absorption spectrum similar to that shown by peridot but with an additional band at 463 nm. The complete spectrum is thus; bands at 493, 475, 463 and 452 nm with general absorption of the violet.

The colour of sinhalite varies from a golden-green to a dark brown, somewhat reminiscent of chrysoberyl; it is found as a contact metamorphic mineral in limestones at granite contacts; most material is found alluvially. It was long thought to be a brown variety of peridot, the mistake being corrected in 1952. The major source is the gem gravels of Sri Lanka although at least one pebble is reported from Burma and pink to brownish-pink crystals with some gem-quality sections have been reported from a skarn in north-east Tanzania. A sinhalite from the Elahera district of Sri Lanka showed RI of (α) 1.669, (γ) 1.704 and DR of 0.035. The SG was about 3.485. Two-phase and negative crystals were included. Recent reports suggest that some green stones described as sinhalite and emanating from Sri Lanka may in fact be peridot. Compared to other green peridot, that from Sri Lanka shows a slightly higher amount of MnO (0.27%) and the RI has been measured at (α) 1.651 (β) 1.660 and (γ) 1.690 with a DR of 0.039 and SG of 3.36. Euhedral inclusions of spinel are reported as well as a smoke-like veiling. Field evidence suggests that the location may be between Pelmadulla and Kahawatta in the Ratnapura District. Interestingly, pebbles (therefore cut stones) tend to be more commonly found in large than in small sizes. Named from an old Sanskrit word for Ceylon.

SMITHSONITE

$ZnCO_3$. Trigonal, forming botryoidal masses and rare rhombohedral crystals with a perfect rhombohedral cleavage. H 4–4.5. SG 4.3–4.5. RI (ϵ) 1.621 (ω) 1.848. Uniaxial negative. DR 0.227. Effervesces in warm acids. May show whitish-blue fluorescence in short-wave UV (Japanese material) but in gem quality smithsonite luminescence is not an important effect.

Smithsonite occurs as a secondary mineral in the oxidized zone of ore deposits, the finest massive material coming from the Kelly mine, Socorro County, New Mexico, USA, where beautiful blue to blue–green material is obtained for carvings. Some crystals from Tsumeb, Namibia, sometimes a yellow but also pink and green, can be faceted. Some fine green smithsonite could be taken for good quality green jadeite and at one time it was marketed under the name Bonamite by the firm of Goodfriend Brothers. Some faceted stones of light colour may rival diamond in fire since they have a dispersion of 0.037. Named for James Smithson, founding benefactor of the Smithsonian Institution, Washington DC, USA.

SOAPSTONE

$Mg_3Si_4O_{10}(OH)_2$, it does not have mineral species status and is best described as an impure talc rock. Monoclinic and triclinic, usually massive with a perfect cleavage in one direction; sectile. H 1. SG 2.20–2.83. RI (α) 1.539–1.550 (β) 1.589–1.594 (γ) 1.589–1.600 (for monoclinic material). Triclinic material has (α) 1.545 (β) 1.584 (γ) 1.584. Biaxial negative. DR 0.050 (monoclinic) 0.039 (triclinic). In practice a shadow edge at about 1.54 would be seen.

Soapstone (or talc) with its characteristic greasy feel, is found in hydrothermally altered ultrabasic rocks and thermally altered siliceous dolomites. Occurrences are world-wide and often form extensive beds. The colours range from white through grey and green to a brownish-yellow. Named from a Greek word meaning fat (perhaps alluding to its colour); talc from an Arabic word for the mineral.

SODALITE

$Na_4Al_3(SiO_4)_3Cl$. Cubic, usually found as masses though some dodeca-hedral crystals are known. H 5.5–6. SG 2.28 for massive material. RI 1.483–1.487. Some sodalite fluoresces orange–red to violet.

Sodalite is a popular cutting material, even faceted stones being fashioned from specimens which are translucent at best. It occurs in nepheline syenites and is a light or dark blue; a sulphur-rich variety is called hackmanite and shows interesting luminescent effects in which it turns raspberry red from its original white in exposure to short-wave UV, then fades back to white. The finest sodalite comes from Ohopoho, Namibia, this material being translucent on occasion: other important sources include Brazil, Dungannon, Ontario, Canada (hackmanite), various places in British Columbia and in the United States. Sodalite is one of the minerals from which lapis-lazuli is composed. It is named for its composition; hackmanite is named for V. A. Hackman, a Finnish petrologist.

SPHALERITE

ZnS, often with some iron and dimorphous with wurtzite. Cubic, forming a variety of shapes, also massive. Perfect dodecahedral cleavage. H 3.5–4.5. SG 3.9–4.1. RI 2.37–2.43. Dispersion 0.156. Absorption bands attributed to cadmium may be seen at 690, 667 and 651 nm. Much material luminesces a bright orange–red in both long-wave and short-wave UV. Triboluminescence, luminescence from rubbing or striking, is well shown by sphalerite from Otavi, Namibia.

The main ore of zinc, sphalerite (alternatively named blende or zinc blende) is found in low-temperature ore deposits, particularly in limestones. Its almost adamantine lustre and high dispersion make it a spectacular though fragile gemstone; colourless stones may suggest diamond but most sphalerite ranges from red through yellow to green. The colourless to pale green material from Franklin, New Jersey, USA has been called cleiophane; fine reddish-orange cleavages come from Santander, Spain and green, zoned transparent material from Cananea, Sonora, Mexico. A deep green sphalerite has been reported from Zaire. The specimen had an SG of 4.18 (another had an SG of 3.98); the red was absorbed to 625 nm and there was a band in the yellow from 595 to 580 nm with absorption in the blue from 470 nm. The material occurs with copper-, gallium- and germanium-bearing minerals. The name is taken from a Greek word meaning treacherous, since the mineral was first thought to be the lead sulphide galena; when smelted, however, no lead was forthcoming. Blende is from a German word meaning to dazzle, referring to the high lustre.

SPHENE

$CaTiSiO_5$. Monoclinic, wedge-shaped and frequently twinned crystals for which the species is named. Weak cleavage in one direction. H 5.5. SG 3.52–3.54. RI (α) 1.843–1.950 (β) 1.870–2.034 (γ) 1.943–2.110. DR 0.105–0.135. Optic sign positive. Dispersion 0.051. Pleochroism greenish-yellow, reddish-yellow and straw, depending on stone colour. May show didymium absorption spectrum.

Although sphene (which has also been named titanite from its composition) is a typical mineral of Alpine granite cavities, most gem varieties come from pegmatites in Baja California, Mexico. Stones may be yellow, greenish-yellow, colourless, brownish-green and emerald green, this last variety, coloured by chromium, coming from the San Quintin area and also from the El Soccorro pegmatite (which also produces tourmaline). An orange sphene from Burma is in the Pain collection at the British Museum (Natural History) and gem quality sphene is also reported from Sri Lanka, where it occurs in wollastonite-bearing gneisses. Lepidocrocite in growth tubes and hexagonal plates of pyrrhotite are among the inclusions. The dispersion and the attractive body colour of sphene make it a desirable gemstone despite its low hardness and the rare emerald-green chromium variety commands high prices. A yellow variety from Brazil is often heavily included giving it a sleepy appearance. Honey-yellow to yellowish-green sphene is found in the Harts Range of Central Australia. Stones in excess of 20 ct can be cut from the crystals. The name titanite should be used for non-gem quality black or reddish-brown material.

SPINEL

$MgAl_2O_4$, forming three series, with magnesiochromite, with gahnite and with hercynite. Cubic, occurring commonly as octahedra or as flat triangular plates with the girdles cleft at the corners. These plates are twinned octahedra, a form so common with spinel that they are often referred to as spinel twins even when encountered in other minerals. Octahedra are often irregularly developed, lending the crystals a similarity to those of ruby. H 8. SG 3.58–3.61 for gem spinel, rising to 4.40 for gahnite, the zinc-rich variety. RI 1.718 for most gem material, rising to 1.805 for gahnite. Dispersion 0.020. The chromium absorption spectrum for red and pink stones shows a broad absorption band at 540 nm with a group of fine fluorescent lines in the red, unlike the emission doublet in ruby; spinel does not show any absorption in the blue as ruby does. Blue spinel shows a band at 458 nm (strong) and a narrower one at 478 nm; other lines are at 443 and 433 nm in the blue, and in the orange to green at 635, 585, 555 and 508 nm. This absorption is due to ferrous iron. Red and pink stones fluoresce crimson under long-wave UV, a weaker crimson under short-wave UV; there is a moderately strong glow under X-rays. Some purple to mauve stones give a red under long-wave UV, are virtually inert under short-wave UV and glow plum to lilac under X-rays; others may give an orange to red glow under long-wave UV, little under short-wave UV and green under X-rays. Some pale mauve stones may show a greenish glow under long-wave UV, as do some pale-blue or violet–blue specimens. Commonest inclusions are octahedra of one of the other spinel varieties such as the iron spinel magnetite, Fe_2O_4; iron-stained films are also seen but rutile crystals are rare. Zircon inclusions surrounded by stress marks (haloes) are also found. Greyish-blue to black stones from Burma may show six-rayed stars when the inclusions forming the rays are oriented parallel to the edges of an octahedral face; otherwise the star will show four rays. Some spinels with an alexandrite-like colour change from greyish-blue to an amethyst colour are known. A spinel which changed colour from pinkish-purple under fluorescent light to purplish-pink under incandescent light was reported in 1984 by the GIA. Under long-wave UV the stone glowed a moderate chalky yellowish-green and was inert to short-wave UV. The stone came from Burma. A spinel with a colour change from dark-blue in daylight to purplish-blue in incandescent light has been reported; absorption bands additional to the customary iron ones were seen at 630, 580 and 540 nm; these are more often seen in cobalt-doped synthetic spinel.

Gahnospinel specimens from Sri Lanka were found to be intermediate members of the spinel–gahnite solid solution with minor components of hercynite. The SG of ten samples tested was 3.60–4.06 and the RI range was 1.716–1.754.

The association of fine red spinel with ruby goes back into history and it is well known that a number of so-called rubies with historical names are in fact red spinels. One example is the uncut Black Prince's ruby in the British crown jewels (it is set in the front of the Imperial State Crown). Another uncut stone in the same collection, the Timur ruby, is engraved with the names of some of its owners. The finest red spinel occurs with ruby in a metamorphosed limestone at the Mogok Stone Tract in Burma and the mineral in general is characteristic of metamorphosed rocks and their weathering products. Fine red spinels are also found in Afghanistan.

Most of the red spinels appear to have kept their crystal form despite weathering and in fact do so better than corundum; blue spinels from the Sri Lanka gem gravels however are water-worn. Recently a fine blue spinel has been reported from a pegmatite near Jemaa in Nigeria and this material has been found to be rich in zinc, thus meriting the name gahnite. It has an SG in the range 4.4–4.59 and RI between 1.793 and 1.794. The absorption spectrum shows two sharp areas with no absorption between areas where absorption is mixed in intensity; these are at 580 and 555 nm; there are also absorption bands at those places in the red–yellow area similar to those in spinel and gahnospinel. Under incandescent lighting the Nigerian material showed markedly red. On heating in an oxidizing environment stones heated to 1000°C for 1 h showed a change from blue to blue–green while stones heated to 1400°C turned dark olive-green. Columbite has been found in the Nigerian spinels, together with beryl and hematite. Spinel crystals of gem quality are found in marble outcropping along the flanks of the Karakoram Range in the far north of Pakistan. The spinel occurs with corundum in marble forming massive intercalations within garnetiferous mica schists and biotite-plagioclase gneisses. The spinel crystals, which are often euhedral, are less common here than in the similar rocks at Mogok, Burma. Predominant colours are red, brownish to plum-red, lilac, violet and blue. The mean RI is 1.716 and the SG is 3.599. A chromium emission line could be seen at 685.5 nm. Long prismatic crystals of a green amphibole, and fine rutile inclusions could be observed.

The magnesium spinels which provide most of the best gem material form a complete series with gahnite, $ZnAl_2O_4$. This can be a fine dark green as well as the blue as in the Nigerian material. An intermediate variety with the composition $(Mg, Zn)Al_2O_4$, coloured blue to dark blue, has been named gahnospinel.

Until recently blue spinels coloured by ferrous iron were the only blue varieties known of natural origin. Although for some years there had been suggestions that a natural blue spinel coloured by cobalt (giving a much brighter and more exciting blue) did exist, no study of the problem was made until that by Shigley and Stockton (1984). The authors examined a number of bright blue stones with refractive index in the natural blue spinel range but showing a cobalt absorption spectrum. They concluded that the presence of an iron-induced absorption band close to 460 nm was proof of natural origin whether or not cobalt was also found in a specimen. The closeness of the iron and cobalt bands in the absorption spectrum can lead to confusion over which element is operating so that it is no longer possible to state that the effect hitherto known as cobalt absorption bands will separate natural from synthetic blue spinels.

Most synthetic spinel is grown by the flame-fusion method and this material has a higher SG and RI than the natural stones from which they can easily be distinguished. The figures are 3.64 and 1.728 respectively;

details of their manufacture are given in the chapter on synthetic stones but at this point it can be said that they are made to imitate other stones rather than to pass for natural spinel. Light blue, light green, yellow and colourless stones seem to be the most popular, the latter being quite dangerous when cut to small standard sizes and set in numbers in a piece of jewellery; in such conditions they make a respectable diamond simulant but can be spotted by their bright blue fluorescence under short-wave UV. Red synthetic spinels are rarely made by the flame-fusion process since the boules are likely to fragment; when seen, cut stones show characteristic curved swathes of colour and numerous gas bubbles; their RI is 1.730 and SG is 3.60; although rich in chromium they do not show the group of emission lines in the red end of the spectrum but rather a single bright line at 685 nm. Verneuil-grown spinels show anomalous double refraction; their included gas bubbles take on interesting shapes, some resembling furled umbrellas. Another characteristic inclusion consists of a flat cavity containing a bubble and either liquid or gas, frequently joined by a thin tube to a similar cavity parallel to and below the original.

Further work on cobalt-coloured natural blue spinel was reported by GIA in 1986: two stones examined had an RI of 1.720 and showed absorption bands at 622, 595, 575, 559, 480 and 434 nm; bands at 480, 460 and 434 nm are not found in the synthetic material. A natural twin blue star spinel was also reported by GIA. The weight was 2.12 ct. The twinning doubles the star, the two images of which are separated by the twin plane.

Spinel is perhaps named from a Latin word for a little thorn, perhaps alluding to the pointed octahedral crystals of the mineral. Gahnite is named for the Swedish chemist J. G. Gahn, the discoverer of manganese.

SPODUMENE

$LiAlSi_2O_6$. Monoclinic, forming prismatic crystals, characteristically flattened with ragged ends and corroded. Perfect cleavage in one direction and brittle. H 6.5–7.5. The SG of gem material usually near 3.18. RI (α) 1.653–1.670 (β) 1.660–1.669 (γ) 1.665–1.682. DR 0.014–0.027. Biaxial positive. Pronounced pleochroism: pink stones have purple–violet to colourless; green crystals show green, blue–green and colourless to pale green. Hiddenite shows a chromium spectrum with a doublet at 690.5 and 686 nm, weaker lines in the orange and a broad absorption centred near 620 nm. Kunzite shows a golden-pink to orange fluorescence in long-wave UV, weaker in short-wave UV and orange with phosphorescence in X-rays, sometimes with an alteration in colour of the stone to blue–green; this colour disappears in sunlight. Yellow–green stones glow orange–yellow in long-wave UV, weaker in short-wave UV but strong in X-rays;

in this case there is no change in body colour. Hiddenite glows orange in X-rays with phosphorescence.

Spodumene occurs in granite pegmatites; the lilac–pink variety kunzite is the best-known but is very hard to cut on account of its easy cleavage, brittleness and sensitivity to heat. Most commercial kunzite comes from Brazil where two pegmatites providing very fine crystals were located in 1962. One is at Corrego do Urucum in the Doce river area of Minas Gerais, the other is at Fazenda Anglo, near Itambacuri in the same state. Green and yellow (coloured by iron) material is also found in Brazil but the green may not be called hiddenite since it contains no chromium. True chromium-bearing hiddenite is found only in North Carolina, USA, at Stony Point and at Hiddenite. Some spodumene in various colours is found in the Malagasy Republic and fine yellows and other colours are found in Pakistan.

Fine crystals of spodumene have been found in the pegmatites at Pala, San Diego County, southern California, where they occur as pale pink to deep bluish-lilac kunzite varieties. Some is colourless to yellow and a small amount of green spodumene has also been reported. The gem variety of spodumene occurs within the cores of the pegmatite, usually in a very coarse-grained quartz–spodumene unit. Fragments of clear material are surrounded by white to deep pink clay. Among the mines in the Pala area the Stewart Lithia mine, the White Queen mine on Hiriart Hill and the Pala Chief mine have produced fine gem spodumene.

Perhaps the finest kunzite known comes from the Nuristan region of Afghanistan. RI is usually (ϵ) 1.659 and (ω) 1.677 and the SG about 3.20. The most intense colour is seen along the long axis of the crystal and at its finest is a deep purple. Three-phase inclusions and growth tubes are reported. Much spodumene is tenebrescent, showing a reversible darkening and lightening of colour with changes in viewing conditions. On mining many spodumene crystals are blue–violet or green; after exposure to the sun for a few days the purple to pink colour appears. Heating to 400°C will bleach this colour which can be restored by artificial irradiation.

Spodumene takes its name from a Greek word meaning burnt to ashes, with reference to the grey colour of the non-gem material. Kunzite is named for George Frederick Kunz, author and gemmologist to the firm of Tiffany & Co. Hiddenite is named for A. E. Hidden, superintendent of the mine in North Carolina where this variety was first found.

STAUROLITE

(Fe^{2+}, Mg, Zn)$_2$Al$_9$(Si, Al)$_4$O$_{22}$(OH)$_2$. Monoclinic, pseudo-orthorhombic, crystals frequently twinned and crossing at nearly right angles or at 60°, the stone taking its name from this habit. H 7–7.5. SG 3.65–3.78. RI (α)

1.739–1.747 (β) 1.744–1.754 (γ) 1.750–1.762. DR 0.011–0.015. Biaxial positive. Pleochroism strong with one ray nearly colourless, the other some shade of red to brown or yellow. A zincian variety with a colour change from red–brown in incandescent light to yellowish green under fluorescent light and with hardness near 7, SG 3.79 and RI 1.721 and 1.731 has been reported. Pleochroic colours are green, yellow and red.

Valued for its cross-like appearance (which it partly shares with the chiastolite variety of andalusite) staurolite occurs in crystalline schists and gneisses, frequently with tourmaline and garnet. It is a typical mineral of the St Gothard and Ticino areas of Switzerland, where it occurs with kyanite to which it is related. Staurolite is also found in Brazil, the USSR and the USA. Rare transparent specimens are a deep brown; the opaque crystals range from brown to yellow.

STEATITE see SOAPSTONE

STIBIOTANTALITE

$SbTaO_4$. Orthorhombic, forming prismatic, striated crystals, often with twinning and with distinct cleavage in one direction. H 5–5.5. SG 7.34–7.46 (the higher value for gem material reported from San Diego County, California). RI (α) 2.37 (β) 2.40 (γ) 2.46, biaxial positive. DR 0.090. Dispersion 0.146. Lustre adamantine to resinous.

Stibiotantalite is a pale brownish-yellow and can resemble sphalerite, though the latter is isotropic and possesses a higher lustre. Stibiotantalite is found in granite pegmatites, most gem quality material coming from Mesa Grande, San Diego County, California. Named for its composition.

STICHTITE

$Mg_6Cr_2(CO_3)(OH)_{16} \cdot 4H_2O$, dimorphous with barbertonite. Trigonal, forming masses with a perfect cleavage in one direction. H 1.5–2.5. SG 2.16–2.22. RI (ε) 1.518 (ω) 1.545. Uniaxial negative. DR 0.027. Pleochroism dark to light red. Chromium absorption spectrum may be seen.

The flexible purple laminae of stichtite, with its characteristic soapy feel, occur in serpentine rocks often associated with chromite. Attractive pink cabochons and large ornaments are fashioned, mostly from material found at Dundas, Tasmania, Australia, Black Lake, Quebec, Canada or the Barberton area of the Transvaal, South Africa. Named for R. Sticht of the Mt Lyell Mining and Railway Co. of Tasmania.

SUGILITE

(K, Na) (Na, Fe^{3+})$_2$ (Li_2Fe^{3+})$Si_{12}O_{30}$, a member of the osumilite group. Hexagonal and first discovered as brownish-yellow veins in an aegirine syenite from Ehemi Prefecture, Japan. H 6. SG 2.74. RI (ϵ) 1.607 (ω) 1.610. Uniaxial negative. The gem variety contains manganese and forms rock-like material of a deep purple. Absorption bands at 437, 419 and 411 nm. The mineral is named for the Japanese petrologist Ken-ichi Sugi who first found it.

The manganoan sugilite first appeared on the market in the late 1970s and has been called Royal Azel or Royal Lavulite. Sugilite from South Africa is described in a paper by Henn (1986). The RI is given as 1.607–1.609 and the SG as 2.76–2.80. Some stones from the Wessels mine, Hotazel, South Africa are translucent but most are opaque with a saturated purple which is quite distinctive. The absorption bands are most easily observed through a blue filter; they are manganese bands.

TAAFFEITE

$BeMg_3Al_8O_{16}$. Hexagonal, well-shaped crystals of any size have yet to be found although some small prismatic ones are reported from Hunan, China. H 8–8.5. SG 3.60–3.61. Cleavage undetermined. RI (ϵ) 1.717–1.720 (ω) 1.721–1.724. DR 0.004. Uniaxial negative. Taaffeite shows some dichroism which helps to distinguish it from otherwise similar spinel. No useful absorption spectrum or luminescence. Contains some crystalline inclusions which may (from their appearance) be phenakite.

One of the rarest known gemstones, taaffeite was discovered (the story goes) in 1945 by Count Taaffe, a member of an old Irish family which for some time had taken Austrian nationality. Living in Dublin, Taaffe collected gemstones among other things, obtaining some of them from combing the junk-boxes invariably kept by jewellers. One small cut stone of 1.42 ct showed doubling of dust specks and scratches, when seen through the stone. This ruled out spinel which has a similar refractive index. At a loss, Taaffe sent the stone to B. W. Anderson of the London Chamber of Commerce gem laboratory. Anderson and his colleagues agreed with Taaffe's findings but could not relate them to any known species. A test at the British Museum (Natural History), using X-rays, finally discredited spinel as a possibility and further tests, including chemical analysis, established the presence of magnesium, aluminium and beryllium (the first reported example of a mineral, let alone a gemstone, in which the combination of beryllium and magnesium occurred). The tests also assigned the stone to a class of the hexagonal system with only one other representative, a form of quartz. In the original stone 5.9% iron was found, accounting for the pale mauve colour. Several more stones have now been found, some with a distinctly red colour and having a very low beryllium content.

Gem quality taaffeite comes from Sri Lanka where it has been found as rolled pebbles, although a crystal in my care at the time of writing is flattened and markedly dichroic. Chinese material has been obtained from a dolomitized limestone. A zinc-bearing taaffeite of a reddish-violet colour, with 4.66% ZnO, an RI of (ϵ) 1.728 and (ω) 1.730 and an SG of 3.71, is reported from Sri Lanka. For this specimen the idealized chemical formula is $Be(Mg, Zn, Fe)_3Al_8O_{16}$ with a greater proportion of magnesium than zinc or iron. The colour is probably due to iron and chromium.

TALC see SOAPSTONE

TANZANITE see EPIDOTE

THOMSONITE

$NaCa_2Al_5Si_5O_{20} \cdot 6H_2O$. Orthorhombic, a member of the zeolite group of minerals, forming radial columns or radiated spherical aggregates with a perfect cleavage in one direction. H 5–5.5. SG 2.25–2.40. RI (α) 1.497–1.530 (β) 1.513–1.533 (γ) 1.518–1.544. Biaxial positive. DR 0.021. Pyroelectric.

Eye-like forms in white, yellow, reddish or brown colours are shown by thomsonite specimens, particularly those found on the shores of Lake Superior and on Isle Royale, USA. Thomsonite is a secondary mineral in lavas and basic igneous rocks; a translucent green variety from Michigan might be mistaken for jade and is locally known as lintonite. Named for Thomas Thomson a Scottish chemist who analysed the mineral. Ozarkite is a local name for a snow-white variety from Arkansas, USA.

TITANITE see SPHENE

TOPAZ

$Al_2SiO_4(F, OH)_2$. Orthorhombic, forming stumpy prismatic crystals, sometimes very large and well-formed; also occurs as water-worn pebbles. Perfect cleavage parallel to the basal pinacoid. H 8. SG (pink stones)

3.50–3.53; (yellows) 3.51–3.54; (colourless) 3.56–3.57; (blue) 3.56–3.57. Both SG and RI are roughly correlated with colour and with the ratio of (OH) to (OH + F) in the chemical composition. Brownish stones from Ouro Preto, Brazil, have SG 3.53 and RI (α) 1.629 (β) 1.631 (γ) 1.637 with DR 0.008; these are rich in (OH) and may contain chromium; fluorine-rich bluish stones from the USSR have SG 3.53 and RI (α) 1.609 (γ) 1.619 with DR 0.010. Pink specimens from Mardan, Pakistan have SG 3.53 and RI (α) 1.632 (β) 1.636 (γ) 1.641 with DR 0.009. Biaxial positive. Pleochroism varies with body colour but is usually varying strengths of this colour. Some pink stones obtained by heating brown or orange material give a chromium emission line at 682 nm but otherwise there is no distinctive absorption spectrum. Characteristic inclusions are planes of minute liquid-filled structures, sometimes with a gas bubble; three-phase inclusions have been reported. Some blue and colourless topaz shows a weak yellowish green luminescence under long-wave UV and a weaker effect under short-wave UV; under X-rays they may show greenish-white to violet-blue; there is a danger that stones may turn brown from the irradiation. Sherry brown and pink stones may glow orange–yellow in long-wave UV, weaker in short-wave UV and brownish-yellow to orange in X-rays. Stones are easily electrified by heat or friction.

Topaz is a characteristic mineral of pegmatites and high-temperature quartz veins; it may also occur in cavities in rhyolite and granite, in contact zones or alluvially as pebbles. Named from a Greek word for a mist-enshrouded island in the Red Sea, the colour of topaz is due to colour centres, apart from the pink to violet and the pink component of some orange stones which are due to chromium. Certain brown stones fade in the light; yellow–brown stones from Utah will lose colour after a few days in sunlight, as will some Mexican stones. A recent report on topaz coloured by electrons in the 10–20 mega-electron volt range shows that different colour centres are present in variable concentrations. The colour centres have been designated as X and Y due to their strong polarization dependency. Their colours are blue and bluish-violet respectively. Irradiated samples show cracks and parting planes parallel to the basal pinacoid. A shell-like structure with a blue rim, a colourless intermediate zone with a high concentration of radiation-induced defects and a colourless core are shown by the crystals. The shell-like structure is due to a temperature gradient from the surface to the centre of the crystals. Pink topaz from Katlang, Pakistan, has well-developed euhedral crystals only rarely; most are broken and highly fractured. Most are up to 3 cm long. Basal pinacoids are not often present. The colour of crystals varies from colourless to very pale beige to light brown, to very pale pink to deep pink (this colour is particularly distinctive; a crystal which I saw at the Gemstone Corporation of Pakistan's headquarters in Peshawar in 1985 was a superb purplish-pink). RI (pink material) is (α) 1.629–1.631 (β) 1.631–1.634 (γ)

1.638–1.642. DR 0.010. SG 3.51–3.53. The commonest inclusions are fissures, often appearing as cleavage cracks running parallel to the basal pinacoid. Two-phase inclusions with the liquid phase dominant are also found in most stones.

Topaz is the archetypal yellow stone and nomenclature problems connected with this still persist in some countries. They centre on citrine which has frequently been called topaz; sometimes true topaz is then called precious topaz. Where there are Trades Descriptions acts such names can no longer be used. The colour of topaz covers a wide range from pale yellow or sherry colour through to a rich orange or reddish-brown to a dark brown. Pink stones occur naturally though some of the pink stones seen in jewellery are derived from heat-treatment of certain brown stones (see the chapter on alteration of colour). Blue topaz is usually pale; a much darker blue is obtained by irradiation. Colourless stones also occur, but cannot compete with most colourless transparent material on account of low dispersion (about 0.014); where colourless topaz comes into its own is in the fine euhedral crystals long associated with Nigerian occurrences such as the Ropp tin mines area. Most topaz crystals are gifts to students since their rhombic cross-section, characteristic terminations of pyramid and dome forms and the flat basal pinacoid characterized by the interference colours indicating incipient cleavage make them easy to recognize; even in the dark a cut topaz is said to be easy to identify from its slippery feel.

Almost all the world's supply of topaz comes from the many pegmatites in Brazil. Traditionally the finest reddish-orange material is known as 'Imperial' topaz and is associated with the mines around Ouro Preto, once the capital of Minas Gerais state. In general fine 'Imperial' topaz from the Ouro Preto area has an SG of 3.53 and RI of 1.63–1.64 (lower and higher figures than topaz from other granitic pegmatites). The colour is a magnificent orange–red from the Capão do Lana mine. The Vermelhão mine is the most important, followed by the Capão mine at Rodrigo Silva; here the topaz is obtained by hydraulic extraction. Pegmatites north of Teófilo Otoni contain paler yellow topaz but the stones are bright and may reach large sizes.

Many topaz pebbles are found in Sri Lanka, the colour varying from pale to dark yellow, colourless and pale green (somewhat resembling beryl). Fine pink crystals are found at Mardan, Pakistan, in a limestone matrix; sherry–coloured crystals are found in the Thomas Range, Utah, USA, in a rhyolite, and crystals of a similar colour occur at San Luis Potosi, Mexico; some of these fade in sunlight. Fine blue crystals occur in the Ural Mountains of the USSR, at Alabashka and Mursinska (a blue crystal with some muscovite adhering is at Harvard; it weighs 4 lb). Topaz from the Sanarka River valley in the southern Urals resembles Brazilian material; a rose-coloured variety is found with chrome mica (fuchsite) and chrome tourmaline in quartz veins or in vugs in a Carboniferous limestone. Some attractive topaz is found in Zimbabwe, a blue variety with quite strong colour occurring at Miami and elsewhere; in Namibia topaz of gem quality is found in the pegmatites at Klein Spitzkopje. At Jos, Nigeria, fine blue crystals of gem quality are found. Fine yellow topaz crystal groups were found at Schneckenstein, Saxony, German Democratic Republic and this was the source of much of the topaz used in eighteenth-century jewellery. This may have higher refractive indices (up to (ϵ) 1.620 and (ω) 1.629) which are higher than those usually attributed to it; the colour is generally a pale yellow. Fine blue crystals are found in the granite rocks at Pike's Peak, Colorado, USA, along with some other colours. One or two locations in Australia also produce gem-quality material; notable crystals come from Tingha, New South Wales and also from locations in Queensland and Tasmania. Lastly the Mogok Stone Tract in Burma (see under ruby) produces the occasional topaz.

TOURMALINE

The name tourmaline is used conveniently to describe a family of minerals (at least six) all of which have similar crystal structures but which vary widely in chemical composition. It is named from a Sinhala word applied to a number of gemstones. They are:

Dravite: $NaMg_3Al_6(BO_3)_3Si_6O_{18}(OH)_4$, forming two series, with schorl and with elbaite
Uvite: $(Ca, Na) (Mg, Fe^{2+})_3Al_5Mg(BO_3)_3Si_6O_{18}(OH, F)_4$
Schorl: $NaFe^{2+}_3Al_6(BO_3)_3Si_6O_{18}(OH)_4$, forming series with dravite
Elbaite: $Na(Li, Al)_3Al_6(BO_3)_3Si_6O_{18}(OH)_4$, forming series with dravite
Liddicoatite: $Ca(Li, Al)_3Al_6(BO_3)_3Si_6O_{18}(O, OH, F)_4$
Buergerite: $NaFe^{3+}_3Al_6(BO_3)_3Si_6O_{21}F$

All members of the tourmaline group crystallize in the trigonal system as striated prismatic crystals with characteristic rounded triangular cross-sections, variously terminated. Only buergerite has a distinct prismatic cleavage. H 7–7.5. SG varies with species but can usefully be summarized under colour: red–pink 3.01–3.06; pale green 3.05; brown 3.06; dark green 3.08–3.11; blue 3.05–3.11; yellow–orange 3.10; black 3.11–3.12. Crystals are piezoelectric and absorb colour parallel to their length, along which they may be colour-zoned; concentric zoning is an alternative. RI is within the range 1.610–1.675 with an average birefringence of 0.018. Uniaxial negative. Some faceted stones may show up to six shadow-edges.

Pleochroism is strong, showing light and dark of the same colour; the ordinary ray is absorbed enough to plane-polarize light and when total absorption of this ray causes only one shadow-edge to appear on the refractometer, the stone might appear to be isotropic. (An apparently black tourmaline showed a fine deep green and deep garnet red through a dichroscope, showing that pleochroism is not always light and dark of the same colour in these minerals.) Some specimens show luminescence: stones from Newry, Maine, USA, may show a chalky to a strong blue in short-wave UV: pink Brazilian stones may glow blue or lavender in this radiation and Tanzanian brown, golden and green stones give a strong yellow. Tourmaline is generally quite heavily included, the commonest features being flat, liquid-filled films that appear black in reflected light. Elongated cavities, sometimes containing two-phase inclusions, are also seen, generally parallel to the length of the crystal. If these cavities are densely packed they may give rise to a chatoyant effect, coarser, however, than that in chrysoberyl. Tourmalines examined for their chatoyancy seemed to indicate that growth tubes were most frequent in green specimens while needles occurred most often in blue and red specimens. The growth tubes were found to be filled with tourmaline but prosopite, cookeite and iron oxides were also found in them. The needles were found to be either tourmaline or epidote. Hornblende and mica crystals are also reported.

Tourmaline is a mineral of crystalline schists, granites and granite pegmatites. Elbaite is most often found in the latter; dravite and uvite occur most frequently in gneiss, marble and other contact metamorphic rocks; quartz (rock crystal) can contain tourmaline inclusions.

The colour varieties shown by tourmaline exceed those displayed by any other gem species; the greatest variety of colours is shown by elbaite; dravite is usually dark, black to brown, sometimes colourless; uvite is black, brown or green, usually dark shades; schorl is black, blue or blue–green; buergerite is brown to black with a bronze-coloured iridescence below the surface; liddicoatite (the calcium analogue of elbaite) shows a variety of colours. The names indicolite (blue tourmaline), rubellite (red) and achroite (colourless) are sometimes used but too many names are confusing. The cause of colour in tourmaline is complex, not all stones of a certain colour having the same origin for that colour. Generally the greens and blues come from iron and so do the yellows and browns, and some pink to red specimens (but some pink–red stones are coloured by manganese). In some cases charge transfer operates. Some green tourmaline is coloured by vanadium or chromium or both. None of the colours is complemented by a really useful absorption spectrum, the red stones showing two narrow bands at 458 and 450 nm in the blue (care should be taken not to confuse these with the absorption bands in the same area in ruby; the ruby bands are much stronger); within the general absorption in the green is a fine line at 537 nm. Green tourmaline usually absorbs all the red part of the spectrum down to about 640 nm; a narrow iron absorption band can be seen at 497 nm. This band can also be seen in blue tourmaline.

The main commercial source for tourmaline is Brazil where a number of states (Bahia, Ceara, Espirito Santo, Goiás, Paraiba and Rio de Janeiro) produce specimens. The best, however, come from the state of Minas Gerais where a tourmaline-rich area is centred on the 17th line of latitude south and the 42nd line of longitude west. In this region there are vast numbers of tourmaline-bearing pegmatites. The main deposits are in the Salinas River area which produces pink and green tourmaline from Porteira, Boqueirão and Lago do Alto; the Araçuai–Itinga region and the Piauí River (the Pirineus mine) which give fine green and blue stones; the Itamarandiba, Malacacheta and Minas Novas region, also giving fine green specimens; the districts north-east of Teófilo Otoni at Marambaya and Ladinha where there are fine green, pink and water-melon stones; and the Cruzeiro deposit, celebrated for red tourmaline which comes close to the pigeon's blood colour of ruby.

In 1977 a lithium-bearing pegmatite was found near Conselheiro Pena on the banks of the Doce River. This contained a geode filled with cleavelandite and pink to pigeon's blood red tourmaline crystals, some the largest ever found. The Araçuai–Itinga area produces mainly green and blue crystals while the Araçuai–Salinas area produces red and multi-coloured stones. The deposits at Virgem da Lapa have produced large green and blue crystals; the Salinas mine specializes in multi-coloured stones and that at Ouro Fino produces cherry-red rubellite. The Virgem da Lapa is a hard rock pegmatite with crystals occurring in unaltered pockets. At Salinas the crystals are recovered from altered pockets and at Ouro Fino the crystals are found in eluvial deposits some way from the primary pegmatite (Proctor, 1985).

In the Ural Mountains of the USSR, cherry-red to rose-red tourmalines are found in the Mursinska area on the east side of the mountains; principal Uralian locations for this colour of tourmaline are Mursinska itself, Alabashka, Lipovskaya, Sarapulskaya and Shaitanka. This last is the most important and tourmaline occurs here with quartz, albite, beryl, green muscovite and lepidolite in druses in a coarse-grained granite. The tourmaline is usually emplanted on the albite and mica or found as loose crystals in a yellow clay-like decomposed granite. Violet–blue colours are also found. Tourmaline is found in the Adun–Chilon region of southern Transbaikalia, particularly around the peak of Soktuj Gora. Here pink–red specimens occur; in museums these are sometimes labelled 'Nerchinsk', which is where most sales take place. Many crystals from the USSR are illustrated in Goldschmidt (1916) and repeated in Scalisi and Cook (1983).

A variety of tourmalines occur in the pegmatites of the Pala district, San Diego County, southern California, USA. Colours include pink, red, green, blue, parti-coloured and colourless and many crystals are clear, the chief flaws being fractures and inclusions. Crystals are found in the Stewart, Tourmaline King and Tourmaline Queen mines. Many crystals from this deposit differ in colour at opposite ends, usually pink and green; some have blue cores with two or more colours surrounding them. They are consistently prismatic and occur as isolated individuals or columnar composites or radiating groups.

On the eastern side of the United States, fine tourmaline has long been known from Maine. In 1895 Augustus Choate Hamlin published *The history of Mount Mica of Maine, USA, and its wonderful deposits of matchless tourmalines,* now a very rare and much-sought book, with its fine coloured pictures of crystals. Hamlin preceded this book with *The tourmaline* published in 1873. Mount Mica is close to the town of Paris in Oxford County and it was here that transparent green crystals of tourmaline were first reported. Fine specimens can be seen in a number of collections. In the 1970s another Maine deposit, also of pegmatitic origin, was worked for gem-quality material: the Dunton pegmatite on Newry Hill, Newry. Most crystals are log-like and are single non-matrix specimens of watermelon tourmaline. Many of the crystals are broken but predominantly prismatic; most have a clear green termination, others have a thin rind of light green over a red core which often shows through the green. The red is a deep burgundy colour and brilliant; the best green is apple-coloured or a soft bluish-green. Blue crystals are usually too dark to cut satisfactorily. Some green stones are fibrous and can make chatoyant stones.

In 1984 a report (Schmetzer and Bank, 1984) showed that a newly-discovered intense yellow tourmaline of gem quality from Zambia was a member of the solid solution elbaite–tsilaisite, the latter being a manganese tourmaline with the composition (taken from the Zambian specimens and idealized) of $Na(Li_{0.5}Mn_{1.0}Al_{1.5})Al_6(BO_3)_3Si_6O_{18}O(OH, F)_3$. MnO content of the Zambian crystals lay between 6.3 and 6.9%; RI was 1.622–1.623 for the extraordinary ray and 1.645–1.648 for the ordinary ray with a birefringence of 0.023–0.025. Specific gravity was 3.13. As there is no absorption band at 1.92×10^{-3} nm^{-1} which is assigned to an Mn^{3+} absorption in rose to rose-red tourmalines (manganese–elbaites) the manganese in the Zambian material is thought to be present only in the divalent state. Absorption bands noted are at 1.58×10^{-3} nm^{-1}, increasing absorption from about 1.7×10^{-3} nm^{-1} to the blue end of the visible area and a weak shoulder at about 2.41×10^{-3} nm^{-1}. A red tourmaline with unusually high constants was reported in the early 1970s from Osarara, Narok, Kenya. This material had specific gravity of 3.07 but a very high RI of 1.623–1.655. These appear to be iron-rich dravites. A red tourmaline from the area of Chipata, Zambia had an intense red to brownish-red colour and RI 1.624 and 1.654 with DR 0.030. SG mean was 3.05. Some two-phase inclusions were observed and the pleochroic colours were bright red and dull brownish-red.

Both Sri Lanka and Burma produce gem tourmaline; the former giving yellow and brown stones from alluvial deposits in the south-east of the country. In Burma, tourmaline of a fine red has been found in an alluvial deposit of decomposed gneiss and granite about 20 miles east of the Mogok Stone Tract. This material may have been the stone used by Chinese mandarins who used it ceremonially.

Liddicoatite in a wide variety of colours comes from the Malagasy Republic and various shades and colours of elbaite are found at Alta Ligonha, Mozambique. Liddicoatite from Madagascar, often beautifully zoned in triangular shapes, shows brown, pink and green. RI is (ϵ) 1.621 (ω) 1.637, with DR of 0.019. Other African sources include Klein Spitzkopje, Namibia where a dark green variety of elbaite is found. This can be altered to an emerald-green on heating, probably to around 650°C (at 725°C essential water is lost and the specimen is destroyed!). A golden yellow tourmaline from the Voi-Taveta area of Kenya was a member of the dravite-uvite series (ratio of 78:22). The RI was (ϵ) 1.619 (ω) 1.642 with a DR of 0.022. Profiled inclusions were observed and negative crystals parallel to the c-axis, curved growth tubes and flat healing fissures. The saturation of the colour was thought to be due to titanium which showed a wide absorption band centred on 440 nm. An unusual East African blue–green dravite had an RI of (ϵ) 1.616 (ω) 1.637, with a DR of 0.021; pleochroic colours were blue–green to light yellow–green. It was found as a stream-worn pebble. Crystals of an emerald-green are reported from Usakos, Namibia and from Tanzania comes a variety of colours, one being a green stone containing chromium or vanadium or both; this will give a strong residual red through the colour filter whereas other green tourmalines remain inert. These are elbaites. Multiple RI in tourmaline has also been noted in rough uvite crystals from Tanzania, the effect due to vicinal faces.

In 1978 a discovery of red tourmaline was made at the Itataia mine, 3 miles east of Conselheiro Pena, near Governador Valadares, Minas Gerais, Brazil. In a pegmatite pocket were very large elbaite crystals associated with albite, quartz, lepidolite and microcline. The colour resembles that of cranberries as a rich deep red and dealers refer to them as cranberry tourmalines. Many crystals show double terminations and reach up to 30 g in weight.

Fine crystals of tourmaline are found in the pegmatites of eastern Afghanistan. Bariand and Poullen (1978) describe three of the gem-bearing areas, Nilaw, Mawi and Korgal. The pegmatite occurs as veins of variable size. Tourmaline of various colours occurs in the veins at Nilaw and multi-coloured specimens at Mawi. The finest tourmalines, however, come from the Korgal deposit where green crystals of virtually complete transparency are found. Pink crystals are found with the green on sheaves of muscovite. Some yellow-brown crystals are also reported and these may come from the Kantiwa field which also provides deep blue specimens. Fine tourmalines are found at a number of places in the pegmatites of the Shingus-Dusso area of Gilgit in the far north of Pakistan. Some tricoloured (black/dark green/pink), green and pink crystals are reported by Kazmi *et al.* (1985). Tourmaline crystals of high quality are reported from two pegmatite mines in the Sankhuwa Sabha district in the Kosi zone of eastern Nepal. Bright, grass-green elbaite, yellow to amber manganese-rich elbaites, pink, lemon-yellow, green, tricoloured and watermelon crystals are reported.

Brown tourmalines from Sri Lanka (the area of Elahera) have been classed as uvites and dravites. RI of the dravites is (ϵ) 1.623 (ω) 1.640 with DR of 0.017 and SG of 2.92; that of the uvites is (ϵ) 1.619–1.631 (ω) 1.635–1.648 with DR 0.016–0.021 and SG 2.85–3.24. Uvite is the calcium–magnesium borosilicate, dravite the sodium–magnesium borosilicate.

Gem quality tourmaline has been found near Peatfold Farm, Glenbuchat, Aberdeenshire, Scotland. The crystals are found in a range of colours; where the matrix is mostly muscovite and lepidolite the tourmaline is usually pink; some water-melon (pink centre, green rind) and some colourless stones are also found as well as green to blue–green. Stones cut from the crystals would be very small; constants are in the usual tourmaline range.

A tourmaline with a yellowish-green colour shown to GIA had a DR of 0.028 which is high for the material. An absorption band at approximately 423.5 nm and a stronger one extending from 420–418 nm, with a cut-off at approximately 410 nm are probably both due to manganese. Many gas and liquid-filled inclusions were observed. It is possible that the stone is of Zambian pegmatitic origin though the colour is not as yellow as that shown by some other reportedly Zambian tourmalines.

An elbaite with an RI of (ϵ) 1.621 and (ω) 1.649, DR of 0.028 and SG of 3.13 has 9% MnO by weight and may approach the theoretical 'tsilaisite' manganese analogue of elbaite.

TREMOLITE

$Ca_2Mg_5Si_8O_{22}(OH)_2$ with some iron, forming a series to actinolite. Monoclinic, forming prismatic crystals or masses. The mineral nephrite (q.v.) is formed from tremolite which is made up of minute fibrous crystals, felted and interlocked. Cleavage in two directions. H 5–6. SG 2.9–3.2 (2.98 for chatoyant material from Ontario; variety hexagonite has 2.98–3.03). RI (α) 1.560–1.562 (β) 1.613 (γ) 1.624–1.643. Biaxial negative. DR 0.017–0.027; 0.019–0.028 for hexagonite. Green transparent crystals from Tanzania have RI of 1.608–1.631 with an SG of 3.02. An emerald green tremolite from Taiwan has an SG of 2.98; RI (α) 1.608 (β) 1.624 (γ) 1.636. Hexagonite shows bluish-red, deep rose and deep red–violet pleochroism; green Tanzanian crystals show light yellowish-green, light green and green. Chrome-bearing material may show a chromium absorption spectrum; some material may show a line at 437 nm (cf jadeite). Hexagonite shows orange to medium-pink luminescence in long-wave and short-wave UV.

Tremolite occurs in contact and regionally metamorphosed dolomites, magnesian limestones and ultrabasic rocks. Hexagonite, a pink transparent variety, is found at Fowler, New York, USA; chatoyant green stones are found in Burma and in Ontario. Transparent green crystals have been found at Lalatema, Tanzania. Tremolite is named from the Tremola valley in Switzerland; hexagonite was first thought to be a hexagonal mineral. Purple–pink chatoyant hexagonite is reported from Fowler, St Lawrence County, New York. The stones are heavily included with needles, parallel cleavage planes and opaque dark brown euhedral crystals. An emerald-green tremolite–actinolite (sometimes known as grammatite) has been reported from Tanzania. SG was found to be 3.24–3.30 and RI (α) 1.609 (β) 1.610 (γ) 1.611 with a DR of 0.002. Needle-like crystals were included and the chromium-content was 0.034%, with 0.09% vanadium and 0.04% iron. The hardness is between 5 and 6. The name smaragdite has sometimes been used for a substance of similar composition.

TUGTUPITE

$Na_4AlBeSi_4O_{12}Cl$. Tetragonal, found as masses. Hardness near 6.5. SG 2.30–2.57, according to impurities. RI (ω) 1.496 (ϵ) 1.502. DR 0.006. Uniaxial positive or negative. Pronounced dichroism, bluish-red and orange-red. Under long-wave UV tugtupite glows orange and salmon red under short-wave UV. Massive specimens are mottled white with dark and light red patches; the lighter patches may lose their colour in the dark but may have it restored on exposure to daylight.

Tugtupite occurs in a nepheline syenite pegmatite in the celebrated Ilimaussaq alkaline intrusion in Greenland where it is named for one of the promontories. The first discovery at Tugtup agtakorfia was made in 1957 but gem material was found in 1965 as angular patches in the syenite which extends over the intrusion. Some blue material has also been found. The name translates as reindeer stone; material has also been reported from Poluostrov Kolskiu (the former Kola Peninsula) in north-west USSR. Some translucent specimens have been faceted and some of these have an SG of 2.38, RI of (ω) 1.492 (ϵ) 1.500 and a birefringence of 0.008.

TURQUOISE

$CuAl_6(PO_4)_4(OH)_8 \cdot 5H_2O$ with some iron. Triclinic, crystals are rare and very small, the stone usually occurring as masses, concretions, veins and

crusts. Turquoise is blue to blue–green with a waxy lustre. H 5–6. SG 2.84
(crystals) 2.6–2.9 (masses); turquoise from Iran is 2.75–2.85. Massive
material gives a shadow–edge at RI 1.62. Crystals have RI (α) 1.61 (β) 1.62
(γ) 1.65. Biaxial positive and DR 0.040. The absorption spectrum shows
distinctive bands (often hard to see) at 460 and 432 nm. The bands are more
prominent in treated stones. Turquoise shows a greenish-yellow luminescence to blue in long-wave UV but is inert in short-wave UV and in
X-rays.

Turquoise is formed by the action of percolating ground-waters in
aluminium-bearing rocks containing copper. Crystals are found at Lynch
Station, Virginia, USA; they are usually very small and only a specimen of
exceptional size could be faceted. The finest turquoise comes from the
Nishapur area of Iran where it occurs in porphyry and trachyte rocks with
brown limonite which occurs as thin veins in the turquoise. Turquoise also
comes from the Sinai Peninsula of Egypt at Serabit el Khadim and
Maharah, mines which have been operating from 1000 BC at least. The
turquoise has been brecciated by earth movements and there is much
limonite associated with it. The colour is blue to greenish-blue and some
specimens are reported to have faded in sunlight. Fine dense turquoise has
been found in Australia.

There are many turquoise locations in the western and south-western
United States. Qualities vary a good deal and some powdery material has
been impregnated with plastic to stabilize it; the treatment also deepening
the colour. Turquoise from the Lavender Blue mine in Nevada shows a
black spider-web pattern with tiny specks of turquoise. Intense dark blue
specimens have been mined at Bisbee, Arizona. Turquoise of good quality
comes from the Chuquicamata area of Chile. A turquoise reported from
China had RI of 1.62–1.64, SG of 2.696–2.698 and H 4.5–5.5. There was a
weak greenish-yellow fluorescence under long-wave and short-wave UV.
In China, turquoise with a hardness over 5 is described as 'porcelain
turquoise' and quality is said to equal the best material found elsewhere.
Turquoise is named from a word meaning Turkish: the material was
originally brought to Europe from Persia via Turkey.

ULEXITE

$NaCaB_5O_9 \cdot 8H_2O$. Triclinic, forming fibrous nodules with a perfect
cleavage in one direction. H 1–2.5. SG 1.65–1.95. RI (α) 1.496 (β) 1.505 (γ)
1.519. Biaxial positive. DR 0.023. May show blue–green fluorescence in
short-wave UV with some phosphorescence.

Ulexite may be cut so that print can be read along the fibres, giving a
natural counterpart of the fibre-optic effect. Cabochons may also be cut
from this white material. 'Television or TV stone' has been used as a

popular name; the mineral is named after G. L. Ulex, a German chemist who performed the first correct chemical analysis of the mineral. It occurs in playa deposits and dry lakes with other borates. Most material comes from California.

VANADINITE

$Pb_5(VO_4)_3$ Cl with some P and As. Hexagonal, forming hexagonal prisms or tabular crystals. H 2.5–3. SG 6.88. RI (ϵ) 2.350 (ω) 2.416. Uniaxial negative. DR 0.066.

Vanadinite is found in the oxidized zones of ore deposits, particularly those of lead. It is a bright orange-red to brown with a resinous lustre, being more commonly cut into cabochons since crystals are collected as mineral specimens. Best facet quality material comes from Mibladen, Morocco and crystals from the Apache and Mammoth mines in Arizona, USA, though small, are of good colour. Named from its composition.

VARISCITE

$AlPO_4 \cdot 2H_2O$ forming a series to strengite. Orthorhombic, usually found in masses. H 3.5–4.5. SG 2.2–2.57. RI: shadow edge near 1.56. Some variscite fluoresces dull green to green in long-wave and short-wave UV.

Pale green to blue–green variscite is made into attractive ornaments and may sometimes imitate turquoise. It occurs by the action of phosphate-bearing waters on aluminous rocks, fine material being found in the USA at Fairfield and Tooele in Utah, where rich green nodules are also found. Named from the old district of Variscia, Germany, the first source of the mineral.

VERDITE

Verdite is the name given to a deep green rock composed of finely intergrown fuchsite or muscovite mica. Material from Zimbabwe may contain ruby and albite feldspar. The hardness is 3 but the corundum–rich material may reach 9. The SG ranges from 2.70 to 2.87; specimens rich in albite are close to 2.65. Two dark green specimens from the Transvaal, South Africa, had an SG of 2.85 and also showed a faint chromium absorption spectrum. Bands of rutile grains have also been observed in material from Zimbabwe. The chief location is Barberton in the Transvaal.

VESUVIANITE

$Ca_{10}Mg_2Al_4(SiO_4)_5(Si_2O_7)_2(OH)_4$. Tetragonal, forming primatic crystals, often euhedral; sometimes found intergrown with grossular garnet which is closely related chemically. Considerable chemical substitution may take place. H 6–7. SG 3.32–3.47. RI (ω) 1.712–1.716 (ϵ) 1.700–1.721. Uniaxial positive or negative. DR may be as low as 0.001. Pleochroism weak giving light and dark of body colour. Absorption bands at 528.5 nm (weak) and 461 nm (strong); in californite the 461 band distinguishes from jadeite. Some Canadian brown stones may give a rare earth absorption spectrum.

Vesuvianite (idocrase is a frequently used alternative name) occurs in serpentinites and in contact metamorphic deposits, particularly in limestones and dolomites. Much vesuvianite is a transparent yellowish to greenish brown; blue masses are known from Telemark, Norway, and fine deep green chromium-bearing material comes from Asbestos, Quebec, Canada, a locality which also produces manganese-bearing pink stones. Fine green crystals, some transparent, are reported from Quetta, Pakistan and green and brown facetable crystals have been found in Kenya. The massive variety known as californite is a mixture of vesuvianite and grossular garnet with a SG of 3.25–3.32; in addition to the Californian localities it is found in Africa and Pakistan. Idocrase is named from two Greek words referring to its mixed appearance, since idocrase resembles crystals of other species; Vesuvianite from Vesuvius.

VILLIAUMITE

NaF. Cubic, forming very small crystals, more commonly masses. Perfect cleavage in one direction. H 2–2.5. SG 2.79. RI 1.327. May show anomalous pleochroism, yellow to carmine-red.

Small crystals, of a deep red colour, have been cut from material found at Los, an island off the coast of Guinea, the occurrence being in a nepheline syenite. Named from a French explorer, Villiaume.

VIVIANITE

$Fe_3(PO_4)_2 \cdot 8H_2O$. Monoclinic, forming prismatic crystals or masses, sometimes bladed. Perfect cleavage in one direction; sectile and flexible in thin pieces. H 1.5–2. SG 2.64–2.68. RI (α) 1.579–1.616 (β) 1.602–1.656 (γ) 1.629–1.675. Biaxial positive. DR 0.040–0.059. Strong pleochroism, blue, pale yellowish-green and pale yellowish-green; or deep blue, pale bluish-green, pale yellowish-green; or indigo, yellowish-green, yellowish olive-green.

Vivianite occurs as a secondary mineral in ore veins and is also found as an alteration product of primary phosphate minerals in granite pegmatites. The colour darkens on exposure to light and is then very rich, so that some faceted stones have been cut despite the cleavage and fragility. Best material for cutting comes from Llallagua and Poopo, Bolivia and from N'gaoundere, Cameroun. Named for J. G. Vivian, the English mineralogist who discovered it.

Odontolite, a fossil material, may be stained by vivianite and then resemble turquoise. The best blues have an SG just over 3 and an RI varying from 1.57–1.63. The hardness is 5. Good quality material comes from Simmore, Gers, southern France. X-ray powder photography shows that odontolite gives a powder pattern typical of apatite, indicating that apatite is an important ingredient in odontolite, which is composed of the fossilized bones and teeth of ancient animals. The name comes from two Greek words meaning tooth and stone.

WARDITE

$NaAl_3(PO_4)_2(OH)_4 \cdot 2H_2O$. Tetragonal, forming pyramidal crystals or crusts and aggregates. Perfect cleavage in one direction. H 5. SG 2.81–2.87. RI (ω) 1.586–1.594 (ϵ) 1.595–1.604. Uniaxial positive; sometimes shows anomalous biaxiality. DR 0.009.

Greenish-white crystals of wardite from Piedras Lavadras, Paraiba, Brazil, have been faceted and cabochons of white wardite mixed with green variscite are fashioned from material found in Utah, USA. Wardite occurs in phosphate masses in sediments, also in pegmatites. Named for H. A. Ward, an American collector.

WAVELLITE

$Al_3(OH)_3(PO_4)_2 \cdot 5H_2O$. Orthorhombic, forming characteristic radial aggregates of acicular crystals with a perfect cleavage in one direction. H 3.5–4. SG 2.36. RI (α) 1.520–1.535 (β) 1.526–1.543 (γ) 1.545–1.561. Biaxial positive. DR 0.025. Some material may glow bluish in longwave UV.

Collected almost exclusively for its fine, radial, crystal aggregates, wavellite provides cabochon material. Colours are always pale shades of white, green or yellow, although some colourless and bluish material is reported. It occurs as a secondary mineral in hydrothermal veins, also occurring in aluminous and phosphate rocks. Best material for fashioning comes from Hot Springs, Arkansas, USA. Named for William Wavell, English physician, who first discovered the mineral.

WHEWELLITE

$CaC_2O_4 \cdot H_2O$. Monoclinic, forming prismatic or twinned crystals with three directions of cleavage. H 2.5–3. SG 2.21–2.25. RI (α) 1.489 (β) 1.553 (γ) 1.649–1.651. Biaxial positive. DR 0.159–0.163.

Small faceted gems with quite a high dispersion (0.034) can be cut from this rare mineral. Most stones are white, sometimes inclining to pale yellow or brown. Crystals are found in coal seams or in concretions. It also occurs as a hydrothermal mineral in ore veins. Crystals are found at Burgk, Germany, and septarian concretions are found at Havre, Montana, USA. Named for William Whewell, English scientist.

WILLEMITE

Zn_2SiO_4. Trigonal, forming prismatic crystals. H 5.5. SG near 4.1. RI (ω) 1.691 (ϵ) 1.719. Uniaxial positive. DR 0.028. Weak absorption bands at 583, 540, 490, 442 and 432 nm with a strong band at 421 nm. Intense yellow–green fluorescence in short-wave UV, less strong in long-wave UV. Material from Franklin, New Jersey, USA, gives this effect most markedly and also exhibits strong phosphorescence of the same colour.

Willemite occurs in zinc ore bodies or zinc-bearing metamorphic deposits. Green crystals are found at Franklin and Sterling Hill, New Jersey, together with orange masses. Small blue crystals are found at Mont Saint Hilaire, Quebec, Canada, and colourless gem quality material is found at Tsumeb, Namibia. Cabochons of willemite with red zincite and black franklinite are fashioned for use under UV light. Named for King William I of the Netherlands.

WITHERITE

$BaCO_3$. Orthorhombic, forming twinned crystals giving hexagonal dipyramids, both botryoidal or globular masses; distinct cleavage in one direction. H 3–3.5. SG 4.27–4.79. RI (α) 1.529 (β) 1.676 (γ) 1.677. Biaxial negative. DR 0.148. Effervesces in acid. May fluoresce greenish or yellowish in short-wave UV (particularly English material); some phosphorescence also shown.

Witherite is found as a low-temperature mineral in hydrothermal vein deposits. Colourless to pale yellow cabochons have been cut from material found at the Minerva mine, Rosiclare, Illinois, USA. Named for the English physician William Withering who first discovered the mineral.

WOLLASTONITE

$CaSiO_3$. Triclinic, forming tabular crystals; more commonly fibrous masses. Perfect cleavage in one direction. H 4.5–5. SG 2.8–3.09. RI (α) 1.616–1.640 (β) 1.628–1.650 (γ) 1.631–1.653. Biaxial negative. DR 0.015. Shows blue–green fluorescence with yellow phosphorescence in short-wave and long-wave UV.

Cabochons have been cut from the reddish wollastonite which occurs at Isle Royale, Lake Superior, Michigan, USA; chatoyant material is also known since the mineral is fibrous. Colourless to grey and pale green material is known; it occurs in metamorphosed limestones and alkalic igneous rocks. Named for the English mineralogist, W. H. Wollaston.

WULFENITE

$PbMoO_4$. Tetragonal, forming square-shaped tabular crystals with a distinct cleavage in one direction. H 2.5–3. SG 6.5–7.0. RI (ω) 2.405 (ϵ) 2.283. Uniaxial negative. DR 0.122.

Bright red to orange wulfenite crystals can be faceted, but most material is too thin. Wulfenite occurs as a secondary mineral in the oxidized zone of ore deposits; the finest specimens come from the Red Cloud mine in Arizona, USA: this state also produces good material from the Mammoth mine and further specimens are found in New Mexico. Yellowish crystals come from Tsumeb, Namibia. Named for the Austrian mineralogist F. X. Wülfen who wrote on the lead ores of Carinthia.

ZEKTZERITE

$LiNaZrSi_6O_{15}$. Orthorhombic with two directions of perfect cleavage. H 6. SG 2.79. RI (α) 1.582 (β) 1.584 (γ) 1.585. Biaxial negative. Colourless to light pink. Light yellow fluorescence in short-wave UV. Named for J. Zektzer of Seattle, Washington, USA.

ZINCITE

ZnO with some manganese. Hexagonal, forming hemimorphic crystals with a perfect but difficult cleavage in one direction. H 4–4.5. SG 5.68. RI (ω) 2.013 (ϵ) 2.029. Uniaxial positive. DR 0.016.

Transparent red manganese-coloured crystals of zincite are found in a metamorphosed limestone with zinc ores at Franklin, New Jersey, USA.

The colour is very attractive but facetable material is extremely rare. Named from the composition.

ZIRCON

$ZrSiO_4$ with some iron, uranium, thorium and hafnium. Tetragonal, forming prismatic or pyramidal crystals, frequently twinned; also found as water-worn pebbles. Very brittle. H 7–7.5, varying somewhat with direction. Over geological time the crystal structure of the zircon may be damaged or even destroyed by the action of α-particles emanating from these elements until the zircon is ultimately transformed into a mixture of quartz and zirconium oxide which is virtually amorphous. These are called low or metamict zircons and intermediate and high zircons also exist; in the latter no decay has yet taken place. There is a complete transition between the high and low types. Constants are affected (see Appendix A, Tables A1 and A2):

low zircon:	SG 3.95–4.20 (4.00 is a common value)
	RI 1.78–1.815 (close to isotropic)
	DR none to 0.008
intermediate zircon:	SG 4.08–4.60
	RI (ω) 1.83–1.93 (ϵ) 1.84–1.970
	DR 0.008–0.043
high zircon:	SG 4.6–4.8 (4.70 is a common value)
	RI (ω) 1.92–1.94 (often 1.925)
	(ϵ) 1.97–2.01 (often 1.984)
	DR 0.036–0.059 (the latter figure is customary)
	Uniaxial positive.

Low zircon is most commonly green but may be brown or orange; intermediate zircon is brownish green to dark red and the high stones are colourless, blue or brownish-orange.

Zircon has a characteristically greasy lustre. It is uniaxial positive and the dispersion is 0.039 for all types. Zircon displays a most beautiful and interesting absorption spectrum which consists of evenly spaced fine lines covering the whole of the spectrum; some 12 of these are stronger than the rest and the line at 653.5 nm in the red is seen in almost all stones, with a weaker companion at 659 nm. With most of the commercial blue, colourless and golden-yellow stones these lines can best be seen by placing the stone table facet down on a black cloth and examining the spectrum by reflected light. The most intense absorption spectrum is shown by dark green or brown Burmese stones in which up to 40 lines can be seen. These stones contain more uranium than those from Sri Lanka; green or greenish stones from that country show a strong ten- or twelve-band spectrum with somewhat indistinct bands. Low zircons may show only a diffuse broad band near 653 nm, but some show a narrow band at 520 nm in the green. When these stones and some others with low properties are heated to about 800°C a series of anomalous bands is developed, Red zircon usually shows no absorption bands.

Zircon may show a mustard-yellow or yellow-orange fluorescence under short-wave UV. Some stones glow a dull yellow with phosphorescence. Under X-rays zircon may show whitish, yellow, greenish or violet–blue. There is also a fluorescence spectrum but the effects of radiation may alter the colour (see below). Blue stones show a distinct pleochroism of deep sky blue to colourless or yellowish-grey; red and brown stones show approximately light and dark of those colours. Inclusions are characteristic; a strong angular zoning and streakiness; rutile crystals intersecting to form a silk-like effect are sometimes seen as well as stress cracks with iron oxide staining. Some metamict stones show bright fissures known as angles; these are stress marks which have been brought about by the disintegration of the zircon structure and correspond to the positions of prism faces and dipyramids; they prove the existence of a former crystal structure. Low zircons may also show fairly large cracks parallel to cleavage planes (zircon has, however, an indistinct cleavage). Cat's-eye zircon is occasionally seen; a stone examined by the GIA was greyish-green with quite a sharp eye, caused by long needle-like inclusions. A fine green zircon cat's-eye and another of brownish yellow are reported by the GIA; the yellow stone could be confused with cat's-eye chrysoberyl but the latter would not show the characteristic zircon absorption spectrum.

Zircon is a mineral of igneous rocks, particularly granites. Fine specimens are found in Thailand, the most important commercial source; Burma provides yellowish and green stones in gravels with ruby; Sri Lankan stones comprise all colours and are also found in gravels; most metamict zircon so far found comes from Sri Lanka. Fine crystals with very sharp edges and an exceptionally bright lustre are found near Miass in the Ilmen Mountains of the southern Urals. They are brown and occur loose or embedded in a feldspar–syenite. They show first and second order prisms. Fine orange gem-quality zircon is found in New South Wales, Australia and a very pale yellow stone from Inverell fluoresces a brilliant golden-yellow in UV.

It is customary to heat zircon to obtain the desirable colours golden-yellow, blue, and colourless from brown material. The stone is first heated for a few hours at about 1000°C, surrounded by charcoal. Some of the stones alter to blue, others become colourless, the remainder developing dark patches. The treatment may be repeated and the off-white stones then heated to about 900°C with air admitted to the furnace. This converts some material to colourless, some to yellow, orange or red. Any stones remaining off-white can be recycled again. Some of the colourless and blue stones may alter with exposure to light or just with time. The original colour may be restored by irradiation and the best material is kept in the sunlight for a few days in any case so that any tendency to discolouration can be noted.

Zircon which was originally blue turned a muddy greyish to greenish brown after exposure to UV in a tanning booth. The material was shown to GIA and the colour restored by placing the test stone on the tip of an 80-watt quartz-halogen fibre-optic light. Two hours sufficed for complete restoration of colour. Exposure to short-wave UV produced a similar alteration of colour. Zircon is named from an Arabic word signifying gold, the word being derived from Persian.

ZOISITE see EPIDOTE

9 Descriptive section: organic materials

CONSERVATION OF ORGANIC MATERIALS

Many substances described in this section involve the use of living creatures and although the materials produced are undeniably attractive, the interests of conservation should be put first; it is far more important to ensure the continuation of a species rather than to use it for merely ornamental purposes. Some recovery techniques are cruel and it is to be hoped that the practices will eventually cease.

AMMONITE

Fossil ammonite shell with a fine play of colour has been found in southern Alberta, Canada. The ammonite concerned is *Placenticeras meeki* and dates from the Upper Cretaceous era (about 70 million years ago). It occurs in a dark grey shale; up to now no complete fossils have been found, only fragments. Only about 5% of the total number of pieces found have any gem use and only 20% of that figure is in any condition to be worked. The ornamental material is the nacreous layer of the ammonite shell, a substance composed of aragonite in the form of overlapping platelets of which mother-of-pearl is made. Most common colours are red and green but the better quality pieces show a complete range of colours. Many pieces show red and orange when the incident light is perpendicular to the surface and green when it is nearly parallel. Under the microscope thin pieces show a dark brown with a very fine-grained texture.

The composition of the shell layer is chiefly calcium carbonate in the form of aragonite, as is the material filling cracks in the shell. This gives a bright yellow fluorescence under long-wave UV, less intense yellow under short-wave UV. SG is between 2.67 and 2.85: RI is 1.512–1.528. The hardness is about 4.

Since most of the material is so thin it is often used in composites, generally as triplets with a quartz cap. Uncapped shell may splinter as it is very brittle. The trade name Korite has been used, although ammolite and calcentine have also been proposed.

AMBER

Amber is a fossil resin from certain pines, notably *Pinus succinifera* which dates from the Oligocene. No tree which produces resin of the amber type exists today so amber is a true fossil. It consists of various substances, each slightly soluble in ether or chloroform (most simulants of amber are decidedly more soluble), together with some bituminous material; the overall composition is near $C_{10}H_{16}O$. It is transparent to translucent and has a characteristic greasy lustre and sticky feel. The colour varies from pale yellow through red to brown; it may show a blue–white fluorescence under UV. The hardness is a little more than 2 and the SG about 1.08. Amber softens at about 150°C and melts at 250–300°C. Between crossed polars amber will display interference colours and the SG allows most amber to remain suspended in a brine. It feels warm to the touch and develops a negative electrical charge on rubbing (this does not distinguish amber from its plastic simulants).

The best-known sources of amber are the area of Gdansk in Poland and the adjoining Baltic shores of the German Democratic Republic. Some amber is mined (pit amber) from deposits of oligocene-period glauconite sand, the rest is washed up on the shore (sea amber). The low density of amber allows it to be carried considerable distances and this is why some pieces can be found on the English North Sea coast.

Pressed amber

Small pieces of amber can be pressed together with a gentle heat to make pieces large enough to be worked, forming a reconstructed amber; pieces are welded together at high pressure, a technique helped by amber's low melting point of 180°C, and can be recognized by the elongation of bubbles and by signs of flow in the structure; the interference colours are not now uniform over so large a field but give a patchwork of contrasting colours. Cloudy amber can be clarified by heating in rape seed oil which enters the air spaces which cause the cloudiness; crack-like markings are often seen in material treated in this way. These marks are known as sun spangles. Some amber contains inclusions resembling nasturtium leaves, probably indicating artificial treatment. Much amber is stained to give a variety of colours, particularly a reddish-brown, thought to give the appearance of age. The reconstructed material made from small pieces shows a brilliant chalky blue fluorescence (modern pieces show this best); some of this material shows no flow marks and no bubbles. SG for this type of amber is lower than for the true material. The name 'ambroid' (or pressed amber) has been used and the first examples of this type of reconstructed material were made in about 1881. Pieces of amber apparently fused together to make irregularly-shaped beads were not apparently produced by the same process that gives the well-known pressed amber. The beads recently examined by GIA did not show a strained internal texture.

Occurrence

Baltic amber contains from 2.5–8% succinic acid, $C_2H_4 (COOH)_2$ and yields this in solid crystalline form on heating at temperatures from 250 to 300°C. Other products of heating are a reddish-brown oil of amber, a watery fluid and a black residue. Strictly the name amber should be reserved for those resin varieties containing succinic acid but the name has such wide circulation that to limit it now would be pedantic. When amber is burnt aromatic and irritating fumes are given off by the succinic acid.

Material with a slightly different composition is found in Burma (burmite); this is much redder than Baltic amber and occurs near Myitkina (close to the jadeite mines) in the Hukong Valley. Amber is mined here by the sinking of shafts. Sicilian amber is reddish-brown and occurs along the Simeto River which gives it the name simetite. Amber from Romania (roumanite) comes largely from the province of Muntenia near the town of Buzau. The amber has less succinic acid and more hydrogen sulphide in its composition; the SG is 1.05–1.12 and the hardness 2–2.5. Colours include yellow, brownish-green, brown, green and blue; a marked fluorescence is noted. Other Romanian amber may be a dark greenish-black or bluish. Material from Transylvania is brown- to honey-yellow.

A lighter variety of amber is known as gedanite from a former name for the present-day town of Gdansk. This is softer and lighter than other types, with a hardness of 1.5–2 and SG near 1.02. It contains very little succinic acid.

Some of the finest amber presently on the market comes from the Dominican Republic where deposits are concentrated in the Monte Cristi Range along the northern border of the Vega Real central plain, north of Santiago between Altamira and Canca. Two particular areas are important, the larger one north-west of Tamboril in the Pena region, the other below Pico Diego de Ocampo near Pedro Garcia in the Palo Alto de la Cumbre region. Mining began at the latter site in 1949 but is no longer carried out on a large scale. The mountains are mostly made up of Tertiary marine sandstones and shales with some eroded conglomerates. Amber is likely to be found wherever carbonaceous material and lignite occur in the sandstones. Although some workers have assigned this amber to the Oligocene period, recent work shows that it occurs in Lower Miocene formations similar to those found in Chiapas, Mexico. Dominican

Republic amber contains no succinite and thus belongs to a group which has been given the name retinite. Most is pale yellow to brown and darker colours may bleach to paler ones in sunlight. Some deep reddish browns and fluorescent blues have been found; one type of blue amber appears clear yellow to yellowish-brown in transmitted light, giving a blue filmy surface by reflected light; this type is expensive. All ambers from the Dominican Republic gives a strong fluorescence in blue or green colours.

Dominican Republic amber from the Altamira formation is found scattered in a 60 km² area to the north and east of Santiago de los Caballeros in the Cibao Valley (Fraquet, 1982). The deposits of Los Cacaos, Palo Quemado and Loma el Penon give good quality amber. The amber-bearing strata are exposed along three faults within the Cordillera Septentrional in the northern part of the island which is included in the Bahamas Arc. Amber is found in a sequence of limestone, sandstone, mudstone and conglomerate and is often associated with lignite. Amber from the Yanigua formation in the east of the island is from the Miocene, strata occurring in small valleys and amber being found in similar conditions to those obtaining in the Altamira deposits. The amber from Yanigua tends to be a brighter yellow and brittle. Blue and green fluorescing examples are common. It is found in larger sizes than that from Altamira and this may be due to its having been less subjected to tectonic disturbance. Inclusions in Dominican amber comprise a wide variety of species, even a complete frog (smaller than the modern frog). A recent development on the island is the coating of recent copal-like resin with an epoxy compound which reduces resin's tendency to craze.

Burmese amber is from the Eocene and contains plant and insect inclusions. It has recently been found that the plant inclusions are related to angiosperms rather than to coniferous trees. Romanian amber dates from the Miocene and Sicilian simetite is found in tertiary lignite. Mexican amber from San Cristobal las Casas in Chiapas is like Baltic amber but is less hard; it was formed during the Miocene or Oligocene epochs of the tertiary period.

Canadian amber (chemawinite or cedarite) comes mainly from Cedar Lake, Manitoba, where it is found as nodules in sand and gravels from the Cretaceous; it is a retinite.

Inclusions

A study (Runge, 1868) found that 174 different species of flies, ants, beetles and moths, 73 species of spiders and many species of centipede had been recorded as inclusions in amber. Flies account for about 54% of all insects trapped by the encroaching resin, and it is worth noting that they comprise a greater proportion of the total insect population today than they did 40 million years ago when amber was formed. One piece of Baltic amber per thousand, on average, contains an insect inclusion, but there are more fauna inclusions in the amber from the Lower Miocene formations of the Dominican Republic, dating from about 30 million years ago; here the proportion of insect inclusions is closer to one per hundred. About 40 different types of beetle are reported from Baltic amber, most from the genus *Anobius* which comprises about 1200 species today.

All true insect inclusions show swirls at the extremities where the insect has tried to evade the encroaching resin. Well-formed insect inclusions should be regarded with suspicion; their host will probably be an amber simulant.

A paper by Koivula (1981) illustrates some internal features of amber, including a variety of bubbles, some of which act as hosts to euhedral crystals which may be of quartz. Some pollen grains have been found and some liquid and gas two-phase inclusions. Dark-field illumination is essential for observing these features, at a magnification of about 40 ×.

Copal

Copal is the oldest and best known of the natural resins likely to be confused with amber. The name is taken from a Spanish word meaning incense and refers to a group of exudates from various tropical trees used mainly in varnishes. These resins are soluble in oils and organic liquids but not in water. Copal can be found in a semi-fossilized form and as raw copal, obtained from the living source. Unlike the source of amber, that of copal lives today. The fossilized copal (true copal) may be anything from 100 to 1000 years old, transparent or translucent and yellow or brown. The material is softer than amber and cracks easily with a marked conchoidal fracture. It will burn with a smoky flame and smell of resin; although the SG and RI are like those of amber, the low hardness and tacky feeling when a drop of ether is placed on it identifies copal. Insect and plant inclusions can be present.

Copal from New Zealand is exuded by the Kauri pine, *Agathis australis* and this Kauri gum has been found up to 100 m below the surface in the north part of the country. The colour is yellow to reddish-yellow or brown with frequent insect inclusions. It contains no succinic acid and its SG is 1.05. The low melting point (187–232°C) causes it to become sticky during polishing. Copal is also found in Zanzibar, Sierra Leone, other countries in west and central Africa. A resin known as Dammar resin, clear, yellowish and pale, has a melting point of about 140°C and is used in the manufacture of lacquers. It may be recent or semi-fossilized and is found in Malaysia and Sumatra. Copal from the Philippines has a melting point of 190°C for the harder varieties and 120°C for softer specimens; the SG is 1.072 and the colour varies, some being very dark to black. It has been known as Manila copal. Dammar from Singapore has SG 1.062 and a similar resin from Borneo called Pontianak has an SG of 1.068; it melts at 135°C.

BONE

Bone is similar in composition to ivory, although its structure is different and its density higher. For ornamental use, it is taken from the long-bone of various animals, particularly the ox, and the mandibles of large whales. There is a greater organic content in bone than in ivory and for this reason bone needs to be de-greased before use. The hardness is near 2.5 and the refractive index near to 1.54. Examination of a peeling shows a series of parallel canals when the peeling is taken from a direction parallel to the length of the bone. Transverse sections of bone are pierced with oval or circular cavities surrounded with dot-like spaces known as lacunae. The canals and the circular units are known as the Haversian system after the anatomist Havers who first described them. Like ivory, bone gives a violet–blue fluorescence under UV (that shown by bone may be somewhat whiter).

Deer-horn is similar in structure to bone; made from the antler of the deer it is sometimes known as stag-horn. It has a hardness of 2.5, SG 1.70–1.85 and RI about 1.56. The Haversian system is present but less prominent than in bone. Deer-horn has been used for inlay work and for knife handles and similar objects.

CORAL

Corals belong to the phylum Coelenterata and include the species *Corallium rubrum*, *C. japonicum* and *C. secundum*. Although all corals are secreted by marine coral polyps, the precious corals are harvested from the horny or calcareous internal skeletons of simple polyps which give a rigid support to the polyp colony on the outside, which itself has a secreted axial skeleton. Specific cells in the precious corals (in that part of the organism called the coenenchyme) secrete the calcareous or horny material which stiffens the central area of the membrane to form a rigid axis. Corals from the order Gorgonacea (at least two different families belong in this order) have a coenenchyme made up of small, specifically-shaped particles of calcium carbonate which supports eight-tentacled, retractable polyps which secrete a rigid or flexible axis to support the colony. If the skeletal axis is formed from the horny material gorgonin then a false black coral results. If the particles (spicules) of calcium carbonate sink through the coenenchyme to consolidate as a solid calcareous axis, true precious coral is formed.

Corals from the genus *Corallium* grow as arborescent colonies branching in one general plane with their marginal twigs turning upwards and forwards. They grow on the sides of fast flowing channels, attached to sediment-free limestone substrates. They have a thick coenenchyme and are pigmented with red spicules which form a consolidated skeleton. Growth rates are about 1 cm a year for *C. secundum* and *C. japonicum*, and 0.3 cm a year for *C. rubrum*. Red coral polyps are photosensitive so this coral is found at depths from 100–500 m or in dark caves in shallower waters. A section of a red precious coral shows a radiating structure formed by dark red bands radiating from the central canal to terminate at convexities on the external surface of the axis. The areas between the radial structures are lighter in colour, fibrous in structure. Some dark circumferential growth rings are superimposed on this pattern.

It should be noted that corallium corals have no structural similarity to reef-building corals: the calcareous axis is an internal rather an external skeleton.

Reef-building coral is the calcareous framework of the coral polyp and is built up into a system of stems and branches which are the reefs encountered in warm seas. Most coral, though not some of the black variety, is composed of calcium carbonate in the form of calcite; the coral is not the skeletal matter of the zoophyte coral polyp since this creature has no bony material but rather the framework upon which the animal lives as part of a colony. The coral used in jewellery is *Corallium nobile* or *C. rubrum*. The animals build up the coral in the form of tubes which fit within each other, the mineral matter being secreted from the sea in the animal's tissues and remaining in place on its death. Crystals of calcite are arranged radially at right angles to the axis of the branch. The hardness is about 3.5 and the SG between 2.6 and 2.7. Refractive indices of 1.65 and 1.49 (figures for calcite) have been obtained on specimens polished for the test. Those corals containing calcium carbonate will effervesce with acids.

Much coral is red or pink, some showing an absorption band at 494 nm in thin sections. There is little response other than a dull red in UV (some uranium glass imitations give a strong green fluorescence and some plastics glow an orange–red). *Corallium rubrum* is white to red and is found in waters where a temperature of 13–16°C can be ensured such as the Mediterranean and the Malaysian and Japanese waters. The best red coral is found off Algiers and Tunis on the African side and off the coasts of Sicily, Calabria, Sardinia and Corsica. Much of the fashioning is carried out at Torre del Greco near Naples.

Coral from the Ryuku Islands south of Japan is blood-red, pink or white; a black coral, *Antipathes spiralis*, is found off the coast of Cameroon and is known as Akabar or King's coral; a blue coral known as Akori coral comes from the same area. Akori coral is *Allopara subirolcea*. A white coral, *Oculinacea vaseuclosa* is also fashioned. A coral from the Agulhas Bank, South Africa is a variety of stylasterine coral *Allopora nobilis*. The stylasterine corals are star-shaped groups of polyps with the central polyp having a prominent stylus; blue and violet varieties (*A. subviolacea*) have been recovered from the West African Gulf since the eighteenth century at least. *A. nobilis* is made up of platelets of orthorhombic aragonite; some specimens are dyed to give an ox-blood colour and some are stabilized by impregnation with a mixture of methyl and butyl methacrylate in a vacuum. The SG of the stabilized coral is 2.41. Parallel growth banding, white star-like spots and white markings resembling the tails of comets have been noted.

Black coral

Black coral from the Indian Ocean and the Mediterranean has been known by various names including Akabar, King's coral and Giojetto. It is an arborescent coelenterate secreting a skeleton to support its family of polyps with its outer surface covered by a thin living tissue. In cross section the coral resembles a tree with characteristic ring patterning. Two types of black coral have recently been distinguished. The 'true' black coral is derived from Antipatharian coral found off Hawaii and the Great Barrier Reef of Australia. 'False' black coral is derived from Gorgonian sea fan coral from shallow tropical or sub-tropical waters. Both types have been formed from concentric lamellae of a horny material deposited round a central canal. The false black coral's axis is derived from a complex tanned collagen called gorgonin, the axis of the true black coral derives from a chemically inert tanned protein, named antipathin by Goldberg, which is neither a collagen nor a keratin.

True black coral shows fine linear structures radiating from the central canal to the external surface. These structures are not found in false black coral. Longitudinal sections have a knotted appearance in true black coral, resulting from the differential deposition of the layers of antipathin over radially arranged spines. Sections of true black coral show strain birefringence between crossed polars. Interference fringes invariably surround the spines of true black coral, never in false material. Black coral from waters around the island of Maui in the Hawaiian group is quite often used in jewellery. Some of the black coral has an SG of 1.34 and a vague RI of 1.56, suggesting that it is composed of organic matter rather than calcite.

IVORY

The chemical composition of ivory is mainly calcium phosphate, close to $(Ca_3OH)_2(PO_4)_6Ca_4$ with some organic matter. It can be softened by nitric and phosphoric acids.

The finest ivory comes from the tusk of the elephant but the tusks of the walrus, hippopotamus, narwhal and sperm whale may also be used. Fossil ivory is a name sometimes used for material taken from the mammoth but this is not a true fossil because no silicification has taken place. All elephant ivory, including that from the mammoth, contains a network of slender tubes extending from the base towards the tip in a generally longitudinal direction in flat spirals of opposite hands. Transverse tusk sections thus show a series of arcs and spirals like the patterns produced with a pair of compasses. No other type of ivory and no ivory simulant shows these patterns which are called lines of Retzius.

The SG falls within the range 1.70 to 1.85 and a vague shadow-edge at a mean of 1.535 can be seen on the refractometer. The hardness is 2.5–2.75. Two types of ivory, hard and soft, are recognized; the best quality material comes from the Cameroons, with a transparent, non-mottled appearance. Other fine material comes from Loango, Congo, Gabon, Ambriz, Angola, Ghana and Sierra Leone. Sudan ivory shows light and dark concentric rings. Ivory from Zanzibar and Mozambique is generally soft. Hard ivory is more difficult to cut and is brighter; soft ivory cracks less easily, having a better temperature resistance. Ivory from Ethiopia is soft with a 'bark or skin'.

Ivory from Africa comes from the African elephant, *Loxodonta africanus*; the Asiatic elephant, *Elephas maximus* is found in Burma, Thailand and India; it gives a denser white ivory which yellows more easily and is easier to work. Tusks of the Asiatic elephant are smaller than those of the African elephant. 'Fossil' ivory from the wooly mammoth (*Elephas primigenius*) is found in Siberia and sometimes in the Yukon River area of Alaska. The latter place gives a rather dark ivory but is used locally. Hippopotamus ivory from *Hippopotamus amphibius* is denser and finer-grained than elephant ivory; material comes from Central African rivers. Walrus ivory from *Odobenus rosmarus* is less dense and coarser than the commoner ivories with a finer texture on the exterior than at the core.

The walrus is found in northern seas. Ivory from the narwhal and sperm whale is coarse and is closer in texture to bone than to commoner ivory. Almost all varieties show a violet–blue fluorescence under UV and in common with most organic matter used for ornament, is inert under X-rays. Ivory will chip with a pen-knife blade where plastics, so often used as imitations, will peel. The teeth of a number of animals have been used in ornament; sharks and whales as well as elephants have played a part in the jewellery of primitive races. Molar teeth of elephants when sliced can be used as knife handles. Warthog canines may be carved. The material known as hornbill ivory comes from the beak of the helmeted hornbill bird *Rhinoplax vigil*, particularly the Indonesian helmeted variety. H 2.5, SG 1.28–1.29 and RI 1.55. Under UV there is a greenish or bluish-white glow. There is no reaction with acid but the hotpoint will produce a smell of burning hair from keratin. It is easily pared. Two colours, carmine and yellowish, are recorded.

Vegetable ivory is not a true ivory but simulates it to some degree. There are several varieties; one is the seed of the ivory palm, *Phytelephas macrocarpa*, found in Colombia. The fruit, which contains a nut called the

Corozo nut, occurs in large drupes (stone-fruits) and the nut may approach a hen's egg in size. Vegetable ivory is close to pure cellulose with the composition $C_6H_{10}O_5$; the hardness is 2.5, the SG 1.40–1.43 and the RI near 1.54. A peeling shows torpedo-shaped cells in roughly parallel lines. The Doom (Doum) palm, *Hyphaene thebaica*, is found in Egypt; its nut has a hardness of just under 2.5, SG 1.38–1.40 and RI 1.54. Peelings show a structure similar to that of the Corozo nut although polygonal-shaped cells may be seen end-on. All types of vegetable ivory take dye well and in some cases may be taken for corals when appropriately coloured; they will not, however, effervesce with acids.

JET

This lustrous black material is a fossil wood similar in origin to brown coal; it is found in the hard shales of the Upper Lias on the Yorkshire coast near Whitby, UK, where a small number of people fashion it today. It is formed from the decomposition of driftwood which sank to the sea-bottom and became emplaced in the rocks into which the fine mud was transformed over geological time. Some mines operate sporadically in the Whitby area and also in some neighbouring dales. Jet has a hardness of 2.5–4 and an SG of 1.30–1.35. It will burn with a sooty flame and a coal-like smell.

Jet is found in horizontal seams in the places where it is mined but most material is found casually on beaches, and also at Villaviciosa, Asturias, Spain (it is worked at Oviedo). The few other world occurrences are of no commercial importance, although a deposit near Colorado Springs, USA, produces material said to be indistinguishable from Whitby jet.

Muller (1980) examined material from England, Turkey, Spain, the USSR and North America. The most recent work shows that jet is a form of fossilized Araucaria. The woody structure of Whitby jet is confirmed by the scanning electron microscope. X-ray emission spectroscopy of the various types showed that the surface of the Whitby jet was high in silicon and significant levels of aluminium, sulphur, potassium, calcium and iron were recorded. The internal surface contained a good deal of aluminium, very little sulphur but some iron and copper. In Spanish jet the internal surface showed both aluminium and sulphur as well as silicon, potassium, calcium, iron and copper. Spanish material has the reputation of breaking up with sudden temperature changes, for which the sulphur content may be responsible. Turkish jet showed significant amounts of aluminium, silicon and sulphur as well as a little potassium and calcium.

SHELL

Many shells have been used for ornament, those showing a play of colour being especially prized. The pearl oysters *Pinctada maxima* and *P. margariti-*

fera from Australian waters and from the Torres Strait provide mother-of-pearl; the latter creature is the principal pearl oyster on the north coast of Australia and can measure up to 20 cm in size. It is valued more for its shell than for its pearl, although in pearl production it ranks second only to *P. vulgaris*. Pearl shell was discovered in 1861 in north-west Australia. Some dark shell gives an effect resembling chatoyancy and may be used as buttons or as more important jewellery. Paua shell which is coloured a bright green and blue is found in New Zealand; in American waters this shell is called abalone, the taxonomic name being *Haliotis*. Some of these shells are large and may produce pearls.

The shell of the sea snail (*Turbo* sp.) produces a material which can be used as a pearl imitation; it has a pearly upper surface and a yellowish non-nacreous back. These are called oil pearls or Antilles pearls. Coque de perle is cut from the central whorl of the nautilus shell and is like a blister pearl but with a thinner skin. Its hollow interior may be filled with wax or cement.

Operculum is the lid or door found on the gastropod *Turbo petholatus*; it is interestingly marked with eye-like patterns. It is calcareous and usually measures 12–25 mm in diameter, being rounded. Its upper surface which is slightly domed, may be green at the apex, turning to yellow and white on one side and reddish-brown to dark brown on the other. On the reverse side are spiral growth lines. The hardness is about 3.5 and the SG 2.70–2.76. The animal is found in seaweed bordering the tropical seas of Oceania, Melanesia and Polynesia.

Cameos are made from two different shells, the most common being the Helmet shell (of *Cassis madagascariensis*) which is found in West Indian waters. Cameos from the Helmet shell are in white relief against a brown background; material from the Malagasy Republic being cut in Italy. Cameos are also cut from the shell of the giant conch (*Strombus gigas*) which produces pink conch pearls. It is found off the West Indies and the Florida coast. The carving may be white on a rose-coloured background or the other way round. Some specimens fade in bright light. Mitchell (1982a) showed that in nearly all cases the brown, orange-brown or pink layer used as the background surrounding the white cameo proper has a fine grain running in virtually parallel lines in one direction and the white cameo material shows fine lines which run at nearly right angles to those in the background. The various layers in fact all have different grain directions and this is no doubt a way in which the strength of the shell is reinforced as the grain represents a direction of weakness. Imitation cameos in such materials as chalcedony do not show graining.

TORTOISESHELL

This material is taken from the Hawk's-bill turtle, *Eretmochelys imbricata*, found in many tropical seas and in particular around the Malay archipelago and the West Indies. (For conservation reasons, its use should now be discontinued.) The finest examples have a rich brown colour on a translucent yellow background, these specimens coming from the Sulawesi area of Indonesia. The dorsal carapace has 13 large scutes or plates but these are fringed by a series of marginal scutes, 26 in number if the nuchal scute is included. The 'shell' also includes ventral plastron and inframarginal scutes. All of them can be used ornamentally after ridges have been smoothed out by grinding and polishing. Some small plates and even chips can be moulded together after softening.

The composition of tortoise shell is mostly keratin (the protein of human hair and nails); the hardness is about 2.5 and the SG fairly constant at 1.29. The RI is 1.55 and the shell is sectile. The mottling of colour can be seen under magnification to be made up of spherical dots; in plastic imitations the colour is found in patches or swathes. Under UV the clear yellow parts of the shell give a bluish-white colour; most imitations give a yellow glow. Most ivory workers will handle tortoiseshell and also re-polish damaged articles.

PEARL

Pearls are produced by various species of bivalve molluscs in which there is a pair of slightly hollow shells, hinged along one edge. These simple organisms consist in most cases of a visceral mass, gills, a foot, a bundle of horny fibres by which they attach themselves to rocks or coral, and a mantle – two lobes or flaps which envelop the mass and have on their surfaces epithelial cells capable of secretion (Fig. 9.1). At the edge of the outer surface of each flap the brown–black scleroprotein conchiolin is secreted. Away from the edge along the outer surface are groups of cells which secrete calcium carbonate in the form of aragonite (the pearl mussel) or calcite (the pearl oyster). Further inwards from the edge are the cells which secrete calcium carbonate as flakes which form nacre or mother-of-pearl. The shell is formed in three stages: as the mantle expands the conchiolin is laid down first, then the second zone of cells deposits a layer of calcium carbonate prisms on the inner side of the conchiolin layer; finally the third zone of cells lines the calcium carbonate layer with nacre.

Any mollusc with a shell lined with nacre should, in theory, be able to produce a pearl, but only a few do so in fact. Only the pearl oyster and the pearl mussel are worth commercial exploitation. Pearl fisheries on the Arabic coast of the Persian Gulf and in the Gulf of Manaar off the north-west coast of Sri Lanka have been established since very early times. Fishing continues in both places although in Sri Lanka during government-designated years only. An important modern source is the north to north-west coast of Australia which produces both pearl and mother-of-pearl.

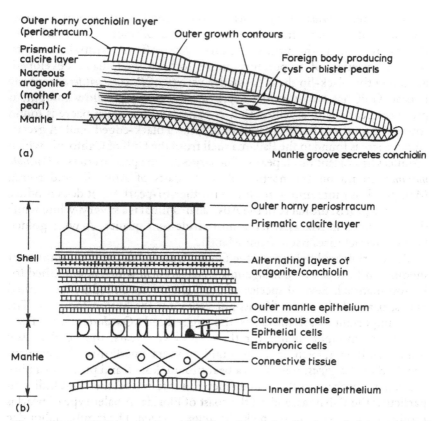

Figure 9.1 (a) Cross-section of shell and mantle. (b) Growth layers of shell. (After J.-P. Poirot.)

Pearls are also found off the coast of Burma, around Tahiti and other Pacific islands, off the coasts of New Guinea and Borneo, in the Gulf of Mexico and off the coasts of Venezuela. They are also found off the west coasts of South America and off Japan; the latter deposits have been neglected in favour of the Japanese cultured pearl industry. River (freshwater) pearls can be found in Great Britain but the only commercial exploitation of river pearl is in North America and Bavaria, West Germany.

Most pearls are produced by the pearl oyster, *Pinctada vulgaris* which is especially common in the Persian Gulf and the Gulf of Manaar; though occurring round Malaysia and off the coasts of Australia and New Guinea it is not the chief pearl oyster in these places. The oyster measures about 6 cm in diameter, the Persian Gulf oysters being a little larger, darker and showing a reddish tint on the shell compared with a pink colour on the shell of the Sri Lanka animal.

The oyster *Pinctada margaritifera* is the second most important pearl producer; it is more important as a source of mother-of-pearl than *P. vulgaris*. This is the principal pearl oyster off the Australian coasts. Several varieties, of importance more for the shell, have developed, including the black-lip shell. A smaller variety of *P. margaritifera* from the Persian Gulf has a rosy tinted nacre with a greenish-yellow margin; it produces pearls with a yellow tint. A variety with a green nacre is found round the southern Pacific islands; it supplies a black-edged shell. A green-edged nacre is found in the Panama shell from the Gulf of California which produces important black pearls. The largest of the pearl oysters is *Pinctada maxima,* found off the north and west coasts of Australia and round Malaysia. It is more important for its mother-of-pearl, but it does produce very large pearls: the nacre of the Australian animal is a silvery-white while shell marketed in Macassar has a more iridescent quality. A wide golden border characterizes nacre from Manila.

The pearl mussel belongs to the family Unionidae, the most important member in Europe being *Margaritifera margaritifera* which is still fished for button material. Several species of pearl mussel are fished in the United States, particularly in the Mississippi valley for the button industry. The most important pearl mussel is the niggerhead, *Quadrula ebena.* Pearl oysters are thought to live for three to eleven years, the pearl mussel somewhat longer, even up to one hundred years if undisturbed. The giant conch, *Strombus gigas,* which has a beautiful pink shell and produces a pink pearl, may reach up to 30 cm in diameter. It is found off the West Indies, in particular the Bahamas, and off the coast of Florida. A paler type of pearl is produced by the common whelk, *Buccinum undatum.* The family Haliotidae which includes the abalones or ear-shells, also yields pearls with a fine quality nacre. Pearls, though small, are found in a variety of colours including green, yellow and a rare blue.

Pearl formation is a process which is even now not completely understood. If the animal is left completely undisturbed and remains free from disease or attack it will not produce a pearl. Some shell distortion is thought to indicate that a mollusc will contain pearl but only about one in forty pearl oysters will produce one. Several pearls may be found in the one oyster (up to 87 have been recorded but one to three is a common number). The pearl builds up from the inside outwards, layer by layer, by secretions from the epithelial tissue. In true pearls the layers are composed of nacre; some pearls have a dark core consisting of organic matter and it was long believed that some form of solid material initiated pearl growth; worms, sand grains and a variety of other potential irritants were suggested. Many pearls when sectioned, however, show no trace of any intruder; perhaps a disease was their 'irritation'.

Pearls have a characteristic lustre due to interference of light from the succession of thin translucent laminae which form the surface, and also to diffraction of light reflected from the grating-like structure made by the closely-packed lines in which the laminae meet the surface. This lustre is known as the orient of the pearl (natural pearls are traditionally known as oriental pearls). The play of colour in mother-of-pearl arises in the same way. There is considerable variation in the translucency of the surface layer; where a pearl is so rich in conchiolin that it shows a brown colour it loses lustre and consequently value. The colour of the pearl as distinct from the surface play of colour is determined by the surface translucency and by the nature of the underlying layers. Conchiolin is yellowish in thin sections, deepening to brown or black whereas calcium carbonate is colourless or white. Impurities in the water in which the pearl grows probably give other tints; after white, the commonest colours are yellowish, salmon-pink, reddish or blackish-grey. Some white pearls are greenish on extraction and lose this colour on drying. So-called blue pearls have a thin crust superimposed on a dark conchiolin-rich kernel; they are lead-grey rather than blue. The colour of black pearl may be due to some impurity in the water. It is sometimes possible to recognize the origin of a pearl from its colour, those from the Persian Gulf appearing creamy, those from Sri Lanka being paler, Australian pearls being white or silvery and those from West Indian waters a bright rose-red with wavy white lines. Pearl from Panama is a golden brown and from Mexico reddish-brown or black. Japanese white pearls may show a greenish tinge. 'Indian' pearls which come in fact from the coast of Sri Lanka may show a faint rose colour. Virtually all pearls regularly seen in commerce show a sky-blue fluorescence of varying intensity.

The shape of the pearl depends upon the position taken in the shell during growth. If the intruder (where there is one) is emplaced between mantle and shell it becomes coated with nacre which adheres to the shell, giving the so-called blister pearl. If the intrusion is through the shell quite a large and even a round blister may be formed: sometimes the blister may contain a round pearl. The best-formed pearls are found within the tissues of the animal; the irritation causes a concave dent to be formed in the outer surface of the mantle. The dent deepens until it closes and takes the shape of a hollow sphere lined with epithelium cells. The cells on the outer surface of the mantle coalesce and the hole made by the dent is covered so that the hollow sphere (pearl-sac) is pressed into the underlying tissues. These pearls are called cyst-pearls. Deposition of nacreous material is usually constant after the first year. Round pearls are the most highly prized; pearls of irregular shape are called baroque and at the time of writing are being used extensively in modern jewellery. Seed pearls are small round pearls which are less than a quarter of a grain in weight (a pearl grain is 0.25 ct).

Large or exceptional pearls are sold individually but the majority come on to the market on strings of silk which are collected in bunches. Price is quoted from a base and the pearls graded for quality. The cost of a bunch is given by:

$$p \cdot \sum \frac{w^2}{n}$$

where n is the number of pearls in each grade, w their total weight in grains, p the base price and Σ the sum. In practice the mean weight is found first and then multiplied by the weight and the product by the price per base. Smith (1972) gives the example: a bunch of pearls consists of 44 strings containing 1529 pearls, grouped into two sizes with a base price of 50p:

Strings	n	ct	w	$\dfrac{w}{n}$	$\dfrac{w^2}{n}$
26	827	143.62	574.48	0.69	396.39
18	702	63.70	254.80	0.36	91.72
44	1529	207.32	829.28		488.11

The total product of 488.11 multiplied by the base price gives £244.05 as the value of the bunch.

Another example: a parcel of 64 button pearls is grouped into four sizes and the base price is 20p:

n	ct	w	$\dfrac{w}{n}$	$\dfrac{w^2}{n}$
5	7.80	31.20	6.24	194.68
10	13.53	54.12	5.41	292.78
17	16.75	67.00	3.94	263.98
32	24.61	98.44	3.07	302.21
64	62.69	250.76		1053.65

the total product of 1053.65 multiplied by the base price gives £240.73 as the value of the parcel.

Most pearls seem to have an approximate composition of 86% calcium carbonate, 12% conchiolin and 2% water. A round pearl sliced through its centre and examined under a microscope, shows a series of more or less continuous concentric layers, something like an onion; the shells differ in character, however, according to the nature of the cells that secreted them. On the refractometer the section gives a shadow-edge corresponding to the greatest of the indices since the arrangement of the crystals is normal to the concentric shells, radiating from the centre; thus, the calcium carbonate, in the form of nearly hexagonal orthorhombic aragonite crystals gives an unvarying ordinary ray reading, the reading for the extraordinary ray varying from this position to that of the lower refractive index.

Pearl has an SG varying from 2.715 for Persian Gulf material to 2.74 for Australian specimens. Venezuelan pearls with their high translucency have an SG in the range 2.65–2.75. Most Japanese pearls are in the range 2.70–2.74, although some go down to 2.66. North American freshwater pearls have a range of 2.66 to over 2.78. Black pearls have an SG in the range 2.61–2.69; pink conch pearls with no nacreous coating, are about 2.85; this indicates a higher proportion of aragonite (SG 2.94) in their composition. Pearl has a hardness of 3.5–4, close to that of aragonite. Their softness and their proneness to damage by acids, and even by the oils of the skin, make careful handling essential.

Production of pearl by inducing a mollusc to cover an alien substance with nacre was known to the Chinese in the thirteenth century. Objects were inserted between mantle and shell, the animal being returned to the water to live another two to three years. All kinds of objects were used, including small lead images of Buddha. Linnaeus discovered the process independently in the mid–eighteenth century. The great developers of the cultured pearl process have been the Japanese, especially K. Mikimoto (1857–1954) who worked on improving it from 1914 onwards. At first, small pellets of mother-of-pearl were cemented to the nacreous lining of the shell to obtain blister pearls, but the process was not economic because the blisters were coated on one side only and had to be cemented on removal to a piece of mother-of-pearl which was then ground to a symmetrical shape. Early Japanese pearls can easily be identified from this join.

Mikimoto introduced a new process. A shell is taken from one oyster and a strip of living tissue cut from the mantle. This was used to wrap a mother-of-pearl bead which was then inserted through an incision into the mantle of another oyster. These traditional methods of pearl cultivation are still used in Japan and elsewhere. Young oysters (spats) are collected as spawn and kept in cages for three years, at which time they are mature. The mother-of-pearl pellets are introduced and the oysters returned to the sea for a further seven years, disease permitting. Under the best conditions perhaps one quarter will produce a pearl. Each year about 0.625 mm of nacre is deposited on the pellet. Mother-of-pearl is used as the pellet since it can easily be pierced for stringing and it is easily shaped. Most cultured pearls range in size from 0.5 to 4 grains; when the larger species *P. margaritifera* is used pearls up to 60 grains may be obtained. A later process dispenses with the bead and makes use of the freshwater mussel *Hyriopsis schlegeli* in the mantle of which up to twenty incisions can be made. Into these incisions are placed prepared mantle tissue from another mussel and this on its own stimulates pearl growth. The pearls are known as non-nucleated. Biwa pearls are harvested in the summer season and there can be at least three crops. It is estimated that up to 90% of the implants in the first crop will yield pearls and 80% in the second crop. Although all have a characteristic external surface with some oiliness and with irregularities, an irregular central cavity surrounded by layers of nacre, the first crop shows a comparatively large cavity with positive X-ray fluorescence of a greenish-yellow; the cavity in the second crop is reduced to a slit and the X-ray picture resembles that for a natural pearl.

Most cultured pearls in which a pellet is present can be identified by X-radiography which will show up the single zone of conchiolin surrounding the nucleus. Conchiolin has a low absorption for X-rays (they pass through to darken a film). This technique is not always recommended because interpretation of X-radiographs is not easy and the process is slow; the results are unambiguous however. A natural pearl consists of concentric shells of a near-spherical shape and a cultured pearl has almost flat layers; the narrow X-ray beam therefore behaves differently in each. The pseudo-hexagonal axes of the aragonite crystals are at right angles to the surface of the shells or of the layers. In natural pearls therefore an infinite number of crystals radiate from the centre and the X-ray beam traverses them in the direction of the pseudo-hexagonal axis, giving a hexagonal spot pattern. In the cultured pearl the hexagonal pattern is only obtained when the beam is perpendicular to the mother-of-pearl layers; in directions at right angles to this, nearly digonal symmetry will be shown, yielding a four-fold spot pattern. Natural hollow pearls have a more opaque centre of nucleation than tissue-nucleated cultured pearls (under X-rays). Successful application of the technique needs very careful alignment of the pearl's nucleus with the X-ray beam. If the pearls are large or irregular and the position of the nucleus cannot easily be determined, the beam may pass well away from it giving diffraction patterns like those shown by natural pearls. Empirical centering and obtaining several photographs of the specimen in a variety of positions is essential.

Cultured blister pearls either have large salt-water nuclei or smaller freshwater nuclei and are described as three-quarter blister pearls. Large nuclei are said to be made from the shell of the mollusc *Pinctada maxima* and there is little conchiolin around the bead when the origin is from salt water. Material comes from the Philippines. Cultured blister pearls and Mabe pearls can be distinguished from each other; Mabe pearls have four parts, a very thin nacreous layer, a mother-of-pearl bead, a filler (?Canada balsam) and a mother-of-pearl base. There may also be a coloured lacquer coating inside the dome. Mabe pearls are made in Japan using *Pteria penguin*, a mollusc not used for culturing pearls. Mabe pearls are known to be fragile. The cultured blister pearls are stronger with their solid nucleus; they also show up easily under X-rays and have characteristic grooves which appear to merge near the girdle. A blister pearl cultivated in the shell of *Pinctada maxima* is marketed at Broome in Western Australia. Plastic of various colours forms the nucleus, giving colour to the whole pearl (Broome pearls). A clam pearl from *Tridacna gigas*, found off Port Moresby, Papua New Guinea, appeared as a white nacreless concretion with a smooth matt surface. It had an SG of 2.80 and a hardness near 3. The surface fluoresces a pale blue under long-wave UV but is inert to short-wave UV. It will effervesce with acids and is opaque to X-rays.

Under X-rays cultured pearls show more fluorescence than natural ones, many of which are virtually inert. The pellet has a strong fluorescence which shows through the whole pearl if the skin is not too thick. Cultured pearls have an SG about 2.75 with a range from 2.72 to over 2.78. Very thick-skinned pearls may be lower than 2.72. Most cultured pearls will sink in a liquid of density 2.74. Recent reports suggest that some pearls formed by tissue implantation (i.e. non-nucleated pearls) do not show the characteristic voids under X-radiography and are thus indistinguishable from natural pearls. If such pearls are together in a piece of jewellery their uniformity should be grounds for suspicion but some items may be classed as natural by default. The name Keshi has been used for such pearls.

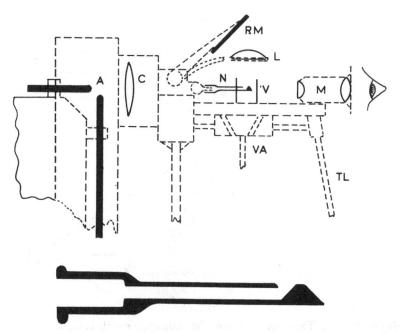

Figure 9.2 Schematic diagram of the endoscope: A, arc-light; C, condenser; RM, reflecting mirror to show surface of pearl; L, lens to enlarge surface of pearl; N, needle; V, vice to hold pearl; M, microscope to view end of needle; VA, vice adjustment; TL, traversing lever to move pearl along needle.

When a cultured pearl is examined under a strong source of light, the beam having passed through it, the kernel may be observed. Natural pearls will appear dark while the surface of the cultured pearl glows. A pearl microscope has been developed on the basis of this phenomenon. Drilled pearls can be tested with the endoscope, a variation of the medical instruments (Figs 9.2 and 9.3). A hollow needle is passed along the drill hole and light from a carbon arc passed along the needle to a sloping mirror which will direct the light around the concentric layers to a further mirror placed back-to-back with the first, thus allowing it to be seen at the other end of the drill-hole. In cultured pearls the light is directed to the surface of the pearl where it appears in a mirror as a flash resembling a momentary

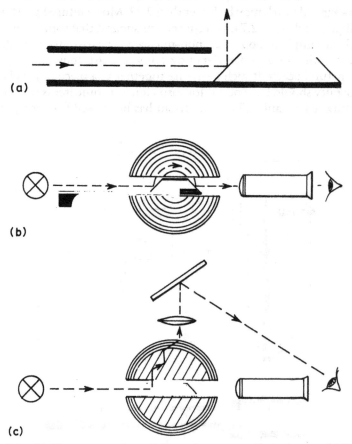

Figure 9.3 (a) The construction of the endoscope needle is shown. (b) Light is channelled round the concentric layers of a natural pearl to produce a flash of light in the viewing microscope. (c) With a cultured nucleated pearl, the light escapes through the mother-of-pearl bead, and can be seen as a line of light on the pearl's surface.

cat's-eye effect. The disadvantages of the endoscope include the high cost of the needles which break easily (and are now only made to order), the slowness of the method compared to X-radiography and the need to test one pearl at a time. All pearls tested must be drilled.

While most natural black cultured pearls glow an indistinct reddish-brown under long-wave UV, treated black pearls are appearing on the market in increasing numbers. These are usually inert to UV radiation but some fluoresce slightly greenish under long-wave UV. Necklaces are usually uniform in fluorescence.

Imitation pearls are glass, either filled or coated with an appropriate substance. Coating is with pearl essence made from the scales of the bleak (*Alburnus lucidus*) or a synthetic substance (essence of orient) giving the same effect. When the coating is dry the interior is filled with wax to give the required solidity and weight. Reflection at the inner shell doubles a pencil spot and they feel smooth to the teeth, compared with the slight roughness of the natural and cultured pearl. This comes from overlapping aragonite plates. The SG of the hollow beads may be 1.55 or even lower while the solid bead types are higher at 2.85–3.18. Plastic beads weighted with small metallic spheres have been reported as centres for imitation pearls. Finely divided particles like those used to give colour play in Slocum stone give iridescence and are set between the metallic core and the surface.

10 Synthetic and imitation stones

Gemstones have been imitated by other substances, particularly glass, for as long as people have used ornament, but the actual manufacture of gem materials with the same chemical composition and physical properties as existing gemstones dates only from the middle of the nineteenth century. There has long been confusion over whether synthetic is the correct adjective to use when describing ornamental man-made substances; some would restrict the term to those substances which have a natural counterpart, others use it in connection with any man-made substances. Inevitably the word has become loosely applied over the years. With the development of Trades Description Acts, the importance of this question is now more than merely academic but I must leave it to readers to apply their terminology sensibly.

10.1 CRYSTAL GROWTH FROM SOLUTION

The various ways in which a crystal can grow account for variation in colour, size, shape and included material, all factors which the gemmologist needs to assess when testing a finished stone.

Crystals grow from a saturated solution, that is one in which no more solid can be dissolved. The concentration at which saturation occurs varies with the solution and the temperature; some solutions, on heating, will accept more solid. When this point of saturation is reached any undissolved solid is removed, the solution is transferred to another container and allowed to cool back to room temperature. If solid added on heating is not rejected on cooling, the solution is described as supersaturated. It is from this supersaturated solution that a crystal can be grown; solid added now will grow until the solution returns to saturation. The supersaturated solution is unstable, there being a critical composition and temperature at which excess solid will be rejected. Sometimes rejection only commences by the addition of a piece of the solid, or even dust particles, upon which the excess can be deposited.

The growth of crystals is still as much an art as a science since it is difficult to obtain the desired size of crystal and to ensure its freedom from dislocations and inclusions. Almost all crystals have defects even though they may appear perfect on fairly rigorous examination. In most cases, (except where colour is influenced) these do not matter very much if the crystal is to be a gemstone; many defective crystals are unacceptable for electronics applications, however. The gemmologist does have trouble with defects such as veil- or smoke-like structures which result from too fast a growth rate; these are especially prominent in synthetic emerald. Supersaturation is low at the centre of a crystal face so that layers spread from the corners and edges of the face and grow over the centre, trapping in their path any undigested growth liquor they encounter. Such inclusions may also be caused by the formation of cracks during growth; these enclose films of solvent which are kept in place by subsequent 'healing' of the crack. By this stage the films often have broken up into small droplets.

10.2   GROWTH BY FLAME-FUSION

A simple way of growing some gem-quality materials, notably corundum and spinel, is by fusion of their constituent ingredients in a hot oxy-hydrogen blowpipe flame. Corundum has a melting point of nearly 2050°C making the use of most crucibles impossible as they will yield some of their substance to the crystal being grown. This problem can be overcome by the use of expensive noble metal crucibles which are also used for growing many gem materials by other methods. The flame-fusion method is cheap and was developed for the growth of corundum for industrial use (watch-bearings in particular). Growth is quick (a one carat stone can easily be obtained in a matter of hours compared with the months needed for some of the other growth methods). Quality is quite acceptable, although most products are not hard to identify.

The use of a downward-pointing blowpipe for the fusion of alumina powder dates back to the work of Gaudin in 1837; the corundum made took the form of tiny hexagonal platelets and it was not until 1877 that the Frémy and Feil paper was published giving an account of the synthesis of ruby in clear though small crystals. Verneuil, Frémy's assistant, continued the work after Feil died in 1876; he obtained very clear rhombohedral crystals but could not resolve the question of size. The so-called 'Geneva' rubies came on to the market in 1885 and were sold as natural stones; they turned out to be made from small boules (shaped like buttons) made by a technique similar to that established later by Verneuil. A three-step method was used and the boule rotated during growth. The name 'reconstructed', to designate a wide variety of synthetic products was and is entirely inappropriate, since it is not possible to fuse together small ruby fragments to make a large single crystal. Early flame-fusion stones of this kind show markedly curved growth lines.

Among those who examined the Geneva product at the time of its introduction was Verneuil who made the important discovery that melting the ingredients needed for ruby does not necessarily result in a glass, as was originally thought. This was the impetus needed for the perfection of the blowpipe process; full publication of the details came in 1904. Simply described, the present-day Verneuil process involves the fusion of alumina powder in an oxy-hydrogen flame, the molten droplets falling upon a ceramic pedestal and building up into a sintered cone (Fig. 10.1). The tip of the cone melts and enlarges while the pedestal is lowered to keep the upper surface of the growing crystal (boule) in the optimum position with respect to the flame. The name boule refers to the characteristic shape of the finished single crystal. The flame is shielded by a ceramic muffle. Growth ceases when the flame is shut off and the propensity of the boule to crack in a longitudinal direction is indulged by nipping the end. This cracking is due to the sudden change in temperature when the flame is shut off. The feed powder used for ruby was and is a mixture of ammonium alum and chromium alum, $(NH_4)Al(SO_4)_2 \cdot 12H_2O$ and $(NH_4)Cr(SO_4)_2 \cdot 12H_2O$, respectively. On firing to give the correct mixture for crystal growth the water is driven off with the ammonium sulphate and sulphur trioxide.

The method of preparation of the feed powder is the same today as in Verneuil's time. Typical boules were about 6 mm in diameter and about 2.5 cm long in the early days but can now attain much larger sizes. The method of delivering the powder to the flame involves the repeated tapping of a hammer mechanism, a vibrating apparatus having been found less satisfactory. The hydrogen gas may be from mains supply though some is obtained from the electrolysis of water, which also gives some oxygen. Liquefaction and fractionation of air may also give oxygen. The particles falling through the flame are only partially melted, complete melting taking place in a very thin layer on the top surface of the boule. The lowering rate is usually about 1 cm per hour; up to 30% of the feed powder may be lost.

The feed powder contains 1 to 3% Cr_2O_3. From half to two-thirds of the chromium in chromium alum used for the growth of ruby is lost in burn-off (the differential volatilization in the flame). A wide range of colours is produced; iron contamination gives the rubies a natural brownish tinge while magnesium gives an orange through a coupled substitution of Al^{3+} by Mg^{2+} and Cr^{4+}. Present-day boules may reach 9 cm in diameter. Colours of corundum produced by the Verneuil method include:

- Orange by adding nickel, chromium and iron.
- Yellow by adding nickel.

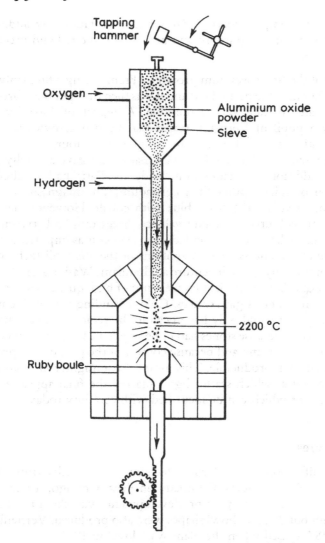

Figure 10.1 Diagram of a Verneuil furnace. Finely divided aluminium oxide powder passes from the hopper through a sieve with the aid of a tapper and then through the flame fed by oxygen and hydrogen. As the boule crystallizes, it is slowly lowered in the furnace.

- Yellow–green by adding nickel, iron and titanium.
- Green by adding cobalt with either vanadium or nickel (strongly reducing conditions are needed for satisfactory growth with included cobalt).
- Blue by adding titanium and iron (see below).
- Purple by adding chromium with titanium and iron.

• A colour-change corundum (alexandrite imitation) by adding vana-dium (the colours are purplish and mauve, not the red and green of true alexandrite).

Some of the first corundum grown by Frémy using a flux (solvent) was violet rather than the desired red and it was then thought that chromium in another valence state was responsible. A report of 1908 by L. Paris, formerly a pupil of Verneuil, stated that he had grown blue sapphire crystals with cobalt and magnesium to give the colour.

Further work by Verneuil showed that cobalt gave a patchy blue and that the addition of magnesium oxide by Paris had resulted in the formation of spinel rather than corundum. The magnesium aluminate spinel does give a satisfactory blue with cobalt. However, natural blue sapphires are coloured by an intervalence charge transfer between iron and titanium and when Verneuil added these elements as impurities he did get the sapphire blue he wanted. Variations of the Verneuil technique were applied in the ruby growth venture at Hoquiam, Washington, USA in the early part of this century. Boules were not of high quality and contained a very large number of gas bubbles. Colour banding is prominent and the largest stone cut could only have been in the region of 2 ct. The venture probably failed because stones had to be shipped a long way (even to New York City) for cutting and because control of the production process was not sufficiently reproducible. This meant that unsightly inclusions such as the gas bubbles (which scatter light) marred the final appearance of the stones, few of which can be identified with certainty today.

Star stones

Later modifications of the Verneuil technique came with its introduction at the Linde Air Products Company in East Chicago, Indiana, USA. Probably the most important development was the growth of star corundum but disc and bowl shapes were also produced. Verneuil growth in the USA ceased when the plant was closed in 1974.

Verneuil star stones are grown in the usual way but the feed powder contains about 0.1 to 0.3% TiO_2. After growth the boule is heated again to about 1300°C at which temperature the rutile (TiO_2) precipitates out in the form of needles which show as a six-rayed star when the stone is cut as a cabochon with the optic axis at right angles to its base. A snag in this method is that the titanium tends to move to the outside of the growing boule so that the finished stones may have rays which are too short. Transparent regions may also be apparent. If, however, the oxygen glow to the blowpipe is varied by a few per cent every few seconds, a thin layer of the growing boule solidifies before deposition of the next layer thus preventing the titanium migrating from the centre to the sides of the boule.

Some Verneuil-type crystals

Corundum Verneuil-corundum has constants similar to those of the natural material, so the best way to distinguish Verneuil-corundum from natural material is by examination of the inclusions. Gas bubbles with bold outlines (due to the marked difference between the RI of the gas and that of the host) are prominent; some resemble tadpoles and others form clots. Some are found enclosed in triangular cavities. Curved growth lines like the grooves on a gramophone record are best seen when the specimen is immersed (methylene iodide with an RI of 1.74 serves well); this effect can be photographed with a stopped-down lens and a narrow beam of light. The lines are best seen at right angles to the table as the table is cut parallel to the sides of the boule. This also accounts for the dichroism so often seen through the table. Colour distribution will be curved rather than angular as in natural stones. Most synthetic Verneuil rubies are more transparent to short-wave UV than natural stones (which often contain some iron) and will phosphoresce after exposure to X-rays.

In blue sapphire the colour banding is more prominent than in ruby; the 450 nm complex of absorption bands is not seen and under short-wave UV the synthetic stones show a rather dusty whitish or greenish fluorescence on the surface, best seen in the dark and when the specimen is viewed on edge. Some natural blue sapphires when free from iron may show this (Sri Lanka stones especially); however, examination of the synthetics while fluorescing will show the curved structure lines on the surface. Unlike the colour banding, the curved lines in synthetic sapphire are as fine as those in its ruby counterpart. Some Verneuil-grown blue sapphires have been found to show polysynthetic twin lamellae when observed between crossed polars. Normally regarded as a characteristic of natural corundum, the presence of the lamellae should be noted and the specimen further tested to see if Plato twinning can also be found before classifying the stone as natural. One stone tested did not show the usual bluish-green glow under short-wave UV as seen in most synthetic blue corundum; it may have been surface diffusion treated. Crack-like markings resembling the patterns made by wavelets on sand can be seen on synthetic corundum, particularly ruby, and are the result of local overheating during polishing. They are found near facet junctions and are known as 'fire' or 'chatter'-marks.

The other colours of corundum may be harder to test than ruby or blue sapphire. Yellow stones may give a red fluorescence under long-wave UV or under X-rays, with some phosphorescence. They will show no absorption spectrum as some natural yellow stones do, nor will they give a yellow fluorescence as other natural yellows may. Curved structure lines are virtually impossible to see. Green synthetic sapphires will not show the absorption spectrum of three broad bands in the blue shown by natural stones. The vanadium–doped corundum intended to imitate alexandrite can easily be spotted by the strong absorption line at 475 nm; curved structure lines are very well shown by this variety whose colour is also quite distinctive to the experienced eye. Some synthetic corundum of an amethyst colour has been made in which curved lines and bubbles are hard to see. Long-wave UV exposure gives a red glow and short-wave UV gives a bluish-white glow, never shown by either natural purple–violet sapphire or by amethyst.

Generally speaking, Verneuil corundum has a rather stark colour; the absence of natural inclusions probably has something to do with this, as might the presence of only one colouring element. Stones are generally cut by completely mechanical processes using such cuts as the 'scissors' which is hardly ever used for natural stones.

Spinel Spinel is $MgAl_2O_4$ but because material with this composition was hard to grow, extra alumina was added, giving a composition including from 1.5 to 3.5 parts Al_2O_3 for each MgO. If up to five parts Al_2O_3 is used, a moonstone imitation is produced. The significance for the gemmologist is that SG and RI are raised from the natural values of 3.60 and 1.718 to 3.64 and 1.728. Stones also show anomalous birefringence between crossed polars and contain a variety of shaped bubbles, some of which resemble hoses, others furled umbrellas or long flasks. They are in fact negative crystals with a hexagonal shape in cross-section. They are occasionally grouped into hexagonal patterns. Two-phase inclusions are sometimes seen and cavities with gas or liquid and a bubble are sometimes joined to each other by thin tubes. Red spinels have prominent swathes of colour and clouds of small gas bubbles; the inclusions sometimes resemble Venetian blinds; these stones have an SG of 3.60 and RI of 1.725. They show a red fluorescence but not the group of emission lines in the red part of the spectrum shown by many natural red spinels. In their place a single emission line may be seen.

Blue spinels show red through the Chelsea colour filter from their cobalt contents and some blues give a reddish fluorescence from chromium. Their cobalt absorption spectrum has bands at 635, 580 and 540 nm, the central band being the widest (in cobalt glass it is the narrowest). In deep blue stones the bands coalesce. Lime green stones which give a strong fluorescence of the same colour show two bands in the violet from manganese.

Synthetic colourless spinel (there is no natural counterpart of this variety) has a dispersion of 0.020 – not high but enough to give a surprisingly good imitation of diamond in small sizes, especially in a piece with other small diamonds. Most of these will show a bluish-white fluorescence under short-wave UV.

Much synthetic spinel imitates aquamarine, blue zircon and peridot, the constants giving it away. A rare dark green variety with red flashes makes a fair alexandrite imitation, better than the vanadium-doped corundum stones.

Rutile Rutile, titanium dioxide, is common in nature, but is too dark for regular gem use (although stones have been cut). Made by the Verneuil process, rutile can be lightened to a yellow (though never a white) and its very high dispersion makes it of some ornamental value though it has a hardness of around 6. The SG is 4.25 and RI (ω) 2.62 (ϵ) 2.90, with a DR of 0.287 and dispersion near to 0.3. It is a member of the tetragonal system and shows a cut-off at the violet end of the spectrum. It is made by a modified technique designed to provide extra oxygen since rutile loses a lot when melted in the usual blowpipe. The modification provides for an extra oxygen-carrying tube, making a tricone burner. One volume of oxygen needs two of hydrogen for effective combustion; in the conventional blowpipe up to three volumes of hydrogen are used, giving a strongly reducing flame. With the tricone burner the oxygen volume can be increased to give a ratio closer to 1.5 volumes of hydrogen per volume of oxygen, giving a strongly oxidizing flame which is larger and hotter than the customary one. Even this flame cannot retain all the oxygen in rutile at its melting point but crystals can be grown. Boules are black (from deficiency of oxygen) but annealing at about 1000°C in pure oxygen will give first a metallic blue, then a paler blue and lastly a pale yellow. Impurities to give red or yellow (the red inclining to brown) have sometimes been added and star rutile has been grown by adding about 0.5% MgO to the TiO_2 and annealing the boule near 1300°C in oxygen.

Strontium titanate The growth of rutile other than for research purposes fell off on the introduction of strontium titanate, grown, as was much rutile, by the National Lead Company in the United States. This material had at least one advantage over rutile in that it is isotropic whereas rutile, a tetragonal mineral, is highly birefringent. Strontium titanate is also colourless; rutile, because of absorption in the deep violet, always has a yellow cast. Unlike rutile, strontium titanate does not exist in nature.

For growth a three-stage torch is needed and the black boules need to be annealed at over 1000°C in oxygen. Occasionally two annealings are needed, one for the removal of strains, the other to improve colour. The dispersion is not as high as that of rutile, though still higher than that of diamond. The stone is both soft (5.5) and brittle. Specific gravity is 5.13 and RI is 2.41, with dispersion of 0.19. It belongs to the cubic system, can be marked by the point of a needle and shows ladder-like surface markings. Impurity additions have given a variety of colours (rarely seen) including yellow, orange, red and blue. It is at its most deceptive when used as the base of a composite whose crown may be a harder colourless material such as synthetic spinel, synthetic corundum or cubic zirconia. Whole stones turn up as melée and are difficult to detect in those small regular sizes.

10.3 HYDROTHERMAL GROWTH

Crystals of a number of materials, especially quartz, have been grown by the hydrothermal method, the only kind of crystal growth echoing any process known in nature. The crystals grow in a closed vessel (autoclave) in which high temperatures ($\sim 400°C$) and pressures (~ 1700 atm) dissolve the nutrient (Fig. 10.2). The autoclave has a liner of noble metal (or of gold or silver) and seed crystals are suspended in the upper portion in a platinum cage or on platinum wires. Nutrient with a mineralizer (sodium carbonate

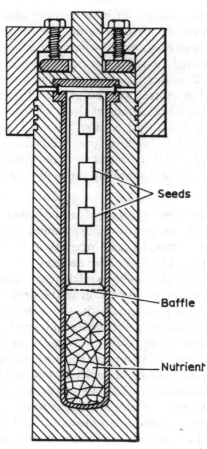

Figure 10.2 A silver-lined laboratory hydrothermal autoclave, about 35 cm (14 in) long.

or sodium hydroxide are regularly used because they increase solubility by an order of magnitude) is placed in the lower portion of the autoclave which is heated from below; molten material is transported by convection to the upper portion where it is deposited on the seeds. These are cut into the orientation desired for the completed crystal, which can be larger than those obtainable by the cooling of a saturated solution. Advantages of the hydrothermal method include the ability to grow materials whose high vapour pressure near their melting point makes them hard to grow by other methods; the sealed autoclave helps to prevent contamination and the close control which can be exerted over the operation can give crystals of good size and purity.

Ruby and emerald have been grown by this method as well as quartz. Ruby production has been largely experimental, with NaOH as a mineralizer and sodium bichromate added to give the red. Corrosion of the container by the solvent is a serious problem and may be responsible for the preferred commercial growth of ruby by other methods. Partially healed fractures with undigested drops of liquid are common in those rubies which have been made. The hydrothermal growth of emerald has been more successful – at least one firm is making them at the time of writing. These crystals (from Vacuum Ventures of New Jersey, USA) have prominent pointed growth tubes emanating from phenakite crystals and sometimes traces of a colourless seed can be seen, possibly with growth tubes leading away from the seed and tapering in the direction of growth.

Growth rates for both ruby and emerald are much slower than those in the Verneuil method: a rate of 0.25 mm per day is fairly fast. Constants for the stones are similar to those of the natural material in the case of ruby; for emerald the Vacuum Ventures production (the Regency emerald) has an SG of 2.67–2.69, a hardness of 7.5–8 and RI 1.570–1.576 with a DR of 0.005–0.006. The stones are not transparent to X-rays nor do they fluoresce but they do show a strong absorption band at 477.4 nm in addition to the characteristic emerald spectrum. The Regency stones are said to be made from a technique originally patented by Union Carbide and at the time of writing are sold as crystals only. In some of the recently released hydrothermal emeralds from the USSR chevron-shaped growth zoning is reported. The RI of the stones is 1.574–1.580 with DR 0.006. The SG is above 2.67 as is customary with hydrothermal products. Fluorescence is doped with iron. Three-phase inclusions have been found in the hydrothermal 'Regency' stones but a report published in 1987 does not believe that the inclusions are artificially induced.

The Austrian manufacturer Johann Lechleitner produces emerald-coated beryl in which the centre is made of pre-cut colourless beryl or pale synthetic emerald and the coating is imposed hydrothermally (Fig. 10.3). Fractures at the junction of seed and overgrowth give a crazy paving effect which is characteristic, as well as parallel lines at right angles to the other set,

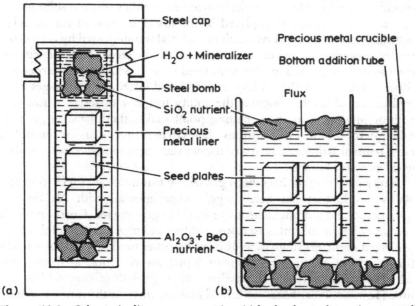

Figure 10.3 Schematic diagrams comparing (a) hydrothermal reaction growth and (b) flux reaction growth for synthetic emerald.

giving a crackled appearance. Tiny dust-like crystals of euclase or phenakite occur at the junction. The SG of these coated emeralds is near 2.695 and the RI is 1.581 and 1.574 with a DR of 0.006. Lechleitner has also attempted a sandwich emerald in which a seed plate of colourless beryl is covered by emerald hydrothermally in his customary way; the stone is then further enlarged by hydrothermal growth which is also colourless. This type of stone needs a closed setting to conceal the various layers. Lechleitner has also grown some solid hydrothermal emeralds. Some of his products were polished only on the upper facets. Linde emeralds (from the Linde Division of Union Carbide) were produced from 1965 till 1970 and were used in especially manufactured jewellery (the Quintessa line). In the growth of hydrothermal emerald a hydrothermal reaction process is used in which the water boils to fill the vessel with one fluid and increases the pressure with increasing temperature until quartz (as an emerald component) dissolves at the top of the autoclave and the other ingredients (Al_2O_3 with BeO) dissolve at the bottom. By convection and diffusion the reagents meet in the centre of the vessel and react to form emerald in solution which then is deposited on the prepared seeds. Too much mixing may give phenakite rather than emerald and other emerald crystals may develop in addition to the ones desired. Temperatures vary from 500 to 600°C and pressures from 691 to 1382 atm. Growth rates of 0.3 mm a day can be attained.

Quartz Quartz made hydrothermally includes the varieties rock crystal, amethyst, citrine, a green quartz and a blue quartz, the latter coloured by cobalt and thus appearing red through the Chelsea colour filter. The blue colour is obtained after doping with cobalt and reduction by heating; brown is obtained by the addition of iron, dark brown by the addition of aluminium and subsequent irradiation. Green quartz also needs iron with reduction, amethyst needs iron with irradiation, citrine needs iron alone and gamma irradiation with heating gives a yellow-green. Brown may be due either to Fe^{3+} ions or to iron silicate particles finely dispersed in the quartz; on heating the brown stones turn green from reduction of Fe^{3+} to Fe^{2+}. Dark brown smoky crystals owe their colour to the operation of colour centres brought about by sodium, lithium or aluminium impurities which activate a colour on irradiation. Some germanium may also be added before irradiation. Constants are as natural quartz; some stones may show breadcrumb-like inclusions but the absence of natural inclusions is probably the best means of identification.

A synthetic amethyst produced in Japan has feather-like inclusions made up of liquid-filled and two-phase structures as well as sharp growth zoning parallel to one rhombohedral face. Twin structures are also seen but can be distinguished from polysynthetic twinning common in natural amethyst. Synthetic amethyst can be identified by a technique devised by Karl Schmetzer in which the stone is examined between crossed polars or by the use of a horizontal immersion microscope. The optic axis must be parallel to the path of light through the microscope or through the crossed polars of the polariscope. In virtually all natural stones Brazil twinning can be identified by the presence of interference colours rather than a succession of broad colour bands shown by the synthetic stones. Schmetzer suggests that a special holder be used but later work has found this to be unnecessary in most cases. In an amethyst made in Japan a characteristic small arrow-head shaped or flame shaped structure has been observed. By the end of 1986 some dealers were reported that up to 25% of their amethyst stock had been shown to be synthetic by the use of this test; it is possible, too, that the same test may be used on the 'ametrine' amethyst–citrine stones, some of which are thought to be artificial. Though signs of hydrothermal origin are rarely seen in synthetic amethyst, GIA has noted acicular crystals and spicules with quartz crystal caps.

10.4 CRYSTAL PULLING

The majority of the better-quality man-made gemstones (particularly ruby, sapphire and emerald) are grown by a liquid–solid equilibrium method which needs less equipment than other methods.

In crystal pulling a long rod-like crystal is pulled from a melt;

Czochralski first used this technique in 1917 and it is often named after him (see Fig. 10.4). The melt is prepared in a crucible of suitable material and a seed is lowered until it just touches the melt surface. It is then slowly drawn upwards taking some of the melt with it. To ensure that growth can begin on a clean surface the temperature of the melt is raised just above the melting point so that the seed (of the same material) can also be partly melted. Diameter is controlled by adjusting the temperature and cooling rate.

Crystals of laser quality are grown in this way, one of the first being scheelite, $CaWO_4$ which has been used as a diamond simulant despite its birefringence. An iridium crucible is needed but the material can be grown in air and appropriate dopants give colour where required. Another birefringent diamond simulant, lithium niobate $LiNbO_3$ (Linobate) is also grown this way and colours are occasionally reported. For ruby the seed is rotated at 10–60 rpm and withdrawn at speeds from 6 to 24.5 mm h^{-1}. The only inclusions found with any frequency in material produced by this

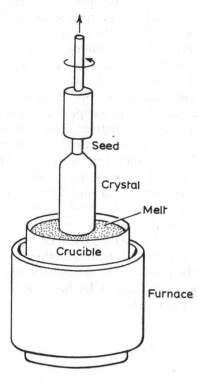

Figure 10.4 Diagram of a Czochralski apparatus. The crucible is filled with powder and melted in the furnace. A rod with a seed crystal attached to the lower end is lowered until it touches the melt and is rotated and slowly withdrawn, 'pulling' the crystal from the melt.

method are elongated bubbles. The oxide group $A_3B_5O_{12}$ is grown by both this method and by the flux-melt technique. For gem use the varieties YAG (yttrium aluminium oxide) and GGG (gadolinium gallium oxide) have been grown; both, but more particularly the former, are used as diamond simulants when colourless. Dopants give colours (see below).

Crystals grown by the Bridgman-Stockbarger process are cooled as feed powder in a cylinder tapered at one end. Once cooled to near the freezing point of the powder the temperature is further lowered at the tapered end to allow a single seed crystal to form there; on cooling this grows into the container until it occupies all of it. Some large fluorides, including fluorite and doped analogues are grown by this technique.

10.5 FLUX-MELT GROWTH

The commonest method of growing ruby and emerald, as well as other gem materials, is from the melt after first being dissolved in a solvent or flux. The advantage of the flux is that growth can take place at temperatures lower than would otherwise be possible and there is no need for crucibles which will withstand high temperatures (platinum or iridium are routinely used). A saturated solution is prepared by keeping the constituents of the flux and of the desired crystal at a temperature slightly above the saturation temperature until a complete solution forms. The crucible is then cooled through a temperature range in which the crystal is known to precipitate. Nucleation (the first growth of the crystal) will usually start in the cooler part of the crucible and when one nucleation has taken place no others should start, so that the crystals from the first nucleation should reach maximum size. When the cooling process is complete, crystals are removed from the crucible by chemical dissolution of the flux or by hand. Two methods are used: 1. growth by slow cooling where the melt precipitates on a seed and 2. growth on a seed placed in a cooler part of the system while the remainder of the solute is in contact with the solvent in a hotter part (this is called growth in a thermal gradient and is similar to aspects of hydrothermal growth). The flux should have a low melting point, be a good solvent for the ingredients required and must not form a compound with the solute.

Emerald

Emerald was grown in this way in the 1930s by IG Farbenindustrie in Germany, the stones being distinguishable from those made hydrother-mally (and from natural stones) by the absence of water which gives a distinctive absorption spectrum in the infra-red. The emerald manufac-turers Chatham in San Francisco and Gilson in France are the two chief commercial manufacturers today. Although many details of the process used are kept secret, the main procedures are fairly well established. Chatham, whose emeralds are marketed under the name Chatham Created Gems, produces crystal groups as well as single crystals. Growth can take up to 12 months; traces of lithium, molybdenum and vanadium show that this mixture is used in the flux.

It is not certain whether Chatham uses purified starting materials or natural beryl or emerald. Natural material has the disadvantage of possible mineral inclusions and time-consuming sorting would be necessary to eliminate such pieces, which might give rise to unwanted nucleation. It is more likely, therefore, that pure chemical starting material is used. The probable flux reaction method is as follows: BeO and Al_2O_3 are in the bottom of the crucible with a platinum screen above them and SiO_2 glass lumps above the screen, all in a flux of $Li_2O–MoO_3$. Growth by diffusion takes place when the SiO_2 dissolves in the flux near the top surface and slowly diffuses downwards while the alumina and beryllia dissolve at the bottom and move upwards. They meet in the centre to form emerald in solution which then crystallizes on to the growing emerald crystals which are just below the platinum screen. This method was used by Espig in the emeralds grown by IG Farbenindustrie who produced the Igmerald which was never placed on the market in the end.

The Gilson emerald is grown by the flux transport method (Fig. 10.5). Growth is first started on a natural colourless beryl seed which is coated with emerald on both sides. The colourless seed is cut away from the emerald sections which are then used as seeds themselves for the run proper. The growth rate is about 1 mm per month. The feed beryl is placed on one side (the hotter side) of the platinum crucible with a screen separating it from the growing crystal on seeds in the other (cooler) side. Mechanical stirrers assist the natural convection currents. To prevent the spontaneous

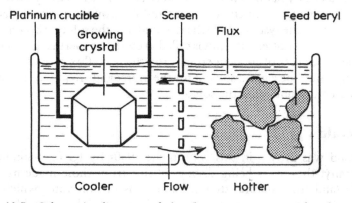

Figure 10.5 Schematic diagram of the flux transport growth of synthetic emerald used by Gilson.

growth of phenakite or of other emerald crystals the supersaturation must be kept to a minimum so that the growth rate is kept low. Both single crystals and point-nucleated clusters are grown. Gilson tried at one time to produce an emerald which would not give the strong red fluorescence characteristic of most synthetic emeralds. He accomplished this by adding a trace of iron which gave an absorption band at 427 nm not seen in natural stones but this so called 'N-series' was only made for a short time.

Many workers have experimented with the flux growth of emerald, particularly with different fluxes, but few of the resulting products could be made commercially successful; such specimens would, however, be of great interest and value to specialist collectors. The latest commercially successful flux-grown emerald is that grown in Japan by the Kyoto Ceramics firm (Kyocera) who call their attractive emerald Crescent Vert. This is only sold in the range of jewellery made for the firm and distributed through their retail outlet in Kyoto.

For historical interest and accuracy it should be said that the emerald grown by Richard Nacken (1884–1971) in Germany from about 1916 to 1927 or 1928 has now been proved to have been made by the flux-melt process through the absence of the infra-red bands of water in the emerald absorption spectrum. His emeralds contained cuneiform nail-like inclusions formed by the nucleation of a small crystal of phenakite followed by an inclusion of flux and the tapering of the whole structure. The stones also contained two-phase inclusions in the form of a fluid-filled cavity containing a gas bubble. These inclusions are seen in many flux-grown emeralds. Modern stones characteristically have SG in the range 2.65–2.67 (lower, generally, than the figures obtaining for hydrothermal stones) with RI (ϵ) 1.560–1.563 and (ω) 1.563–1.566 with DR 0.003–0.005. The most characteristic inclusion is made up of a smoke- or veil-like structure which can occur profusely in lower-quality stones and which can be seen very well in the production by Zerfass of Idar-Oberstein, West Germany. This emerald was the descendent of the Igmerald (see above) and grown for a short time only so specimens are rare. Gilson grades his cut emeralds on a clarity basis, the top quality showing scarcely any included material while the lower grade has quite a lot of 'smoke'. Other inclusions may be phenakite crystals and flakes of platinum from the crucible. Stones show a bright red through the Chesea colour filter. It should be remembered that most synthetic emeralds transmit short-wave UV down to about 230 nm. Natural stones transmit only down to about 300 nm.

Russian flux-grown synthetic emeralds described in *Gems and gemology* **21**, 2, 1985, have RI 1.559–1.563 with DR 0.004. The SG is near 2.65. Flux inclusions may show reddish; platinum crystals have been seen as well as healed fractures. The emeralds are in most ways characteristic of other flux-grown material. Seiko emeralds made in Japan have RI 1.560–1.564 and SG 2.655. The stones show colour zoning in green and colourless layers parallel to the table and sub-rectangular flux pieces oriented in one general direction, contained in a plane between the colour zones. They also show included crystals situated just above a plane between the colour zones and a dust-like mass composed of tiny flux particles. A report in *Gems and gemology* **21**, 3, 1985, describes the Biron synthetic emerald. Made in Australia, it is thought hydrothermally, it has a notably transparent appearance with RI 1.569 and 1.573 with DR 0.004–0.005. The SG is 2.68–2.71 and the stones do not luminesce. Fingerprints, veils and fractures are noted with some nail-head spicules with liquid and gas phases. White particles form 'comet-tails'.

A beryl with a pink core and green skin, resembling a water-melon tourmaline, is made by the Japanese firm of Adachi Shin Industrial Co. of Osaka. The RI is 1.559 (ϵ), 1.564 (ω) and the SG is 2.66. The green portion shows absorption lines at 660, 620 and 477 nm; no absorption is shown by the pink section. The stones are said to be grown by a new method in which fluorine and oxygen react at higher temperatures than is customary in beryl growth with crystalline or amorphous beryllium oxide, silicon dioxide and aluminium oxide. Dopants are added to give the colour. The mixture then migrates to a cooler zone of the apparatus and thence on to seeds.

Corundum

Growth of corundum by methods other than flame-fusion did not really get under way until the 1950s when laser work became important, necessitating the development of better quality crystals. In a flux-melt process patented in 1963 a flux of lead oxide was used with some boron oxide added. Corundum ingredients were dissolved in the flux at 1300°C in a small platinum crucible and cooled at a rate of 2°C per hour over a period of eight days. When the temperature had reached 915°C the flux, still molten, was poured off and the ruby crystals recovered by dissolving the rest of the flux in dilute nitric acid.

Today the most common fluxes are probably either a combination of PbF_2 with PbO or a combination of PbO with B_2O_3. Platinum crucibles may be up to 20 cm in diameter and 20 cm in height. Powder contracts on melting and the crucible may need to be replenished as growth proceeds. Heat is applied from the sides of the crucible and the crystals grow at the bottom. The crystals vary from flat hexagonal shapes to virtually equidimensional rhombohedra. The flux chosen has a great influence on the final shape, flat crystals occurring most often when there is a high PbF_2 content and lower temperatures; equidimensional crystals are obtained with higher temperatures and high PbO, MoO_3 or Bi_2O_3 content. Growth near the bottom of the crucible also favours equidimensional growth.

Chatham grows ruby and crystal groups of orange and blue sapphires, the individual crystals of which are usually too thin to cut. In recent years the Kashan ruby has been prominent on the market. This was first reported in 1972, the manufacturer being Ardon Associates of Dallas, Texas. The stones showed several interesting inclusions, notably flattish, elongated forms with parallel sides and rounded ends which have been called 'paint-splash' inclusions or 'moccasin prints'. A report by Henn and Schrader (1985) grouped inclusions in Kashan rubies into four types: feathers, fingerprints, strings of pearl-like bodies and comet- or hairpin-like inclusions. All are melt residues. Microprobe analysis was successful in finding inclusions in stones of apparently near-perfect clarity. Later Kashan stones which have turned up in parcels of rough natural ruby in Bangkok have shown dust-like inclusions with some traces of the 'smoke' or twisted veil structures. Some Kashan stones may contain iron as they do not always transmit short-wave UV: iron absorption bands seem to bear this out. A recent evaluation of the Kashan ruby has shown that inclusions of cryolite are a useful diagnostic feature. At one time Kashan rubies were sold as crystals, crystal groups and cut stones, and this may still be the case.

In 1983 a new synthetic ruby, the Ramaura ruby, was marketed by Overland Gems, Inc. of Los Angeles. The stones were grown by Judith Osmer. Growth is by a high temperature flux method with spontaneous nucleation, according to the recent report in *Gems & gemology* **19**, 3, 1983. Some crystals show virtually equidimensional rhombohedral form and in the majority the rhombohedron is the prominent form with the basal form subordinate. Other crystals grow as thin plates with the basal form predominating. Stones are noticeably transparent and this may be due to the seedless growth method adopted, making forced growth unnecessary. RI was measured at (ϵ) 1.762 (ω) 1.770 (other stones had slightly lower indices) with a DR of 0.008. Specific gravity was found to be 3.96–4.00. There is nothing diagnostic in the absorption spectrum but some minute amounts of dopant are said to be added to the rubies to cause a fluorescence shift towards the orange–yellow; stones so far examined, however, do not bear this out, nor does EDS-XRF (energy dispersive X-ray fluorescence) show any trace of a dopant. Exposure to long-wave UV shows up a chalky red to orange–red fluorescence which may at times be a pure red similar to that shown by Verneuil rubies. Most crystals showed fluorescence zoning. Short-wave UV and X-rays showed variable responses but there was no perceptible phosphorescence after exposure to X-rays. The best indicators of Ramaura ruby as far as fluorescence is concerned are chalky yellow and bluish-white zones seen under long-wave or short-wave UV.

Inclusions are the best indicators of this product. Among them are orange–yellow and white flux fingerprints, inclusions resembling comet tails and fractures. Features of colour zoning also provide a clue; the use of dark-field illumination is recommended.

Very fine ruby crystals with unusual forms have been grown since the late 1970s by Professor Paul Otto Knischka of the University of Steyr,

Austria. Details of the growth process are unknown but crystals are produced from a melt using a technique involving a temperature gradient with supercooling and supersaturation. Crystals of a superb colour with many faces are produced as individuals, groups or twins. Some of the crystals have a pseudo-cubic habit and one crystal was found by Dr Eduard Gübelin to have 42 faces (most natural ruby crystals have no more than 20 and even this number is unusual). Other details are given in *Gems & gemology* **18**, 3, 1982. There are cloudy veils of unknown origin, together with healing fissures inside the stones (which are sold only as crystals at the time of writing). Some negative crystals characteristic of the growth method have been observed and many platelets of platinum or silver from the crucible are present in earlier products, although the manufacturer claims now to be able to suppress them. Large gas bubbles, probably the gas phase of two-phase inclusions, can be seen under low magnification. The SG is measured at an average of 3.976, RI (ϵ) 1.760–1.761 (ω) 1.768–1.769 with a birefringence of 0.008. A clear strong carmine red fluorescent colour could be seen under crossed filters and both types of UV. After X-ray exposure, stones showed a weak phosphorescence lasting about 7 s. Other features of the Knischka product are similar to those shown by other synthetic rubies.

Other colours of corundum The flux-melt growth of blue and orange sapphire by the firm of Chatham in San Francisco has been briefly mentioned above. Most of the production is sold as groups of bladed crystals but a few cut stones do exist (see *Gems & gemology* **18**, 3, 1982). Strong colour zoning is easily seen and the groups have a ceramic glaze on the base in which gas bubbles can be seen. Both blue and orange stones vary in shade within the stone – the orange groups showing a chalky yellow fluorescence in zones; the blue stones showing a variable reaction to UV, some being inert while others respond strongly giving a chalky greenish-yellow to reddish-orange in long-wave UV. Neither blue nor orange material phosphoresced after exposure to X-rays.

The orange stones show three sharp narrow lines in the blue part of the absorption spectrum, at 476.5, 475 and 468.5 nm. The first two form a doublet in which the line at 475 nm is very hard to see. Faint lines characteristic of chromium can be seen in the red and orange and there is a broad band of absorption blocking out a small part of the orange, all of the yellow and green with some of the blue. All the violet is absorbed. The blue stones show a faint diffuse band on the green side of 450 nm but neither this nor the spectra shown by the orange stones are any help with identification. Many angular platelets of platinum and wispy veils of flux inside the stones are good indications of artificial origin.

A specific gravity of 3.702 was recorded from a crystal cluster of Chatham orange sapphire, the low value being provisionally attributed to the ceramic glaze coating on the bottom of the cluster. Included gas bubbles may also have lowered the SG. The material showed a strong fluorescence under long-wave UV radiations, the colour being reddish to yellowish-orange. Natural orange sapphire recorded from Tanzania is practically inert as were samples manufactured by Djevahirdjan. The Japanese product gave similar results to the Chatham material. Inclusions of flux and platinum were noted. The electron microprobe revealed that fillings reaching the surface of a faceted blue sapphire were aluminosilicate glass. The glass appeared to be man-made.

Ruby and sapphire synthesized by Johann Lechleitner of emerald overgrowth fame are reported in *Gems & gemology* **21**, 1. Lechleitner has produced both solid synthetic emeralds and the better-known coated beryl; the ruby and sapphire recently examined are solid stones and other experimental colours including an orange padparadschah, alexandrite imitation, colourless and pink are reported. Some stones have been sold in Japan.

The Lechleitner ruby is described as a purplish red with medium tone and strong saturation. The blue sapphire is also medium tone with moderately strong saturation. Both stones are transparent with hazy areas and are relatively free from inclusions. Under magnification flux inclusions, the cause of the haziness, can be seen. The optic axis direction is nearly parallel to the table in the ruby and in the blue sapphire the axis is oriented about 20–30° away from the plane of the table. The ruby has an SG of 4.00 and an RI of 1.760–1.768 with a DR of 0.008. The pleochroism is a strong purplish-red parallel to the c-axis and a pale orange–pink at right angles to this direction. There is a strong red fluorescence but no phosphorescence under long-wave UV; under short-wave UV there is a moderate red fluorescence with slightly chalky white overtones and no phosphorescence. Under X-rays the fluorescence is a moderate chalky red and again no phosphorescence is observed. The absorption spectrum of the ruby has lines at 694.2, 692.8, 668, 659.2, 468.5, 476.5 and 475 nm with a broad band blocking all of the violet and some of the blue, all of the green and yellow and a small area in the orange. Flux fingerprint inclusions can be seen with wispy veils, ranging from transparent to opaque and from near-colourless to white. Their arrangement varies from thin tight patterns to loose mesh-like structures. Curved striae are reported.

The blue sapphire has an SG of 4.00 and RI of 1.760–1.768 with DR of 0.008. Pleochroism is strong violet–blue parallel to the c-axis with pale greenish-blue with some grey at right angles to this direction. Under long-wave UV there is a very weak chalky-whitish blue fluorescence and no phosphorescence and the effect under short-wave UV is similar with the fluorescence being opaque. There are no visible absorption lines or bands; there is a moderately strong broad absorption in part of the violet and in the far red area of the spectrum. There are more inclusions than in the ruby but they are of the same type. They appear in moderate to high relief and there is curved colour banding. Growth could have been effected by placing a seed grown by flame-fusion in the flux crucible; this would account for the curved striae observed. Alternatively a large Verneuil crystal with dopant added to give the required colour could have been placed in the flux crucible until a flux-grown corundum enveloped it.

Lechleitner is also producing various combinations of overgrowths, including synthetic pink corundum over synthetic Verneuil colourless corundum, synthetic ruby over Verneuil ruby and synthetic ruby over natural corundum.

10.6   MANUFACTURE OF IMPORTANT STONES

Garnet-type crystals

Flux growth of the so-called synthetic garnets, already mentioned briefly in the discussion of the Czochralski technique, dates from the early 1960s but the Czochralski method is much more satisfactory for these materials. The crystals are not garnets in the true sense since they contain no silica but they do have a structural similarity to garnet. The general formula is $A_3B_5O_{12}$ (or sometimes $A_3B_2C_3O_{12}$). The complex and frequent substitution of elements in natural garnets is echoed by the man-made products, many of which contain rare earths. Impetus was given to the Czochralski growth of synthetic garnet by the development of the neodymium laser, crystals for which could easily be grown by pulling. This with the related development of colourless crystals for diamond simulation took place in the 1960s, the composition being $Y_3Al_5O_{12}$ (YAG). This has an RI of 1.83 and dispersion of 0.028 (compared to the 0.044 of diamond); stones are somewhat lifeless, but they are hard at H 8.5 and have an SG of 4.55. The most common trade name is probably 'Diamonair' but many others have been used. A large production of characteristic crystals, 5 cm in diameter and 20 cm in length, ensured that supplies were plentiful and stones cheap. The advent of cubic zirconia with its higher dispersion really put paid to YAG as a diamond simulant. Coloured varieties continue to be made, however, although many are still experimental. These have been doped with chromium to give a fine green and rare earths give the following colours:

- Terbium   – pale yellow
- Dysprosium   – yellow-green
- Holmium   – golden yellow
- Erbium   – yellowish-pink
- Thulium   – pale green
- Ytterbium   – pale yellow
- Lutecium   – pale yellow

The fine green chromium–doped stones may also contain neodymium and show an absorption spectrum which is a combination of that of both elements. Blue stones very similar in appearance to aquamarine have also been seen on the market. Their constants are, of course, higher than those of beryl. Some YAG gives a yellowish fluorescence in long-wave UV with a bright mauve under X-rays with a discrete band in the yellow in the fluorescence spectrum. In general the synthetic garnets are not heavily included, although angular flux particles have been recorded for those specimens grown from the melt (in a technique paralleling the flux growth of ruby).

The analagous substance GGG ($Gd_3Ga_5O_{12}$) was for a time offered as a diamond simulant under the name 'Galliant' but the stones are softer than YAG and much denser (H 7, RI 1.97, dispersion 0.045, SG 7.02; figures from D. L. Wood of Bell Laboratories). Stones have a tendency to turn brown if trace elements are present; the UV component in sunlight may be sufficient to accomplish this. Trace elements also give some colourless GGG an interesting fluorescence spectrum under both types of UV and under X-rays. This is accompanied by a pale straw fluorescence under long-wave UV and a peach colour under short-wave UV with a lilac colour under X-rays; first reported by the author, these phenomena presumably arise from impurities. As with YAG, GGG can be doped; trivalent ions can occupy any of the three sites, depending on size. Chromium gives green, praseodymium yellow and neodymium a lilac colour. Elements with divalent ions need a coupled substitution with a four-valent ion (e.g. silicon). Blue stones can be obtained in this way by replacing $2Al^{3+}$ with $Co^{2+} + Si^{4+}$. A pink to red colour can be obtained from manganese coupled with silicon. Cut stones with these colours are rare but GGG crystals resembling those of natural garnet in form may be seen from time to time; I have red, green and blue–green specimens in my own collection.

Alexandrite

In 1973 a synthetic alexandrite was produced by Creative Crystals, Inc. of San Ramon, California, the process being patented in 1975. Both flux and pulling techniques have been used but the stones offered commercially show characteristic inclusions of flux. About 0.7% chromium and 0.3% iron are used to obtain the alexandrite effect. Similar stones are produced by Seiko and Kyocera in Japan. Chrysoberyl with a low chromium content (giving a very weak alexandrite effect) is routinely used as a laser material. The American stones show dust-like inclusions and smoke-like structures; they have an SG of 3.73 and RI of 1.746–1.755. A report of 1975 on a Czochralski-grown alexandrite gave SG 3.715 and RI 1.740 to 1.749 with a strong red fluorescence under long-wave and short-wave UV, and X-rays. Randomly oriented needles and lath-shaped crystals were included. The colour change was from greenish-blue in daylight to purple to violet in incandescent light.

Spinel

Spinel has been grown by the flux-melt process but, although a variety of attractive crystals and colours is obtainable, they have as yet no commercial importance. Red from chromium, blue from cobalt and yellow from nickel or titanium are among the available colours. The gahnite variety, rich in zinc, has also been grown in a variety of colours.

Cubic zirconia

For the manufacture of the latest and most successful diamond simulant, cubic zirconia (CZ, ZrO_2), a new method of growth had to be found since the melting point of CZ is 2750°C. The method used has been called 'skull-melting', the term referring to the shape of the framework of water-filled copper tubes within which the melting takes place (Fig. 10.6). The 'skull' is placed inside a copper coil connected to a radio-frequency source (4 MHz and 100 kW). Energy from the source penetrates between the 'fingers' made by the copper tubes and enters a block of zirconia with some zirconium metal added to ensure the heating up of the powder which contains either calcium or yttrium as a stabilizer. Although zirconia exists in nature as the monoclinic mineral, baddeleyite, one of these elements

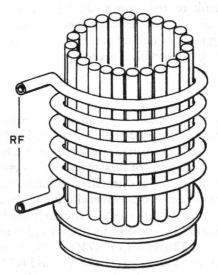

Figure 10.6 Schematic drawing of one form of a skull-melting apparatus.

needs to be added to it to ensure growth of the cubic form. The heated zirconium metal heats the powder near to it which then conducts the electricity from the radio-frequency source and melts. The metal produces more zirconia by reacting with oxygen in the air. The zirconia gradually melts leaving only a thin layer next to the cooling tubes, thus the melt is confined in a container of its own substance, preventing contamination from the copper. On reducing the heat, a skin forms over the contents of the skull and crystal growth begins at the bottom. Columns of parallel crystals nucleate near the bottom and grow upwards, stopping when the melt is exhausted. Crystals can be removed by separating the columns. A charge of 1 kg has been said to give a maximum of 500 g of facetable material. Various colours are obtainable though no really fine reds, greens or blues have been reported so far. An attractive orange–red results from the addition of cerium oxide; several different elements give yellow and nickel gives brown. Identification of mounted zirconia is probably best carried out with a reflectivity meter or thermal conductivity probe. The calcium-stabilized material appears to be inclusion-free, but some inclusions do occur in the yttrium-stabilized material. Some feather-like inclusions, resembling those seen in a variety of natural stones, may be seen from time to time. During 1986 there were several reports of octahedra of CZ with artifically etched trigons on their faces.

Yttrium-stabilized zirconia: H 8.25
                                SG 5.95
                                RI 2.171
                                Dispersion 0.059
                                some greenish yellow or reddish
                                    fluorescence
                                parallel rows of small semi-transparent
                                    isometric crystal-like cavities extending
                                    into hazy stripes of tiny particles
                                stones may show roundish facet edges,
                                    percussion marks or small chips
Calcium-stabilized zirconia: H 8.5
                                SG 5.65
                                RI 2.177
                                Dispersion 0.065
                                some distinct yellow fluorescence
                                    virtually free from inclusions
                                similar external characteristics to Y-
                                    stabilized stones

As for all diamond simulants, the thermal conductivity of CZ is much lower than that of diamond; a simple test is to put an ink blot on the surface of the material: diamond's surface maintains the ink blot in a coherent shape, whereas the surface of simulants breaks the ink up into beads. When the table of a stone is placed over a narrow source of light the pavilion facets of a CZ will show monochromatic colours red to yellow.

Nassau (1981) lists the colouring agents used to dope CZ:

- yellow–orange–red: CeO_2, Ce_2O_3
- Yellow–amber–brown: CuO, Fe_2O_3, NiO, Pr_2O_3, TiO_2
- Pink: Er_2O_3, Eu_2O_3, Ho_2O_3
- Olive-green: Cr_2O_3, Tm_2O_3, V_2O_3
- Lilac–violet: Co_2O_3, MnO_2, Nd_2O_3.

Small advertisements in the lapidary magazines have recently been offering a material called hafnia, with an SG higher than that of CZ, as a diamond simulant.

Diamond

The synthesis of diamond has a colourful and often doubtful history. The synthesis of gem quality diamond took a long time to develop but in 1970 General Electric in the United States announced that the problem was solved and that they had made diamonds up to 1 ct in weight and in a number of colours.

Accounts of the exact process vary and there may in any case be a number of modifications and new processes under development. However, the main feature of the apparatus is a ring of tungsten carbide, with the centre forming a cylinder with flared ends. Pistons, hydraulically powered, drive into opposite ends of the cylinder and a pyrophyllite container filled with graphite or some other form of carbon with nickel or tantalum, is placed in the hole between the pistons (Fig. 10.7). The advantage of pyrophyllite is that it has a melting point which rises with a rise in pressure and can in fact reach 2720°C. Pressure is applied by the pistons and heat by passing an electric current through the container. The pyrophyllite flows, allowing the pistons to compress the metal at pressures of up to or over 102 000 atmospheres and temperatures over 2000°C with transient peaks in the region of 3000°C. A small mass of synthetic diamond crystals is placed in the centre of the cylinder with a bath of the nickel or iron catalyst on either side. The crystals melt, allowing free carbon atoms to traverse the bath. In the hotter part of the cylinder more carbon is dissolved so that free atoms tend to crystallize at the ends. The diamond seed crystals prevent the formation of graphite instead of diamond. Growth rate is said to be about 2–3 mg h^{-1}, the rate increasing with the distance between the centre and ends of the pressure chamber, increasing this distance can increase the temperature by as much as 28–33°C. Temperatures are held for

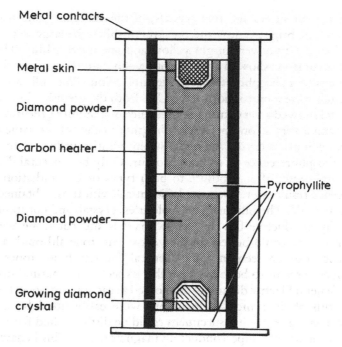

Figure 10.7 The pyrophyllite tube arrangement used for growing large synthetic diamond crystals. A thin skin of molten nickel is moving through the diamond powder, converting it to single crystal.

several days at a pressure of about 61 250 atmospheres. The better quality crystals grow at the bottom of the melt, dirt and poor quality crystals moving upwards. Impurities can be added to influence properties, including colour.

The first crystals made were colourless, canary yellow and pale blue and faceted stones were cut from them. Polishers were said to find no difference between them and natural diamonds. It was later found that if nitrogen was excluded crystals of a good white could be made and that the addition of boron produced blue. The first crystals emerged from the apparatus as truncated octahedra with modified cube faces (white and blue stones); the yellow crystals were well-shaped octahedra with one point diminished. Some stones were virtually clear, others had round or plate-like nickel inclusions; all stones showed very fine dust-like white particles under high magnification. Some blue stones showed a whitish cross under high magnification which may have been due to a particular formation of the dust-like particles. All diamonds except the yellows were semiconductors; in nature this property is shown only by blue stones. No absorption spectra were shown. All stones appear to be inert to long-wave UV but differed in their reactions to short-wave UV. Eight stones were tested by the GIA in

1984: three cut stones and five crystals; of the cut stones one was near-colourless, one bright yellow and one greyish-blue; the largest crystal was near-colourless, two were bright yellow and one greyish-blue. The near-colourless cut stone showed a very strong yellow fluorescence with a very strong persistent phosphorescence of the same colour. The yellow cut stone and crystals were inert to short-wave UV. Both the greyish-blue cut stone and crystal showed a very strong fluorescence of a slightly greenish-yellow colour with a very strong persistent phosphorescence of the same colour. The near colourless crystal showed a strong whitish yellow with a long-lasting phosphorescence of the same colour. Only blue natural diamonds will phosphoresce after exposure to both types of UV radiation. X-ray fluorescence results were found to be identical with those obtained under short-wave UV. The blue and near-colourless crystals and cut stones were electrically conductive but the yellow crystals and cut stone were not. Growth zoning could be seen in the yellow cut stone although a similar appearance can be seen in some General Electric blue stones. Strain birefringence seems to be lower than that expected in a natural diamond.

The General Electric diamonds all showed a reaction to a pocket magnet, stones with obvious metallic inclusions were easily lifted by a pocket 'horseshoe' magnet. Other specimens could be distinguished from natural stones by the use of a superconducting magnetometer. This behaviour has not so far been reported for natural diamond.

The GIA report (*Gems & gemology* **20**, 3, 1984) gives a useful summary of how to spot a synthetic diamond (although it is hoped that there are not too many around!).

- Near-colourless stones conducting electricity.
- Near-colourless stones fluorescing and phosphorescing strongly under short-wave UV but inert to long-wave UV.
- Fancy yellow stones with no absorption spectrum and inert to long-wave UV.
- Near-colourless stones with no hint of blue, grey or brown and with no absorption line at 415.5 nm.
- Stones with a strong yellow fluorescence under X-rays with persistent strong yellow phosphorescence.

Most natural diamonds fluoresce blue under X-rays, occasionally orange. Yellow fluorescence under X-rays is very rare and thus suspicious.

The Japanese firm of Sumitomo is manufacturing synthetic gem-quality yellow diamonds for industrial purposes, though some crystals have been and no doubt will continue to be cut. Sizes are up to 0.40 ct, about 8 mm in maximum dimension. Typical crystals are distorted octahedra with cube and dodecahedral faces as modifications. There is no fluorescence under long-wave UV but under short-wave UV there is a moderate to intense green or greenish yellow glow. There is no phosphorescence but there is a weak to moderate X-ray fluorescence of a bluish–white colour. There are no sharp absorption bands to be seen. Stones are Type Ib, as are the General Electric yellow stones. The stones are non–conductive and their thermal conductivity is similar to that shown by natural diamonds. Stones contain vein–like colourless areas which are characteristic and some contain opaque black inclusions of flux material. Graining can be seen when light is reflected from the surface, the effect persisting in some cases even after cutting. A distinctive cross-shaped interference pattern is shown, the four arms of the cross-shaped pattern either coinciding with or at 45 degrees to the direction of the radiating internal grain lines. Stones faceted from the crystals did not show this pattern. The report is fully given in the Winter 1986 issue of *Gems & gemology*.

Coral

An imitation of coral is made by Gilson who offers a variety of shades of reddish–pink made of calcite from a French source. The Gilson material has an SG of 2.44 (natural coral is 2.6–2.7). Both imitation and natural are about 3.5 on Mohs' scale and both will effervesce in acids. The imitation material has an RI of about 1.55 against 1.49 and 1.65 for natural coral. Gilson material has a brecciated structure while natural coral has a structure like the grain of wood. It has been found to give a reddish–brown streak when drawn across an unglazed porcelain plate; natural coral gives only a whitish mark.

Opal

Successful synthesis of opal could not be completed until the cause of its play of colour was understood in the 1950s (see chapter on colour).

Three stages are needed for synthesis: 1. the production of uniformly-sized spheres (if materials other than silica are used it is better to describe the product as an opal imitation); 2. the spheres are settled in an orderly close-packed arrangement; 3. the structure is consolidated, i.e. the spaces between the spheres are filled with a substance that will enable the opal to be made hard enough to fashion.

The silica spheres can be prepared by dispersing an organic silicon compound in fine droplets in an alcohol–water mixture. A compound often used is tetraethyl orthosilicate, $(C_2H_5O)_4Si$. When ammonia or other mild alkali is added to the mixture, stirring carefully, silica spheres containing some water are formed:

$$(C_2H_5O)_4Si + 2H_2O \rightarrow SiO_2 + 4C_2H_5OH$$

The spheres must all be the same size.

Settling may take more than one year. When complete the array of spheres is fragile and would easily dry out with consequent loss of colour. Some heating at fairly low temperatures sinters the mass and sometimes a little pressure is applied, taking care to preserve the ordering. The exact details of the techniques used are a trade secret but extra silica may be added at some stage to fill up the voids during consolidation in the form of silica gel.

Pierre Gilson has been producing varieties of opal since 1974, the latest product being a fine orange fire opal with play of colour. He has also made white and black stones and water opal, selling both rough and cut material and composites. Under the microscope the Gilson opal shows columns perpendicular to the cabochon base and within the colour patches a lizard-skin effect which is diagnostic. Some examples have a chalky bluish-white body glow with a stronger surface glow of the same colour, with some phosphorescence. Long-wave UV is more effective than short-wave UV in producing these reactions. Those samples which fluoresce also show some thermoluminescence. Affinity for water seems to be less than in natural opal.

In general, the Gilson opal products have silica spheres to give the play of colour. Three of the products have a composition different from that of natural opal and one sample examined (*Journal of gemmology* **19**, 1, 1984) has about 0.5% of crystalline ZrO_2. Some stones with a yellowish-brown body colour contained many organic compounds.

Several manufacturers are now making opal or opal imitations, some with spheres of plastic; latex has also been used. Hardening is by impregnation with another plastic with a different refractive index. The imitation opal is softer than natural opal and has an SG which may be as low as 1.0, in appearance, however, it is very successful, a particularly good example being made in Japan. Infra-red spectroscopic study of one variety of plastic opal showed that it consisted entirely of a co-polymer of styrene and methyl methacrylate. The RI of this example was 1.465; another plastic imitation gave an RI of 1.48 and an SG of 1.17.

Many synthetic opals and some plastic imitations give a blush, not necessarily of red or pink, when the stone is moved under a single source of light. This is not a sure sign as natural opals may give the same effect but it is sufficiently common to arouse suspicion where none may exist before. An early Gilson black opal in my own collection shows the very attractive play of colour in long thin bands along the long direction of the oval-shaped stone. This is the only example I have seen of this pattern; the same stone shows marks on the surface which may be from some kind of exudation.

An interesting imitation of opal is achieved by 'opal-essence', formerly Slocum stone, made by John S. Slocum of Rochester, Michigan, USA. This is a glass made by a controlled precipitation process, with an SG of 2.4–2.5 and RI of 1.49–1.50. The play of colour seems to have been achieved by the incorporation of flakes of tinsel-like laminated material which diffracts light from the roughly 0.3 μm spacing between the laminations. Bubbles and swirl marks prove its glassy nature. A re-examination of Slocum stone reviewed in the *Journal of gemmology* **19**, 7, 1985 suggests that some at least of the colour flakes are the disrupted remnants of what were probably continuous sheets produced by a sedimentation process. Distortion of the profuse gas bubbles may be due to mixing and agitation in a viscous liquid medium, the glass matrix.

Further notes on opal imitations made from plastic are given in the *Journal of gemmology* **18**, 8, 1983. In one type the lizard-skin effect could not been seen and in the other, where the colour domains were scattered rather than compact the opal closely resembled the natural material. The compact type transmitted wavelengths at about 430 nm; this effect has been noted with natural opal. The main component of the plastic opal is colourless polystyrene but in both types an absorption region at 590–565 nm could be seen. This may arise from some structural feature. The RI was 1.485 and the SG 1.18. Under long-wave UV a strong bluish-white fluorescence could be seen with a similar but weaker effect under short-wave. No phosphorescence was observed.

Lapis-lazuli

Lapis-lazuli has been made by Gilson using a ceramic technique, the product being issued both with and without pyrite. It is more porous than natural lapis and thus has a lower SG. It reacts more strongly with acids. The hardness is 4.5 against 5.5 for natural lapis and it leaves a strong blue streak on an unglazed porcelain plate. The RI is a vague edge at 1.50 and the SG about 2.46 (natural lapis 2.81). X-ray analysis showed traces of quartz and calcite with some iron.

Lapis can be imitated by stained howlite which, after removal of the dye, may show a bright orange fluorescence. A polycrystalline material containing cobalt and showing red through the Chelsea colour filter also shows a granular structure when examined with a lens in a good light. The synthetic sintered spinel imitation with gold flecks added to simulate pyrite has an SG near 3.52 and RI near 1.725. Stained jasper ('Swiss lapis') has an SG of 2.58.

A report in the *Journal of gemmology* **19**, 7, 1985, holds that the Gilson lapis should be regarded as an imitation though it is sold as synthetic. It consists of ultramarine and hydrous zinc phosphates as the main components with small pyrite crystals as accessories. Some dyed lapis–lazuli is unaffected by acetone but will give a dark blue stain on a cotton swab when tested with a solution of 10% HCl. A dyed blue quartzite has been used as a lapis imitation. The dye penetrated about 1.5 mm into the stone.

A string of beads purporting to be dyed howlite imitating lapis-lazuli were in fact dolomite from its RI of 1.50 and 1.68 and SG of 2.85.

Turquoise

Turquoise has been imitated by a variety of substances, but only the Gilson turquoise, marketed in 1972, can be described as synthetic. It is thought to be made by precipitation, grinding and pressing with some heating but details have never been released. The absence of iron impurity usually found in natural turquoise suggests that pure chemicals are used rather than ground-up natural pieces. Gilson makes both a matrix-free material and a turquoise in which 'veins' are prominent. The colour is stable and the porosity slight. Examination of the surface under the microscope shows a mass of angular dark blue particles against a whitish ground-mass. The SG of Gilson stones is near 2.74 and the RI averages 1.60. The absorption band at 432 nm is faint. Constants, therefore are not conclusive and magnification of 30–40 × provides the only sure and easy test, in the absence of facilities for infra-red spectroscopy or X-ray diffraction testing.

Recent examination of Gilson material suggests that there are two different types of product. One is turquoise with one or two additional phases, the other is a substitute consisting mainly of calcite. These later materials show absorption lines not seen in natural turquoise. Gilson has named his medium blue material 'Cleopatra' and a darker blue product 'Farah'.

Williams and Nassau (1977) discussed Gilson turquoise, a simulated turquoise made by the Syntho Gem Company of Reseda, California, a so-called 'reconstituted' turquoise from Adco Products of Buena Park, California and 'Turquite' from Turquite Minerals, Deming, New Mexico. Adco and Syntho material were similar to the Gilson stones; Turquite had little aluminium but a good deal of sulphur, silicon and calcium. The Gilson came closest in composition to the natural stone but had less iron. Only the Gilson product had the same crystal structure as natural turquoise.

Many turquoise imitations can be detected by placing a drop of Thoulet's solution (potassium and mercuric iodides) on the specimen, when a brown spot will show; natural materials show a white spot. Some plastic-covered imitations will show thread-like marks and have an SG below 2.62. Stained howlite has an SG of 2.53–2.59 and an RI of 1.59. Stained ivory has an SG of 1.80.

Dyed blue magnesite has been used as a substitute for turquoise. The SG is about 3.0 and the RI 1.51 (ϵ) and 1.70 (ω). The magnesite will effervesce when touched with warm dilute hydrochloric acid.

Jadeite

In late 1984 R. C. DeVries and J. F. Fleischer of the Inorganic Materials

Laboratory of General Electric reported the synthesis of jadeite for jewellery. The jadeite was formed by a high-pressure process and various colours are possible, although not the Imperial green quality. Most results were obtained from glasses of the desired composition made by melting Na_2CO_3, Al_2O_3 and SiO_2 in appropriate proportions to obtain $NaAlSi_2O_6$. The melting was carried out in platinum crucibles in air at about 1550°C. The molten liquid was cooled and quenched glass was broken out and crushed and refined as many times as were necessary to get a homogeneous product; gels were used later on in the study as the original liquid was viscous even at 1550°C. Oxides were added to give the required colour.

Other possible gem materials

Many crystalline materials are made for research or industrial purposes; some are attractively coloured and hard enough to use ornamentally. Many of these when cut pose problems for the gem tester although those examples doped with rare earths often show a complex and characteristic absorption spectrum. Some of the more interesting examples are:

Bismuth germanate ($Bi_{12}GeO_{20}$ or $Bi_4Ge_3O_{12}$); a fine orange with RI of 2.07 and hardness 4.5; SG 7.12.

Bismuth silicate ($Bi_{12}SiO_{20}$) has been grown by the pulling method and is brown to orange. Colourless crystals can be obtained by annealing.

Bromellite (BeO) with a hardness of 9, RI 1.720 and 1.735 and SG about 3.01 is a colourless transparent material which might often be cut were it not for the toxicity of the beryllium dust.

Fluorite (CaF_2) doped with a variety of elements to give interesting colours is often grown. Examples include a colourless crystal which gave a green fluorescence under UV and a brilliant green with exceptionally long phosphorescence under X-rays. Indium has been suggested as the dopant. Red fluorite crystals are also known.

Germanates with rare-earth dopants have been grown. Germanium has similar properties to silicon and is important for this reason.

Periclase (MgO) is a colourless cubic material with RI 1.73 and SG 3.55–3.60. Some stones have been marketed under the name Lavernite and some specimens show a whitish glow under UV.

Phenakite (Be_2SiO_4); some phenakite has been doped with vanadium to give a most attractive light blue.

Scheelite (CaWO$_4$) is grown and may be offered as a diamond simulant. It glows a bright bluish white under short-wave UV; various dopants and colours are recorded.

Silicon carbide (SiC) has RI 2.648 and 2.691 with DR 0.043. The SG is 3.20 and the hardness 9.5. It has been suggested as a simulant for green diamond but the birefringence is against it in this respect. It shows a mustard-coloured fluorescence under long-wave UV. The dispersion is about twice that of diamond.

Yttrium orthoaluminate (YAlO$_3$) was suggested as a diamond simulant but is frequently discoloured by iron impurities. The hardness is over 8, SG about 5.35 and RI 1.938. Yttrium aluminium gallium garnet had an RI of 1.885 and an SG of 5.05. The deep green of the stone recalled vanadium-coloured grossular garnet and lines at 690, 670 and 660 nm with a general absorption 620–570 nm were seen in the absorption spectrum. The stone showed a red transmission and fluorescence.

Yttralox is the name given to a ceramic material with the composition Y$_2$O$_3$. The RI is 1.92, SG 4.84, dispersion 0.039 and hardness 7.5–8. It is isotropic.

Zincite (ZnO) has been doped with various elements to give a number of colours.

Zircon (ZrSiO$_4$) doped by vanadium to give a strongly dichroic purple crystal is found in comprehensive collections. Small crystals of zircon doped with terbium gave a yellow colour.

10.7 COMPOSITES

Composite stones, doublets or triplets, have been known since early times. Generally they are joined at the girdle but there are cases of more subtle manufacture. Despite modern synthetic stones, it seems that it is still worthwhile to produce composites, as some very complicated examples have been marketed recently. A number of examples are shown in Figs 10.8–10.10.

Composites may not always be designed to deceive. In opal doublets and triplets the preservation of a fine quality slice of opal too thin to make a stone by itself is the main reason for manufacture and glass imitation stones can be made harder and more resistant to wear by fusing a garnet slice on the top.

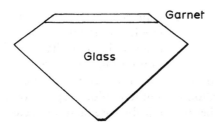

Figure 10.8 A doublet consisting of a garnet top fused to a lower portion of glass.

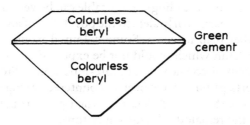

Figure 10.9 A triplet consisting of two pieces of colourless beryl joined by a layer of green cement.

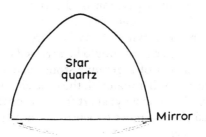

Figure 10.10 A star quartz foil back made to imitate star sapphire.

Doublets

True doublets consist of two parts of the same material but these are rare. Semi-genuine doublets have a genuine crown cemented to a pavilion of a different material; examples include diamond cemented to rock crystal or to synthetic spinel; these are encountered rarely. Immersion when the stone is unset easily shows up the two parts (note that immersion for too long may weaken the cement). Many diamond doublets are set in pieces with rubbed-over settings so that the join is not visible. If a diamond doublet is held so that a dark border can be seen on the opposite edge of the table its nature will become apparent; similarly a reflection of the table seen a little way down inside the stone indicates a diamond doublet. Prismatic colours from interference caused by air penetrating between the two parts of the composite are also clues. Some Indian lasque diamonds (thin platy crystals with parallel sides) may also show the dark shadow of the table edge and there is no question of a composite in these cases.

Semi-genuine doublets with a crown of natural greenish-yellow sapphire on a pavilion of synthetic blue sapphire or synthetic ruby have been seen; remember to check the base of all suspected stones for the signs of Verneuil growth.

Doublets in which the crown is made of a colourless stone like rock crystal and the pavilion of a coloured glass can be called false doublets; these can be deceptive and RI readings on the table can be very dangerous so that the mind has to be particularly alert to the possibility of an apparently 'easy' stone being other than it seems. Some coloured pavilions may show an absorption spectrum which would not be expected in the simulated stone (e.g. the spectrum of cobalt in a sapphire-blue stone). Stones with a glass crown and quartz pavilion have been encountered, strange as it may seem. Webster reports a stone which had only the tip of the pavilion of natural blue sapphire, the remainder being white topaz.

Hollow doublets have a crown of rock crystal or glass which is hollowed out below with highly polished walls. The hollow is filled with a coloured liquid which is held in by a pavilion of the same material as the crown. Viewing from the side should identify such specimens, which are very rare. Doublets in which both crown and pavilion are made of colourless glass cemented by a colourless cement should be revealed by immersion; triplets in which there is a crown of a genuine stone such as rock crystal and a pavilion of the same or similar material, the central portion being coloured glass can be tested in the same way; apart from the opal triplet in which the cover is rock crystal (for the more expensive stones), glass or plastic; these stones are not often seen.

The garnet-topped doublet presents a much greater threat as it is more common than the types mentioned above and it also occurs in small sizes which tend to be suspected less and are harder to test. They are made by the fusion of a thin slice of almandine garnet to a blob of coloured glass which is then cut. Many different kinds are made (even red and colourless stones). All show a red ring round the edge when placed table facet down on a white background and they will give odd and unexpected red flashes when moved in a strong light. Rutile needles and an almandine absorption spectrum may be seen in the crown, with the usual gas bubbles and swirls characteristic of glass in the pavilion. The stones will probably alert the tester when he has failed to get an expected RI reading.

The name soudé has been given to emerald imitations more than to other stones; here the stone has crown and pavilion of rock crystal with a green coloured gelatine to give adhesion and colour. Early green stones show red through the Chelsea filter but the RI will of course be that of quartz. Over a length of time the gelatine may turn yellowish which will catch the eye. Alexandrite and blue sapphire types have been reported. The more modern soudé stones (from the 1920s onwards) have a layer of coloured glass replacing the gelatine. The SG is about 2.8, probably from the heavy lead glass layer. These stones show green through the Chelsea filter. In 1951 a green soudé of colourless spinel for both crown and pavilion was made (some other colours have also been reported). The SG is well above that for emerald, at 3.66–3.70, with RI 1.73. Immersion is the easiest test and it should be remembered that colourless synthetic spinel gives a bluish-white glow under short-wave UV. The stones show green through the Chelsea filter.

An imitation emerald made from a colourless synthetic spinel crown and base with a centre section of green glass was manufactured by Jos Roland of Sannois, France from 1951. GIA tested a stone of 11.66 ct which gave multiple readings (strong edge at 1.724 and weak edge at 1.682 for the spinel and the glass respectively). Viewed nearly perpendicular to the girdle the crown shows a strong chalky-yellow-white fluorescence with the glass layer inert. The pavilion glows a strong pure yellow fluorescence. When the culet was positioned nearest to the light source an opposite effect was noted.

Beryl doublets with crown and pavilion of colourless beryl and an emerald-coloured duroplastic cement were made in 1966 and marketed under the trade name Smaryll. Inclusions typical of aquamarine can be seen where poor quality aquamarine was used; the stones show green through the Chelsea filter and give a dyestuff absorption spectrum with woolly lines in the red. Later products have crown and pavilion of emerald, so it is important to spot the dyestuff spectrum. Try to get the stone free from its setting in all cases of doubt.

Doublets in which the crown is made of a synthetic colourless corundum or spinel and the pavilion of strontium titanate give a good display of fire, although the RI can be measured satisfactorily in both cases.

Star doublets are made by capping a natural star corundum base with a crown of synthetic corundum or spinel of varying colours; a star rose quartz doublet is made by cementing a blue mirror to the base of a rose quartz cabochon containing a star. When examined under a single bulb an image of the bulb can be seen at the centre of the star. As the stone is brought closer to the lamp the image enlarges.

Opal doublets and triplets

Opal doublets are simply slices of fine quality opal cemented to a backing of dark common opal (potch). Sometimes black onyx or black glass (particularly a variety called opalite from Belgium) is used instead of potch. Triplets provide a cover for the doublets; this may be (in recent stones) synthetic spinel or corundum. Identification is not always easy and the stone may need to be free from its setting. Many solid opals have part of the seam wall as their backing and these can look like doublets. Where colour goes right down to the base of the stone when viewed from the side the opal must be solid; doublets look flat and when a strong light is passed through will show flattened bubbles in the junction layer. Opal triplets were patented in Australia in 1958 and in general need a thinner layer of opal; very expensive ones may be capped with synthetic colourless corundum or spinel. Triplets using a layer of fish skin (Schnapper-skin) show colours unlike those seen in opal and also show grey at the girdle. Some triplets are known as Triplex opals. An imitation opal made of a rock crystal cabochon with a slice of mother-of-pearl cemented to the back turns up from time to time. A hollow rock crystal cabochon filled with fragments of opal has been reported.

Jadeite triplet

The very deceptive jadeite triplet was first reported in 1958. It consists of a hollow cabochon of fine translucent white jadeite about 0.5 mm thick, a cabochon of smaller size of the same material made to fit into the larger cabochon and a flat oval section made to close the base. The centre is coloured by a green jelly-like dye to simulate Imperial jade; the dyed centre piece is fitted in and the back closed. If the stone is unset the join at the back may be visible; otherwise the woolly dyestuff absorption bands are the best test.

Other composites

Composites imitating true cameos are usually made entirely of glass, with the upper piece moulded to give the intaglio. The top section is cemented to a plate with a reddish-brown cement to give a cornelian effect. The cement often crazes and the colour is unstable. Where the back of the imitation is a true cornelian the effect and life are much improved. Moss agate doublets are made by placing a chemical (possibly MnO_2) on a glass plate with gelatine and allowing a dendrite to form. The dendrite is covered by another glass plate and the whole thing fashioned. These pieces look unnaturally transparent for true moss agate and if they are unset the construction will be apparent. Turquoise doublets with a base of stained blue chalcedony and a top of turquoise coloured opacified glass are known.

A new type of triplet was developed in 1972: the colour is produced from a transparent film containing a mosaic of three colours with white patches, the whole being enclosed in synthetic spinel. A doublet with a crown of synthetic green spinel and a pavilion of Verneuil synthetic blue sapphire is reported by the GIA. Fusion was probably not used to join the two portions as the melting points of the two substances are 2135 and 2050°C respectively; joining was probably accomplished by the use of cyanoacrylic cement such as 'Super Glue'.

Among other composites reported from time to time are: topaz on a seed of beryl manufactured by Lechleitner; a triplet with glass between two pieces of synthetic spinel and imitating peridot; quartz on glass to simulate ruby. GIA reported a heat-crackled ruby cabochon drilled at the back. A natural ruby had been inserted into the drill-hole, the whole piece being polished to simulate a natural stone.

Natural sapphire on synthetic sapphire is fairly common; a more interesting example reported consisted of a natural sapphire crown on a base of synthetic ruby. The upper layer gave a bluish colour and rutile needles were used to make the stone appear wholly natural; the red in the base was strong enough to colour the whole stone red. Under crossed-filters only the base glowed red. A star sapphire doublet was made by joining a natural top and natural base. The star was in the top portion and the colour in the base. The star was sharp and the stone gave a cobalt absorption spectrum. The colour came from a dye and the base was crackled. Another cobalt-bearing star imitation consisted of two pieces of natural greyish corundum joined by a cement in which there were arborescent patterns.

An odd-looking opal imitation consists of a white opal crown cemented to a fire opal pavilion. Coloured cements (some of which may fluoresce) have been used to give imitations of amethyst, emerald, topaz, ruby or sapphire with a crown and pavilion of synthetic spinel or quartz. At least one stone with a crown of synthetic sapphire and a pavilion of synthetic rutile has been reported.

A jadeite composite examined by GIA apparently consisted of two parts, one extremely thin (0.1 mm) at the top with a thicker part (2.2–2.3 mm) below. The top layer was green and the lower one white. A yellowish-green cement with gas bubbles joined the two. The top layer had a mottled colour with small near-colourless veining while the lower showed signs of crystalline structure. The green portion gave RI readings of 1.64–1.74 and the absorption spectrum suggested chromium. X-ray diffraction revealed that both portions were jadeite.

10.8 GLASS

Glass is without doubt the oldest gemstone imitation. Faience was used by the Egyptians as far back as 5000 BC to imitate a variety of gem materials, turquoise in particular. Faience is a glazed ceramic material. Pliny records the use and value of glass at which time the confusing name 'crystal' was first used.

Glass has often been described as a super-cooled liquid. It results from the cooling of a melt which becomes too viscous to allow crystallization to take place; most inorganic materials crystallize on cooling after melting. The glassy or vitreous state is metastable in that a disorder of the constituent atoms is provoked; this may later be released by the process known as devitrification, seen in some glass gem imitations and sometimes mistaken for natural inclusions (in a sense, the products of devitrification are just that but they arise in an epigenetic way). Most glasses used as ornament are silica glasses. Crown glass is a soda–lime glass (the common window glass). Lead oxide added gives flint glass whose optical properties are more suited to ornament, the dispersion being higher, though if too much lead is added the glass becomes too soft, may take a yellowish colour and tarnish. Dispersion can be increased by the addition of thallium oxide. Colouring agents may be added or the glass made translucent or opaque by the addition of opacifiers. Glass is made on a large scale by melting in ceramic crucibles in kilns and the cheaper varieties attain the desired final shape by being moulded; the more expensive glass gemstones are faceted in the usual way.

The names used for glass in gemstones vary and are used loosely today so that while paste and glass originally differed slightly in significance they are now used interchangeably; the term strass has largely disappeared. Glass is usually soft, about 5–6.5 and can easily be scratched. Conchoidal fracture marks can be seen at edges and gas bubbles are common, being either very round or characteristically elongated. They show a bold outline. Swirl marks arising from incomplete mixing of the ingredients are easily seen. Glass is a poor conductor of heat and feels warm compared to crystalline substances. It usually shows strain birefringence between crossed polars. Gemstone imitation glasses generally have an RI in the range 1.50–1.70; only the natural isotropic gems pollucite and rhodizite occur in this area. The composition of the majority of gem imitation glasses is $34SiO_2 \cdot 62PbO \cdot 3K_2O$.

Some unusual applications of glass include the production of chatoyancy in the so-called Cathay stone. This is a very good imitation of chrysoberyl cat's-eye and is made by heating and drawing bundles of glass rods to form parallel hollow tubes; these reflect light in the same way as mineral inclusions in such stones as tiger's-eye; sharper eyes for the Cathay stone are made from fused mosaics of optical fibres of several different glasses. Star stones have been imitated by glass with rays scratched on the back or impressed on a foil which is stuck to the back. A white opaque glass can be pressed to form a cabochon with six ridges in star formation and then covered with a thin layer of a deep blue glaze which conceals the ridges so that the star seems to be on or just below the surface. Low quality opal or even pearl imitations are made from translucent lime glass with fluorides or phosphates added. Insoluble calcium compounds precipitate to give the effect.

A partly crystallized glass imitating jade is 'Victoria stone' made by Iimori Laboratory of Tokyo (the name meta-jade has also been used for this product). A chatoyant effect can be obtained by using fibrous inclusions in parallel bundles. Iimori stone is clear to translucent material also imitating jade. Goldstone is made by including cuprous oxide with the glass ingredients and reducing during annealing; this gives small precipitated crystals of metallic copper which imitate sunstone feldspar. If the glass is blue a lapis imitation of doubtful quality may be obtained. A green glass said to be made from ash from the 1980 eruption of Mount Saint Helens in Washington, USA, had at most 10% of this ash; much had none at all.

10.9 METALS

The use of metals to imitate gem materials is limited, but there are several substitutes for hematite on record. One is a titanium dioxide of steel-grey colour with a yellow-brown streak (the streak of hematite is reddish-brown). The SG was about 4 and the hardness 5.5. Another hematite imitation appeared to have been made from powdered lead sulphide, perhaps with some silver added. This material had a hardness of 2.5–3 and SG of 6.5–7. It was brittle and easily fusible. Hemetine is a sintered product of several constituents, including lead sulphide (the mineral galena). This has an SG of 7.0 and a black streak.

10.10 CERAMICS

Ceramics are made from finely ground powders of inorganic substances which are sintered, heated or fired, sometimes with the use of pressure. A fine grained polycrystalline solid results which may be completely transparent. Gilson turquoise, lapis-lazuli and the Adco, Syntho and Turquite mentioned above must all be made in this way. Yttralox, made since 1966 by General Electric, has been used as a diamond simulant; it is a stabilized yttrium oxide (Y_2O_3) which takes a cubic form on the addition of 10% ThO_2; this material could be radioactive. It is hot pressed to give its final form; the hardness is 6.5, RI 1.92 and dispersion 0.039. The SG is 4.84.

10.11 PLASTICS

Many of the plastics in common use have been used as gem imitations at some time. Most have a hardness of 1.5–3.0, SG 1.05–1.55 and RI 1.5–1.6.

Dense fillers raise the constants. Most gem imitations are made by the injection moulding process: they are sectile and give off a variety of odours, mostly unpleasant, when touched with the hotpoint. Celluloid was one of the earliest plastics to be manufactured. It is a mixture of the lower nitrates of cellulose and camphor heated under pressure to 110°C. This is a very inflammable material but for many years has been treated with acetic acid to reduce the risk. The old type has an SG of 1.35 rising to 1.80 with the addition of fillers; its RI is between 1.495 and 1.520 and the hardness is 2. The safety celluloid (cellulose acetate) has an SG of 1.29 and a range of RI from 1.49 to 1.51. It burns with the small of vinegar. Both types are sectile.

The protein part of milk (casein) can be made into a hard plastic by the addition of formaldehyde. This has an SG in the range 1.32–1.39 (usually 1.33) and RI of 1.55. Concentrated nitric acid will turn casein yellow; it is sectile and chars with a burnt milk smell.

Bakelite is a phenolic resin and can take transparent variously-coloured forms though it is not very tough and yellows with age. It has an SG of 1.25–1.30 and a somewhat high RI for plastic of 1.61–1.66. Webster gives a useful test for chips of bakelite; they should be placed in a small test tube and covered by distilled water. After boiling, a pinch of 2.6 dibromo-quinonechlorimide is added. The liquid cools and a drop of a very dilute alkali solution is added. If a blue colour forms phenol is indicated and the piece must be bakelite. A modification of bakelite, the amino plastic, is translucent and can be dyed to give different colours. The hardness is near 2, the SG about 1.50 and the RI 1.55–1.62.

The acrylic resin Perspex is a clear, glass-like material with a low SG of 1.18 and RI 1.50. It is used for the manufacture of cheap beads and for the cores of imitation pearls. Polystyrene resins have been made to resemble faceted stones; they have an SG of 1.05 and RI of 1.59. As well as being sectile the polystyrene resins are easily dissolved by such liquids as toluene, methylene iodide and bromoform.

Appendix A Identification tables

These tables summarize information useful in the identification of gemstones. Many of the less common gemstones are not included in the tables. They occur infrequently, so that the need to identify them will arise rarely. Figures quoted in the tables may not agree with those in the body of the text in every case. This is because main text references cover some examples of species with unusual properties, whereas the tables give the constants most likely to be found.

A1 SPECIFIC GRAVITY

Species	Specific gravity	Species	Specific gravity
Gadolinium gallium garnet	7.05	Smithsonite	4.3–4.5
		Rutile	4.2–5.6
Cassiterite	6.8–7.0	Spessartine	4.12–4.20
Cerussite	6.45–6.57	Sphalerite	3.9–4.1
Anglesite	6.30–6.39	Celestine	3.97–4.00
Scheelite	5.90–6.10	Almandine	3.95–4.30
Cuprite	5.85–6.15	Zircon (low type)	3.95–4.20
Cubic zirconia	5.6–5.9	Corundum	3.98–4.00
Zincite	5.66–5.68	Willemite	3.89–4.18
Proustite	5.57–5.64	Siderite	3.83–3.96
Strontium titanate	5.13	Demantoid garnet	3.82–3.85
Hematite	4.95–5.16	Azurite	3.77–3.89
Pyrite	4.95–5.10	Malachite	3.60–4.05
Bornite	4.9–5.4	Chrysoberyl	3.71–3.72
Marcasite	4.85–4.92	Pyrope-almandine garnet	3.70–3.95
Zircon (high type)	4.60–4.80		
Lithium niobate	4.64	Staurolite	3.65–3.78
YAG	4.57–4.60	Pyrope garnet	3.65–3.70
Baryte	4.3–4.6	Kyanite	3.65–3.68

Species	Specific gravity	Species	Specific gravity
Grossular garnet	3.65	Brazilianite	2.98–2.99
Benitoite	3.64–3.68	Boracite	2.95
Spinel (synthetic)	3.61–3.65	Phenakite	2.93–2.97
Taaffeite	3.60–3.61	Aragonite	2.93–2.95
Spinel (natural)	3.58–3.61	Nephrite	2.90–3.02
Topaz	3.53–3.56	Datolite	2.90–3.00
Sphene	3.52–3.54	Prehnite	2.88–2.94
Diamond	3.514–3.518	Pollucite	2.85–2.94
Rhodochrosite	3.50–3.65	Dolomite	2.85–2.93
Sinhalite	3.47–3.50	Beryllonite	2.80–2.85
Rhodonite	3.40–3.70	Pectolite	2.74–2.88
Rhodizite	3.44	Sugilite	2.74
Clinozoisite	3.37	Bytownite feldspar	2.72–2.75
Hydrogrossular		Calcite	2.71–2.94
garnet	3.36–3.55	Lapis lazuli	2.70–2.90
Zoisite	3.35	Emerald (natural)	2.69–2.76
	(Tanzanite	Beryl	2.68–2.90
	variety)	Labradorite	2.68–2.72
Odontolite	near 3.00	Charoite	near 2.68
Peridot	3.30–3.40	Pearl	2.67–2.78
Diaspore	3.30–3.50	Emerald (synthetic)	2.66–2.68
Vesuvianite	3.32–3.47	Quartz	2.651
Jadeite	3.30–3.86	Scapolite	2.63–2.71
Kornerupine	3.28–3.34	Oligoclase feldspar	2.62–2.65
Axinite	3.26–3.36	Albite feldspar	2.62
Dumortierite	3.26–3.41	Serpentine	
Enstatite	3.26–3.28	(williamsite)	2.61
Epidote	3.26–3.50	Turquoise	2.60–2.90
Californite	3.25–3.32	Coral	2.60–2.70
Sillimanite	3.23–3.27	Serpentine	
Diopside	3.22–3.38	(bowenite)	2.58–2.62
Fluorite	3.18	Howlite	near 2.58
Apatite	3.17–3.35	Sanidine feldspar	2.52–2.62
Spodumene	3.17–3.19	Cordierite	2.57–2.61
Andalusite	3.13–3.17	Moonstone	2.55–2.63
Euclase	3.10	Variscite	2.20–2.57
Lazulite	3.08–3.10	Hauyne	near 2.4
Actinolite	3.05	Petalite	2.30–2.50
Amblygonite	3.00–3.10	Tugtupite	2.30–2.57
Saussurite	3.00–3.40	Soapstone	2.20–2.83
Tourmaline	3.00–3.12	Tektite	2.20–2.50
Montebrasite	2.98	Hambergite	2.35
Grandidierite	2.97–3.00	Moldavite	2.33–2.47
Danburite	2.97–3.03	Obsidian	2.33–2.47
Herderite	2.95–3.02	Thomsonite	2.25–2.40

Species	Specific gravity	Species	Specific gravity
Sodalite	2.25–2.30	Celluloid	1.36–1.42
Silica glass	2.21	Black coral	1.34
Chrysocolla	2.00–2.45	Tortoiseshell	1.26–1.35
Bone	2.00	Bakelite	1.25–1.30
Ulexite	1.99	Jet	1.20–1.30
Opal	1.98–2.20	Vulcanite	1.15–1.20
Ivory	1.70–1.98	Copal resin	1.03–1.10
Vegetable ivory	1.38–1.42	Amber	1.03–1.10

A2   REFRACTIVE INDEX

The refractive indices of some species are omitted as readings are not available via normal gem testing methods. Birefringence is given where appropriate in A3.

Species	Refractive index	Species	Refractive index
Cuprite	2.85	Pyrope-almandine garnet	1.75–1.78
Rutile	2.62–2.90	Chrysoberyl	1.75–1.76
Diamond	2.417	Staurolite	1.74–1.75
Strontium titanate	2.41	Grossular garnet	1.73–1.74
Sphalerite	2.37–2.43	Epidote	1.736–1.770
Lithium niobate	2.21–2.30	Azurite	1.73–1.84
Cubic zirconia	2.17	Pyrope garnet	1.73–1.75
Gadolinium gallium garnet	2.03	Rhodonite	1.73–1.74
Zincite	2.013–2.029	Clinozoisite	1.72–1.73
Cassiterite	2.006–2.101	Spinel (synthetic)	1.728
Zircon (high type)	1.925–1.984	Hydrogrossular garnet	1.72
Scheelite	1.918–1.937	Taaffeite	1.718–1.722
Sphene	1.94–2.10	Kyanite	1.715–1.732
Demantoid garnet	1.89	Spinel (natural)	1.718
Anglesite	1.877–1.894	Vesuvianite	1.700–1.721
YAG	1.83	Zoisite (Tanzanite)	1.692–1.701
Cerussite	1.804–2.079	Willemite	1.69–1.72
Spessartine	1.79–1.81	Rhodizite	1.694
Zircon (low type)	1.78–1.81	Dumortierite	1.68 (mean)
Almandine	1.78–1.81	Axinite	1.674–1.704
Corundum	1.76–1.77	Sinhalite	1.66–1.71
Benitoite	1.757–1.804		

Species	Refractive index	Species	Refractive index
Diopside	1.67–1.70	Williamsite	1.57
Kornerupine	1.67–1.68	Emerald (synthetic)	1.560–1.563
Enstatite	1.65–1.67	Labradorite	1.56–1.57
Spodumene	1.66–1.67	Bowenite	1.56
Jadeite	1.66	Variscite	1.56 (mean)
Malachite	1.65–1.90	Black coral	1.56
Sillimanite	1.65–1.68	Hambergite	1.55–1.62
Peridot	1.65–1.69	Beryllonite	1.55–1.56
Phenakite	1.65–1.67	Scapolite	1.55–1.56
Euclase	1.65–1.67	Quartz	1.544–1.553
Siderite	1.65–1.87	Oligoclase feldspar	1.542–1.549
Andalusite	1.64–1.65	Soapstone	1.54 (mean)
Jet	1.64–1.68	Albite feldspar	1.54 (mean)
Baryte	1.636–1.648	Tortoiseshell	1.54
Apatite	1.63–1.64	Copal resin	1.54
Topaz (brown, pink)	1.630–1.638	Amber	1.54
Danburite	1.630–1.636	Ivory	1.54
Smithsonite	1.62–1.85	Vegetable ivory	1.54
Datolite	1.625–1.669	Bone	1.54
Celestine	1.622–1.635	Aragonite	1.530–1.685
Actinolite	1.620–1.642	Cordierite	1.53–1.54
Tourmaline	1.62–1.64 (mean)	Dolomite	1.502–1.681
		Thomsonite	1.52–1.54
Bakelite	1.61–1.66	Microcline	1.52–1.53
Lazulite	1.61–1.64	Orthoclase feldspar	1.52–1.53
Amblygonite	1.578–1.692	Sanidine feldspar	1.518–1.524
Turquoise	1.61–1.65	Pollucite	1.517–1.525
Prehnite	1.61–1.64	Petalite	1.504–1.516
Topaz (white and blue)	1.610–1.620	Lapis lazuli	1.50 (mean)
Herderite	1.61 (mean)	Chrysocolla	1.50 (mean)
Nephrite	1.61 (mean)	Hauyne	1.496
Rhodochrosite	1.60–1.82	Tektite	1.49–1.53
Brazilianite	1.60–1.62	Tugtupite	1.496–1.502
Grandidierite	1.60–1.63	Celluloid	1.50 (mean)
Pectolite	1.59–1.64	Calcite	1.486–1.658
Montebrasite	1.59–1.63	Obsidian	1.48–1.51
Howlite	1.59 (mean)	Moldavite	1.48–1.50
Aquamarine	1.57–1.58	Sodalite	1.48
Emerald (natural)	1.57–1.58	Silica glass	1.46
Bytownite feldspar	1.57–1.58	Opal	1.44–1.46
		Fluorite	1.43

A3   BIREFRINGENCE

Only the larger values are included as they are of particular importance for gem testing. Other values can be found in the descriptive section.

Species	Birefringence	Species	Birefringence
Rutile	0.287	Brazilianite	0.019
Smithsonite	0.228	Tourmaline	0.018
Rhodochrosite	0.220	Spodumene	0.015
Calcite	0.172	Chrysoberyl	0.009
Aragonite	0.155	Quartz	0.009
Sphene	0.120	Corundum	0.008
Cassiterite	0.096	Topaz	0.008
Zircon (high)	0.059	Orthoclase feldspar	0.008
Benitoite	0.047	Beryl	0.007
Peridot	0.038	Danburite	0.006
Sinhalite	0.038	Taaffeite	0.004

A4   DISPERSION

Dispersion is in general only important with colourless transparent stones, though coloured stones with high dispersion, such as demantoid garnet, show a distinct liveliness. Only figures of significance are quoted here.

Species	Dispersion	Species	Dispersion
Rutile (synthetic)	0.28	Gadolinium gallium	
Strontium titanate	0.19	garnet	0.045
Sphalerite	0.156	Diamond	0.044
Lithium niobate	0.13	Benitoite	0.044
Cassiterite	0.071	Zircon	0.038
Cubic zirconia	0.060	YAG	0.028
Demantoid garnet	0.057	Spinel (natural and	
Sphene	0.051	synthetic)	0.020

The most highly dispersive glasses have a dispersion near 0.016.

A5   HARDNESS

Hardness testing is not appropriate for fashioned stones but can be used on crystals. Some idea of hardness can be gained from examining the facet edges of stones. Only the more important species are listed.

Hardness Species

10	Diamond
9	Corundum
8.5	Chrysoberyl
8	Spinel, taaffeite, topaz
7.5	Andalusite, beryl, euclase, almandine garnet, phenakite, sillimanite
7.25	Grossular garnet, pyrope garnet, spessartine garnet, high zircon
7	Axinite, danburite, jadeite, quartz, spodumene, tourmaline
6.5	Benitoite, cassiterite, epidote, demantoid garnet, kornerupine, nephrite, peridot, sinhalite, vesuvianite
6	Amblygonite, feldspars, lapis-lazuli, rutile, scapolite, strontium titanate, turquoise, low zircon, zoisite
5.5	Brazilianite, enstatite, opal, rhodonite
5	Apatite, beryllonite, bowenite, datolite, diopside, dioptase, lazulite, obsidian, smithsonite, sphene
4	Fluorite, kyanite, malachite, rhodochrosite, zincite
3.5	Aragonite, azurite, sphalerite
3	Calcite, jet
2.5	Serpentine, coral, pearl
2	Chrysocolla, gypsum
1.5	Stichtite
1	Soapstone

A6   CLEAVAGE

Cleavage	*Species*
Octahedral	diamond, fluorite
Dodecahedral	sodalite, sphalerite
Rhombohedral	calcite, dioptase, rhodochrosite, smithsonite
Basal	topaz, willemite, zincite
Pinacoidal	amblygonite, beryllonite, brazilianite, cordierite, epidote, euclase, feldspar, gypsum, hambergite, kyanite, peridot, sillimanite, zoisite
Prismatic	Diopside, enstatite, kornerupine, rhodonite, scapolite, sphene, spodumene

Octahedral cleavage gives 4 cleavage planes, dodecahedral 6, rhombohedral 3 and basal 1. Corundum has a basal parting rather than cleavage. Enstatite and diopside have pinacoidal parting and sphene has pyramidal parting.

A7   LESS COMMON AND TRADE NAMES

A very large number of trade names have been used for a variety of gem materials; many of the names are now out of fashion and many had a very short currency. From a mineralogical point of view the use of any of these names is undesirable but some are now so well-known that inclusion here is indicated. Other lists can be found in Webster (1983) and in Embrey and Fuller, *A manual of new mineral names, 1892–1978*, British Museum (Natural History) 1980. So far as the United Kingdom is concerned the 1968 Trade Descriptions Act has ruled out the use of all names other than those which actually indicate fact (i.e. Burma ruby must be a ruby which can be shown to have come from Burma). A list of suggested permissible names can be obtained from the Gemmological Association of Great Britain. Some provisions may strike the reader as unusual – for example a green beryl which can be shown to contain chromium can be named emerald even though it may not look very like that stone.

Name	*Description*
Ametrine	quartz containing both amethyst and citrine components
Astrilite	lithium niobate
Balas ruby	red spinel
Bone turquoise	odontolite
Cacholong	a porous type of common opal
Cape ruby	pyrope garnet
Chrysolite	an undesirable name formerly used for several different gemstones (chrysoberyl, peridot)
Cymophane	cat's-eye chrysoberyl
Diamonair	YAG
Djevalite	cubic zirconia
Fabulite	strontium titanate
Galalith	German name for a casein plastic
Galliant	GGG
Goldfluss	glass imitation of sunstone
Goldstone	glass imitation of sunstone
Hematine	imitation of hematite
Hyacinth	name formerly used for orange-brown zircon and other stones
Iimori stone	glassy man-made jade imitation
Imperial jade	emerald-green jadeite, translucent and with no trace of white streaks
Jacinth	name formerly used for reddish-brown zircon or hessonite garnet
Lactoid	casein

Name	Description
Maori stone	nephrite
Mass aqua	glass of beryl-like composition
Matar (Matura) diamond	colourless zircon, especially from Sri Lanka
Mtorolite	chrome green chalcedony
Meta-jade	Japanese glass jade imitation
Morion	dark brown quartz
Muller's glass	a type of glassy opal
New Zealand greenstone	nephrite
Padparadschah	name used for a type of pink to orange sapphire, now better abandoned
Paste	used interchangeably with glass though at one time having a specific meaning
Rose de France	name for a pinkish-violet colour sometimes seen in amethyst but more frequently used for similarly-coloured synthetic corundum
Ruby spinel	highly undesirable name for red spinel
Sagenite	rock crystal with acicular inclusions
Titania	synthetic rutile
Victoria stone	Glassy man-made jade imitation

Appendix B Useful sources of information

Bulletin d'information, 1965–1975 Continued as *Revue de gemmologie* a.f.g, 1975–	Association française de gemmologie, 162 rue St Honoré, 75001 Paris, France.
Australian gemmologist	Gemmological Association of Australia, PO Box 35, South Yarra, Victoria 3141, Australia.
Bernstein-Forschungen (1929–1939)	Berlin.
Bulletin	Canadian Gemmological Association, PO Box 1106, Station Q, Toronto, Canada M4T 2P2.
Zeitschrift, 1969–	Deutsche Gemmologische Gesellschaft (formerly *Zeitschrift,* Deutsche Gesellschaft für Edelsteinkunde.) Postfach 12 22 60, D-6580 Idar-Oberstein, FRG
Gem instrument digest	PSTS c/o P. G. Read, 68 Forest House, Russell-Cotes Road, Bournemouth, Dorset BH1 3UB, UK.
Gemologia	Associação Brasileira de Gemologia e Mineralogia. Previously published by the Departmento de Mineralogia e Petrografia São Paulo University, for the Associação.
Gemmological news (1935–1943) Continued as *Journal of gemmology*	National Trade Press, London.
Gemmological newsletter	Michael O'Donoghue, 7 Hillingdon Avenue, Sevenoaks, Kent TN13 3RB, UK.

Gemmologist Now incorporated with *Retail jeweller*	Vols 1–31 nos 1–377. NAG Press, London, 1931–1963. Vols 1–4 were the official journal of the Gemmological Association.
Gemmology (Also entitled *Jemoroji*)	Zenkoku Hoseīgaku Kyokai, Tokyo.
Gems & gemology (1934–)	Gemological Institute of America, 1660 Stewart St, Santa Monica, CA 90404, USA.
Gems and Mineral Realm Formerly *Gems*	The Randal Press, 9 Kennet Road, Crayford, Kent, DA1 4QN, UK.
Indiaqua	De Beers Industrial Diamond Division Ltd, 35 Ely Place, London EC1N 6TD, UK.
Jewelers' circular-keystone	Radnor, PA 19089, USA.
Journal of gemmology Continuation of *Gemmological news*, 1944–1946	Gemmological Association of Great Britain, Carey Lane, London EC2V 8AB, UK.
Journal of gemmology and *Proceedings of Gemmological* *Association of Great Britain*	Carey Lane, London EC2V 8AB. 1947–
Journal of the Gemmological *Society of Japan* (1974– (Also entitled *Hoseki* *Gakkaishi*)	The Society, Sendai (Tohoku University).
Lapidary Journal	PO Box 80937, San Diego, CA 92138, USA.
Mineralogical abstracts	Mineralogical Society, 41 Queen's Gate, London SW7 5HR, UK.
Mineralogical record	6349 N. Orange Tree Dr., Tucson, AZ 85740, USA.
Retail jeweller	Knightway House, 20 Soho Square, London W1V 6DT, UK.
Synthetic crystals newsletter	Michael O'Donoghue, 7 Hillingdon Avenue, Sevenoaks, Kent TN13 3RB, UK.

Appendix C   Birthstones

Month of birth	Gemstone	Colour
January	Garnet	Dark red
February	Amethyst	Purple
March	Aquamarine or bloodstone	Pale blue
April	Diamond or rock crystal	Colourless
May	Emerald or chrysoprase	Bright green
June	Pearl or moonstone	Cream
July	Ruby or cornelian	Red
August	Peridot or sardonyx	Pale green
September	Sapphire or lapis lazuli	Deep blue
October	Opal	Variegated
November	Topaz	Yellow
December	Turquoise	Sky blue

Bibliography

Achard, F. C. (1779) *Bestimmung der Bestandtheile einiger Edelgesteine*. Berlin.

Aloisi, P. (1932) *Le gemme*. Firenze.

Anderson, B. W. (1983) *Gem testing*, 9th edn, Butterworths, London. (First edition appeared in 1943 and there have been several translations.)

Arem, J. E. (1973) *Man-made crystals*, Smithsonian Institution Press, Washington, DC, Second edition, 1987.

Arem, J. E. (1977) *Color encyclopedia of gemstones*, Van Nostrand Reinhold, New York.

Atkinson, D. (1983) Kashmir sapphire. *Gems & gemology*, **19**, 2, 64–76.

Bacci, A. (1609) *De gemmis et lapidibus pretiosis*, Francofurti.

Bancroft, P. (1984) *Gem and crystal treasures*. Western Enterprises/Minera-logical Record, Fallbrook, California.

Bariand, P. and Poullen, J. F. (1978) The pegmatites of Laghman, Nuristan, Afghanistan. *Mineralogical record*, **9**, 5, 301–8.

Bauer, M. (1896) *Edelsteinkunde*, Leipzig. (Translation by L. J. Spencer published in 1904 and republished with additions in 1968 by Dover Press, New York.)

Baumers, J. W. (1774) *Naturgeschichte aller Edelsteine wie auch der Erden und Steine*, Wien.

Becker, V. (1980) *Antique and twentieth-century jewellery*. NAG, London.

Belyaev, L. M. (1980) *Ruby and sapphire*, Amerind, New Delhi. (Translation of *Rubin i sapfir*, Moscow, 1974).

Bishop, H. R. (1906) *Investigations and studies in jade*, privately printed, New York.

Blakemore, K. (1983) *The retail jeweller's guide*, 4th edn, Butterworths, London.

Boardman, J. (1968a) *Archaic Greek gems*, Thames & Hudson, London.

Boardman, J. (1968b) *Engraved gems*, Thames & Hudson, London.

Boismenu, E. De (1913) *Fabrication synthétique du diamant*, Paris, 145–60.

Boodt, A. B. De (1609) *Gemmarum et lapidum historia*, Hanoviae.

Boodt, A. B. De (1644) *Le parfaict ioallier, ou histoire des pierreries . . . par A. Toll*, Lyon.

Bosshart, G. (1982) Distinction of natural and synthetic rubies by ultraviolet spectrophotometry. *Journal of gemmology*, **18**, 2.

Boutan, L. (1925) *La perle*, Paris.

Boyer, J. (1909) *La synthèse des pierres précieuses*, Paris.

Boyle, R. (1672) *An essay about the origine and virtues of gems*, London.

Brard, C. P. (1808) *Traité des pierres précieuses, des porphyres, granitis, marbres, albâtres et autres roches*, Paris.

Brückmann, U. F. R. (1757) *Abhandlungen von Edelsteinen*, Braunschweig.

Brückmann, U. F. R. (1778; 1783) *Gesammlete und eigene Beyträge zu seiner Abhandlung von Edelsteinen*, (2 parts) Braunschweig.

Bruton, E. (1978) *Diamonds*, 2nd edn, NAG Press, London.

Calmbach, W. F. Von (1938) *Handbuch Brasilienische Edelsteine und ihre Vorkommen*, Rio de Janeiro.

Cassedanne, J.-P. and Sauer, D. A. (1984) The Santa Terezinha de Goiás emerald deposit. *Gems & gemology*, **20**(1) 4–13.

Cattelle, W. R. (1903) *Precious Stones*, Lippincott, Philadelphia.

Cattelle, W. R. (1907) *The pearl, its story, charm and value*, Lippincott, Philadelphia.

Cavenago-Bignami, S. (1980) *Gemmologia*, 4th edn, Hoepli, Milan.

Chalmers, R. O. (1967) *Australian rocks, minerals and gemstones*, Angus and Robertson, Sydney.

Chikayama, A. (1973) *Gem identification by the inclusion.* (In Japanese: vernacular title reads *Inkurūjon ni yoru hōseki no kanbetsu: kenbikyō shashin to sono hanbetsuhō*), Tokyo.

Chudoba, K. F. J. (1939) *Bezeichnungsübersicht und Bestimmung der Schmucksteine*, Leipzig.

Chudoba, K. F. J. and Gübelin, E. J. (1953a) *Schmuck- und edelsteinkundliches Taschenbuch*, Bonn.

Chudoba, K. F. J. and Gübelin, E. J. (1953b) *Edelsteinkundliches Handbuch*, (2nd, revised edn of the above). Bonn, 1966. (A further edition was published in 1974.)

Chudoba, K. F. J. and Gübelin, E. J. (1956) *Echt oder synthetisch?* Rühle-Diebener-Verlag, Stuttgart.

Church, A., Sir (1883) *Precious stones*, London.

Claremont, L. (1906) *The gem-cutter's craft*, Bell, London.

Copeland, L. L. (1960) *The diamond dictionary*, GIA, Los Angeles.

Copeland, L. L. (1966) *Diamonds, famous, notable and unique*, GIA, Los Angeles.

Crookes, W., Sir (1909) *Diamonds*, Harper, London.

Crowningshield, G. R. (1983) Padparadscha: what's in a name? *Gems & gemology*, **19**(1), 30–6.

Dake, H. C., Fleener, F. L. and Wilson, B. H. (1938) *Quartz family minerals*, Whittlesey House, New York.

Daneu, A. (1964) *L'arte trapanese del corallo*, Fondazione Ignazio Mormino, Banco di Sicilia, Palermo.

Davies, G. (1984) *Diamond*, Hilger, Bristol.

Deer, W. A., Howie, R. A. and Zussman, J. (1978) *Rock-forming minerals*, 2nd edn, Longmans, London.

Doelter, C. (1893) *Edelsteinkunde*, Leipzig.

Dolce, L. (1617) *Trattato delle gemme che produce la natura*, Venetia. (1597 on colophon.)

Doughty, O. (1963) *Early diamond days*, Longmans, London.

Dragsted, A. (1933) *De aedle stene og deres mystik*, Kobenhavn. (Reprinted 1967.)

Dunn, P. J. (1977) The use of the electron microprobe in gemmology. *Journal of gemmology*, **15**(5), 248–58.

Dutens, L. (1776) *Des pierres précieuses et des pierres fines*, Paris.

Elwell, D. (1979) *Man-made gemstones*, Horwood, Chichester.

Emanuel, H. (1865) *Diamonds and precious stones*, Hotten, London.

Embrey, P. G. and Fuller, J. (1980) *A manual of new mineral names, 1892–1978*, British Museum (Natural History) London.

Eppler, A. and Eppler, W. F. (1934) *Edelsteine und Schmucksteine*, Verlag Wilhelm Diebener, Leipzig.

Eppler, W. F. (1933) *Der Diamant und seine Bearbeitung*, Leipzig.

Eppler, W. F. (1966) *Journal of gemmology*, **10**(2) 49–56.

Eppler, W. F. (1973) *Praktische Gemmologie*, Rühle-Diebener-Verlag, Stuttgart.

Evans, J. (1922) *Magical jewels of the Middle Ages*, Clarendon Press, Oxford.

Evans, J. and Serjeantson, M. S. (1933) *English mediaeval lapidaries*, (Early English Text Society; Original series, 190.) EETS, Oxford.

Evans, J. (1953) *A history of jewellery, 1100–1870*, Faber, London.

Eyles, W. C. (1964) *The book of opals*, Tuttle, Rutland, Vermont.

Farrington, O. C. (1903) *Gems and gem minerals*, Mumford, Chicago.

Farrington, O. C. (1923) *Amber; its physical properties and geological occurrence*, Chicago.

Farrington, O. C. and Laufer, E. (1927) *Agate: physical properties and origin. Geological leaflets*, no. 8, Chicago Field Museum of Natural History.

Fersman, A. E. (1952) *Stories of self-coloured precious stones*. (In Russian.) Moscow.

Fersman, A. E. and Goldschmidt, V. (1911) *Der Diamant*, Heidelberg.

Feuchtwanger, L. (1838) *A treatise on gems*, New York. (Later editions were issued.)

First International Gemological Symposium: proceedings (1982) GIA, Santa Monica, California.

Fladung, J. A. F. (1819) *Versuch über die Kennzeichen der Edelsteine und deren vortheilhaftesten, Schnitt*, Pesth.

Fontanieu, P. E. (1778) *L'art de faire les cristaux colorés imitans les pierres précieuses*, Paris. (English translation, 1787.)

Fraquet, H. R. (1982) Amber from the Dominican Republic. *Journal of gemmology*, **18**(4) 321–33.

Frémy, E. (1891) *Synthèse du rubis*, Dunod, Paris.

Gill, J. O. (1978) *Gill's index to journals, articles and books relating to gems and jewelry*, GIA, Santa Monica.

Goette, J. (1937) *Jade lore*, Kelly and Walsh, New York. (Originally published Shanghai, 1936.)

Goldschmidt, V. (1919) *Atlas der Krystallformen*, Heidelberg.

Goodchild, W. (1908) *Precious stones*, Constable, London.

Grigor'ev, D. P. (1965) *Ontogeny of minerals*, Israel Program for Scientific Translations, Jerusalem.

Grodzinski, P. (1936) *Diamant-Werkzeuge*, Berlin.

Grodzinski, P. (1942) *Diamond and gem stone industrial production*, NAG Press, London.

Grodzinski, P. (1953) *Diamond technology*, NAG Press, London.

Grossman, H. and Neuburger, A. (1918) *Die synthetischen Edelsteine*, Berlin.

Groth, P. H., Von (1887) *Grundriss der Edelsteinkunde*, Leipzig.

Gübelin, E. J. (1952) *Edelsteine*, Hallwag, Berne.

Gübelin, E. J. (1953) *Inclusions as a means of gemstone identification*, GIA, Los Angeles.

Gübelin, E. J. (1969) *Edelsteine*, Silva-Verlag, Zürich.

Gübelin, E. J. (1979) *Internal world of gemstones*, 2nd edn. Newnes-Butterworths, London. (Translation of *Innenwelt der Edelsteine*.)

Gübelin, E. J. (1981a) Zabargad: the ancient peridot island in the Red Sea. *Gems & gemology*, **17**(1) 2–8.

Gübelin, E. J. (1981b) Recent observations on an apparently new internal paragenesis of beryls. *Journal of gemmology*, **17**(8) 545–54.

Gübelin, E. J. (1982a) Mineral inclusions contribute towards elucidating the genesis of the diamond. *Journal of gemmology*, **18**(4) 297–320.

Gübelin, E. J. (1982b) Gemstones of Pakistan: emerald, ruby and spinel. *Gems & gemology*, **18**(3) 123–39.

Gübelin, E. J. and Koivula, J. (1986) *Photoatlas of inclusions in gemstones*, ABC, Zürich.

Hansford, S. H. (1950) *Chinese jade carving*, London.

Hansford, S. H. (1968) *Chinese carved jades*, Faber, London.

Hardinge, C., Sir (1961) *Jade, fact and fable*. School of Oriental Studies, University of Durham.

Hatch, F. H. (1961) *Petrology of the igneous rocks*, 12th edn, Murby, London.

Hatch, F. H. (1971) *Petrology of the sedimentary rocks*, 5th edn, Murby, London.

Hauy, R.-J. (1817) *Traité des caractères physiques des pierres précieuses*, Paris.

Henn, U. (1986) *Zeitschrift, Deutsche Gemmologische Gesellschaft*, **35**(1/2) 65–7.

Hinks, P. (1975) *Nineteenth century jewellery*, Faber, London.

Holmes, M. J. B. *The crown jewels*, HMSO, London.

Holmes, A. (1978) *Holmes' principles of physical geology*, 3rd edn, Nelson, London.

Hughes, G. (1964) *Modern jewelry 1890–1964*, 2nd edn, Studio Vista, London.

Hurlbut, C. F. and Klein, C. (1977) *Manual of mineralogy* (after J. D. Dana), 19th edn, Wiley, New York.

Hurlbut, C. F. and Switzer, G. S. (1979) *Gemology*, Wiley, New York.

Imperial Institute (1934) *Gemstones*, 2nd edn, Imperial Institute, London.

Iyer, L. A. N. (1953) The geology and gem-stones of the Mogok Stone Tract, Burma. Calcutta. (*Memoirs of the Geological Survey of India*, vol. 82.)

Jackson, B. (1984) Sapphire from Loch Roag, Isle of Lewis, Scotland. *Journal of gemmology*, **19**(4) 336–42.

Jacobs, H. and Chatrian, N. (1880) *Monographie du diamant*, Anvers.

Jacobs, H. and Chatrian, N. (1884) *Le diamant*. 2nd edn, Paris.

Jannettaz, E. (1880–81) *Diamant et pierres précieuses*, Paris.

Jeffries, D. (1750) *Treatise on diamonds and pearls*. London. [Many later editions.]

Jobbins, E. A. (1980) Opal from Piauí State, Brazil. *Zeitschrift der Deutschen Gemmologischen Gesellschaft*, **29**(1/2).

Jobbins, E. A. and Berrangé, J. P. (1981) The Pailin ruby and sapphire gemfield, Cambodia. *Journal of gemmology*, **18**(8) 555–67.

Kazmi *et al.*, (1985) *Mineralogical record*, **16**, 5.

Keller, P. C. (1981) Emeralds of Colombia. *Gems & gemology*, **17**(2) 80–92.

Keller, P. C. (1982) The Chanthaburi-Trat gem field, Thailand. *Gems & gemology*, **18**(4) 186–96.

Keller, P. C. (1983) The rubies of Burma. *Gems & gemology*, **19**(4) 209–19.

Keller, P. C., Koivula, J. I. and Jara, G. (1985) Sapphire from the Mercaderes-Rio Mayo area, Cauca, Colombia. *Gems & gemology*, **21**(1) 20–5.

Kievlenko, E. Ia. *et al.* (1974) *Geologiia mestorozhdenii dragotsennykh kamne*, Nedra, Moscow.

King, C. W. (1866a) *The handbook of engraved gems*, Bell and Daldy, London.

King, C. W. (1866b) *Antique gems*, London.

King, C. W. (1867a) *The natural history of gems or decorative stones*, London.

King, C. W. (1867b) *The natural history of precious stones and of the precious metals*, London.

Klobius, J. F. (1666) *Ambrae historiam . . . exhibet J. F. K. Wittenbergae.*

Koivula, J. A. (1981) The hidden beauty of amber. *Gems & gemology*, **17**(1) 393–412.

Kraus, E. H. F. and Slawson, C. B. (1947) *Gems and gem materials*, 5th edn, McGraw-Hill, New York. (First edn by Kraus and E. F. Holden, published in 1925.)

Krauss, F. (1929) *Synthetische Edelsteine*, Berlin.

Kunz, G. F. and Stevenson, C. H. (1908) *The book of the pearl*, Macmillan, London.

Kunz, G. F. (1915) *The magic of jewels and charms*, Lippincott, Philadelphia.

Kunz, G. F. (1971) *The curious lore of precious stones*, reprinted by Dover Press, New York. (First published 1913.)

Kunz, G. F. (1968) *Gems and precious stones of North America*, 2nd edn reprinted by Dover Press, New York. (First published 1892.)

Laufer, B. (1912) *Jade: a study in Chinese archaeology and religion*. Field Columbian Museum Publications. Anthropological series. New York.

Laufer, B. (1915) *The diamond*, Field Museum of Natural History. Anthropological series, **15**(1).

Launay, L. A. A. De (1897) *Les diamants du Cap*, Longmans, Paris.

Larsson, S. G. (1978) *Baltic amber, a paleobiological study*, (Entomonograph, vol. 1.) Scandinavian Science Press, Klampenberg.

Leechman, G. F. (1961) *The opal book*, Ure Smith, Sydney. (Reprinted 1978.)

Lenzen, G. (1984) *Edelsteinbestimmung mit gemmologischen Geräten*, Lenzen, Kirschweiler.

Lenzen, G. (1966) *The history of diamond production and the diamond trade* Duncker and Humblot. (Translation of *Produktions- und Handelsgeschichte des Diamanten.*)

Lenzen, G. (1983) *Diamonds and diamond grading*, Butterworths, London. (Translation of *Diamantenkunde*, 1979.)

Leonardus, C. (1502) *Speculum lapidum*, Venetiis. (Translated as *The mirror of stones*, London, 1750.)

Lewis, M. D. S. (1970) *Antique paste jewellery*, Faber, London.

Lewis, H. C. and Bonney, T. G. (1897) *Papers and notes on the genesis and matrix of the diamond*, Longmans, London.

Liddicoat, R. T. (1981) *A handbook of gem identification*, 11th edn, GIA, Santa Monica, California.

Lieber, W. (1972) *Kristalle unter der Lupe*, Ott Verlag, Thun.

Liesegang, R. E. (1915) *Die Achate*, Dresden.

MacCallien, W. J. (1937) *Scottish gem stones*, Blackie, London. (Reprinted 1965.)

MacInnes, D. (1973) *Synthetic gem and allied crystal manufacture*. Noyes Data Corporation, Park Ridge, New Jersey. (See Yaverbaum, L. H. for later material.)

MacLintock, W. F. P. (1912) *A guide to the collection of gemstones in the Geological Museum*, London. (Several later editions, the most recent revised by P. M. Statham, 1983, published by the Geological Museum.)

Marbodus, Bishop of Rennes (1531) *Marbodei Galli poetae vetustissimi de lapidibus pretiosis enchiridion, cum scholiis pictorii villingensis*, Paris.

Marbodus, Bishop of Rennes (1799) *Liber lapidum seu de gemmis varietate lectionis et perpetua annotatione illustratus a J. Beckmanno*, Göttingae.

Mawe, J. (1813) *A treatise on diamonds and precious stones*, London. (Reprinted as *An old English book on diamonds* with notes by P. Grodzinski, London, 1950.)

Meen, V. B. and Tushingham, A. D. (1968) *Crown jewels of Iran*, University of Toronto Press, Toronto.

Merrill, G. P., Moodey, M. W. and Wherry, E. T. (1922) Handbook and descriptive catalogue of the collection of gems and precious stones in the United States National Museum. Washington. (*Bulletin*. 118.)

Metz, R. (1964) *Precious stones and other crystals*, Thames & Hudson, London.

Michel, H. (1926a) *Nachahmungen und Verfälschungen der Edelsteine und Perlen und ihre Erkennung*, Graz.

Michel, H. (1926b) *Die kunstlichen Edelsteine*, 2nd edn, Diebener, Leipzig. (First edn 1914.)

Michel, H. (1940) *Perlen und Kulturperlen*, Leipzig.

Mikimoto, K. (1907) *Japanese culture pearl*, Tokyo.

Mikimoto, K. (1920) *The story of the pearl*, Tokyo.

Mitchell, R. K. (1982a) The structure of cameo shell. *Journal of gemmology*, **18**(4) 334–8.

Mitchell, R. K. (1982b) Oiled opals, *Journal of gemmology*, **18**(4) 339–41.

Mitchell, R. S. (1979) *Mineral names: what do they mean?* Van Nostrand Reinhold, New York.

Mobius, C. (1857) *Die echten Perlen*, Hamburg.

Monnickendam, A. (1941) *Secrets of the diamond*, Muller, London.

Monnickendam, A. (1955) *The magic of diamonds*, Hammond, London.

Muller, H. (1980) A note on the composition of jet. *Journal of gemmology*, **17**(1) 10–18.

Mumme, I. A. (1982) *The emerald*, Mumme Publications, Port Hacking, NSW, Australia.

Murray, J. (1839) *A memoir on the diamond*, 2nd edn, London.

Nassau, K. (1980) *Gems made by man*, Chilton, Radnor, PA.

Nassau, K. (1981) Raman spectroscopy as a gemstone test. *Journal of gemmology*, **17**(5) 306–20.

Nassau, K. (1984) *Gemstone enhancement*, Butterworths, London.

Natter, L. (1754) *A treatise on the ancient method of engraving precious stones compared with the modern*, London.

Nicols, T. (1652) *A lapidary; the history of precious stones*, Cambridge University Press, Cambridge.

Nott, S. C. (1936). *Chinese jade throughout the ages*, Batsford, London. (Reprinted 1966 by Tuttle, Rutland, VT.)

O'Donoghue, M. (1976a) *The encyclopedia of minerals and gemstones*, Orbis, London.

O'Donoghue, M. (1976b) *Synthetic gem materials*. Worshipful Company of Goldsmiths, London.

O'Donoghue, M. (1976c) *Synthetic gems: the case for caution*. Worshipful Company of Goldsmiths, London. (Technical Advisory Committee. Special report. no. 30.)

O'Donoghue, M. (1983) *Identifying man-made gemstones*, NAG Press, London.

O'Donoghue, M. (1986a) *The literature of gemstones*, British Library, London.

O'Donoghue, M. (1986b) *The literature of mineralogy*, British Library, London.

Ogden, J. (1976) *Jewellery of the ancient world*, Trefoil Books, London.

Orlov, Iu. L. (1973) *The mineralogy of the diamond*, Wiley, New York. (Translation of *Mineralogiia alamaza*. 'Nauka', Moscow, 1973.)

Pagel-Thiessen, V. (1973) *Handbook of diamond grading*, 4th edn, Pagel-Thiessen, Frankfurt.

Palache, C., Berman, H. and Frondel, C. (1944) *Dana's system of mineralogy*, 7th edn, Wiley, New York. (Silicates not yet covered.)

Palmer, J. P. (1967) *Jade*, Spring Books, London.

Pearl, R. M. (1965) *Popular gemology*, Wiley, New York.

Pogue, J. E. (1915) The turquoise. *Memoirs*, National Academy of Science 12(2), Washington.

Poynder, M. (1981) *The price guide to jewellery*. Antique Collectors Club, Woodbridge. (Kept up-to-date by supplements.)

Proctor, K. (1984) Gem pegmatites of Minas Gerais, Brazil: exploration, occurrence and aquamarine deposits. *Gems & gemology*, 20(2) 78–100.

Proctor, K. (1985a) Gem pegmatites of Minas Gerais, Brazil: the tourmalines of the Araçuai districts. *Gems & gemology*, 21(1) 3–19.

Proctor, K. (1985b) Gem pegmatites of Minas Gerais, Brazil: the Governador Valadares district. *Gems & gemology* 21(2) 86–104.

Quick, L. H. and Leiper, H. (1960) *Gemcraft*, London.

Quick, L. H. and Leiper, H. (1963) *The book of agates*, Pitman, London.

Rainier, P. W. (1943) *Green fire*, Murray, London.

Read, P. G. (1981) Test report on the Riplus ER 602 refractometer. *Journal of gemmology*, 17(5) 321–4.

Read, P. G. (1982) *Dictionary of gemmology*, Butterworths Scientific, London.

Read, P. G. (1983) *Gemmological instruments*, 2nd edn, Butterworths, London.

Read, P. G. (1984) *Gem instrument digest*, 7(1).

Reece, N. C. (1958) *The cultured pearl*, Tokyo.

Rice, P. C. (1980) *Amber, the golden gem of the ages*. Van Nostrand Reinhold, New York.

Richter, G. M. A. (1968) *Engraved gems of the Greeks, Etruscans and Romans*, Phaidon, London.

Roberts, W. L., Rapp, G. R. and Weber, J. (1974) *Encyclopedia of minerals*, Van Nostrand Reinhold, New York.

Rojo, J. B. (1747) *Theurgia general, y especifica de las graves calcedades de las mas preciosas piedras del universo*, Madrid.

Rueus, F. (1547) *De gemmis aliquot*, etc. Paris.

Ruff, E. (1950) *Jade of the Maori*, Gemmological Association of Great Britain, London.

Rutland, E. H. (1969) Corundum from Malawi. *Journal of gemmology*, **11**(8) 320–3.

Sakikawa, N. (1968) *Jade*, Japan Publications, Tokyo.

Santos Munsuri, A. (1968) *La esmeralda*, Instituto Gemológico Español, Madrid.

Sauer, J. R. (1982) *Brazil, paradise of gemstones*. Sauer, Rio de Janeiro.

Scalisi, P. and Cook, D. (1983) *Classic mineral localities of the world. Asia and Australia*, Van Nostrand Reinhold, New York.

Scandinavian Diamond Nomenclature and Grading Standards (1970) Scandinavian Jewellers' Association, Helsinki.

Schlee, D. (1980) *Bernstein-Raritäten*. Staatliche Museum für Naturkunde, Stuttgart.

Schlossmacher, K. (1950) *Leitfaden für die exacte Edelsteinbestimmung*, Stuttgart.

Schlossmacher, K. (1969) *Edelsteine und Perlen*, 5th edn, E. Schweizerbart'sche Verlagsbuchhandlung, Stuttgart. (First published 1954.)

Schmetzer, K. and Bank, H. (1984) Intensive yellow tourmaline (manganese tourmaline) of gem quality from Zambia. *Journal of gemmology*, **19**(3) 218–23.

Schubnel, H.-J. (1969) *World map of gemstone deposits, scale 1 : 40 000 000*, Association francaise de gemmologie, Orleans.

Shigley, J. E. and Foord, E. E. (1984) Gem quality red beryl from the Wah Wah Mountains, Utah. *Gems & gemology*, **20**(4) 208–21.

Shigley, J. E. and Stockton, C. M. (1984) 'Cobalt-blue' gem spinels. *Gems & gemology*, **20**(1) 34–41.

Shipley, R. M. (1948a) *Famous diamonds of the world*, 5th edn, GIA, New York. (First published 1939.)

Shipley, R. M. (1948b) *Dictionary of gems and gemology*, 4th edn, GIA, Los Angeles. (First published 1945.)

Sinkankas, J. (1959; 1976) *Gemstones of North America*, Van Nostrand Reinhold, Princeton and New York.

Sinkankas, J. (1968) *Van Nostrand's standard catalog of gems*, Van Nostrand Reinhold, New York.

Sinkankas, J. (1970) *Prospecting for gemstones and minerals*, Van Nostrand Reinhold, New York.

Sinkankas, J. (1972) *Gemstone and mineral data book*, Winchester Press, New York.

Sinkankas, J. (1981) *Emerald and other beryls*, Chilton Press, Radnor, PA.

Sinkankas, J. (1985) *Gem cutting*, 3rd edn, Van Nostrand Reinhold, New York.

Sitwell, H. D. W. (1953) *The crown jewels*, Dropmore Press, London.

Spencer, L. J. (1946) *A key to precious stones*, 2nd edn, Blackie, London. (First published 1936.)

Sperisen, F. J. (1950) *The art of the lapidary*, Milwaukee.

Steno, N. (1671) *The prodromus to a dissertation concerning solids naturally contained within solids . . . English'd by H. O.*, London.

Stopford, F. (1920) *The romance of the jewel*, Mappin and Webb, London.

Strack, E. (1982) *Perlenfibel*, Rühle-Diebener-Verlag, Stuttgart.

Streeter, E. W. (1882) *The great diamonds of the world*, Bell, London.

Streeter, E. W. (1886) *Pearls and pearling*, Bell, London.

Streeter, E. W. (1892) *Precious stones and gems*, 5th edn, Chapman and Hall, London. (First published 1877.)

Sutton, J. R. (1928) *Diamond*, Murby, London.

Tavernier, J.-B. (1676; 1682) *Les six voyages de Jean-Baptiste Tavernier . . .* Paris. (Translations by V. Ball in 1889 and W. Crooke in 1925.)

Tescione, G. (1968) *Italiani alla Pesca del Corallo ed Egemonie Marittime nel Mediterraneo*, Fausto Fiorentino, 2nd edn, Naples.

Theophrastus (1774) *History of stones*, London. (Translation by J. Hill.)

Tolansky, S. (1955) *The microstructure of diamond surfaces*, NAG Press, London.

Tolansky, S. (1962) *The history and use of diamond*. Methuen, London.

Tolansky, S. (1968) *The strategic diamond*, Oliver and Boyd, Edinburgh.

Tolkowsky, M. (1919) *Diamond design*, Spon, London.

Twining, E. F. (Baron Twining) (1960) *A history of the crown jewels of Europe*, Batsford, London.

Vargas, G. and Vargas, M. (1972) *Descriptions of gem materials*, Thermal, CA.

Vargas, G. and Vargas, M. (1979) *Faceting for amateurs*, 2nd edn. Thermal, CA.

Wade, F. B. (1918) *A text-book of precious stones*, G. P. Putnam's Sons, New York.

Wagner, P. A. (1914) *The diamond fields of Southern Africa*, Transvaal Leader, Johannesburg.

Walton, J., Sir (1952) *Physical gemmology*, Pitman, London.

Watermeyer, B. (1980) *Diamond cutting*, Purnell, Cape Town.

Webster, R. (1976) *Practical gemmology*, 5th edn (i.e. 6th). NAG Press, London. (First published 1943.)

Webster, R. (1979) *The gemmologist's compendium*, 6th edn, revised by E. A. Jobbins. NAG Press, London. (First published as *The gemmologist's pocket compendium*, 1938.)

Webster, R. (1983) *Gems*, 4th edn, revised by B. W. Anderson. Butterworths, London. (First published 1962.)

Weinstein, M. (1944) *Precious and semi-precious stones*, 4th edn, Pitman, London.

Westropp, H. M. (1874) *A manual of precious stones and antique gems.* Sampson Low, London.

Whitlock, H. P. (1945) *The story of the gems.* Emerson, New York. (First published 1936.)

Wild, G. O. (1936) *Praktikum der Edelsteinkunde,* Stuttgart.

Williams, G. F. (1906) *The diamond mines of South Africa,* Buck, New York. (First published 1902.)

Williams, A. F. (1923) *The genesis of the diamond,* Benn, London.

Williamson, G. C. (1932) *The book of amber,* Benn, London.

Williamson, G. C. (1938) *The book of ivory,* Muller, London.

Wills, G. (1972) *Jade of the East.* Weatherhill Orientations, New York.

Wilson, A. N. (1982) *Diamonds from birth to eternity,* GIA, Santa Monica, California.

Wodiska, J. (1909) *A book of precious stones,* Putnam, New York.

Wollaston, T. C. (1924) *Opal, the gem of the never-never,* Murby, London.

Wright, R. V. and Chadbourne, R. L. (1970) *Gems and minerals of the Bible,* Harper and Row, New York.

Wyart, J., *et al.* (1981) Lapis-lazuli from Sar-e-Sang, Badakshan, Afghanistan. *Gems & gemology,* **17**(4) 184–90.

Yaverbaum, L. H. (1986) *Synthetic gems production techniques,* Noyes Data Corporation, Park Ridge, NJ.

Zwaan, P. C. (1982) Sri Lanka: the gem island. *Gems & gemology,* **18**(2) 62–71.

Index

Bold page numbers indicate the principal reference to a mineral species in the text. For convenience gemstones are listed alphabetically in the sections on absorption spectra (pp. 85–91) and alteration of colour (pp. 95–107). In these cases they are not included in the main index.